CREATIVE BIOLOGY TEACHING

Delma E. Harding

Roger P. Volker

David L. Fagle

THE IOWA STATE UNIVERSITY PRESS, AMES, IOWA

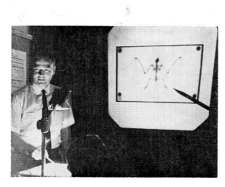

CREATIVE

BIOLOGY

TEACHING

DELMA E. HARDING, professor in the Department of Zoology at Iowa State University, holds the B.A. and M.S. degrees from the State University of Iowa and the Ph.D. degree from Iowa State University. She has taught science in various high schools and universities and has directed National Science Foundation Summer Institutes for high school biology teachers. At the present time she is supervisor of student teachers in the biological and physical sciences. Besides this book, Dr. Harding is author and coauthor of numerous journal articles and laboratory manuals.

ROGER P. VOLKER, director of the Instructional Resources Center, Iowa State University, holds the B.S. and M.S. degrees from Iowa State University. He has taught physical and biological sciences in high school and junior college. For several years he was managing editor of the Iowa Science Teacher's Journal, and has published a series of newspaper and magazine articles on science and science teaching. He served as a botanical technician during the production of the Iowa State Plant Science Studies single-concept loop films. He has conducted workshops for elementary and secondary teachers in science and received the Webster City, Iowa, Junior Chamber of Commerce Outstanding Teacher Award in 1965.

DAVID L. FAGLE, head of the Science and Mathematics Division at Marshalltown Community College, holds the B.A. and M.A. degrees from the University of Northern Iowa. He has taught high school and university sciences. While at UNI Mr. Fagle received the Outstanding Chemistry Student Award and the Old Gold Award for outstanding scholarship in science. In 1965 he received Iowa's Teacher of the Year Award and was first runner-up for *Look's* National Teacher of the Year Award. Besides this book he has published articles in numerous journals and is coauthor of *Design for Life*.

© 1969 The Iowa State University Press
Ames, Iowa, U.S.A. All rights reserved
Composed and printed by
The Iowa State University Press
First edition, 1969
Second printing, 1970
International Standard Book Number:
   0-8138-0539-4
Library of Congress Catalog Card
   Number: 68-17496

ACKNOWLEDGMENTS

WE OWE a great deal of credit to Richard Trump for critically analyzing the manuscript and to Larry Barr, El Gless, Charles Greiner, Robert Haupt, and Richard Kraemer, Sr., for illustrations.

We are grateful to the many persons who contributed in other ways to make this book possible. We express our thanks and appreciation to Benton Buttrey, Aalt G. Boon, Kenneth D. Carlander, John Dodd, Willard F. Hollander, Nels R. Lersten, William R. Shobe, David J. Staszak, Joan Sturtevant, Lois Tiffany, Milton Weller, and faculty members and students of various schools who have contributed their time.

We also owe our families a great deal—especially Carol, Paul, Christopher, Timothy, and Opal.

DELMA E. HARDING
ROGER P. VOLKER
DAVID L. FAGLE

Preface

THIS TEXT was written to help make teachers aware of changes and problems that confront today's biology instructors—to help them in their constant effort to keep abreast of new knowledge. It was written because we believe teachers need help—creative help—in solving these teaching problems.

What happens when a biology teacher is confronted with new problems? How do new teachers, as well as those who have been in the classroom several years, integrate new knowledge into their teaching? What do they do to keep the student aware of the dynamic nature of biology—in a sense keeping him on the living edge of the discipline?

Some teachers do little or nothing. They resign themselves to going over the same ideas, in the same way, year after year. But others try to keep pace with current trends in biology by reading, attending conferences, returning to school, and continually revising their teaching.

Contemporary biology can be reflected in teaching in several ways: by the use of new teaching tools as well as by the use of more traditional materials; even in the layout and arrangement of materials in the classroom itself. In this book we have shown how to achieve these and have also given practical suggestions for using living and preserved specimens; brought out teaching techniques, such as proper voice modulation and correct use of the chalkboard, overhead projector, and other educational media. We have discussed artistry in the classroom—an aspect of biology teaching that is vital for the effective presentation of well-chosen subject matter.

We have tried to achieve a balance between traditional biology and molecular biology. Broad-based preparation in the natural sciences is important for biology teachers, as is a working knowledge of the biochemical basis of life.

We believe both the beginning teacher and the practicing professional will find the book helpful. The text portion will be most valuable to the student preparing to teach, although many career teachers will find new ideas in it. The appendices will be useful to all biology teachers, regardless of their experience. New laboratory exercises, references on stains, cultures of living materials, and a number of other resources are outlined in the appendices.

DELMA E. HARDING
ROGER P. VOLKER
DAVID L. FAGLE

Part 4
APPENDICES, 183

A. Teaching Aids, 184
  Leaders and Scientists Who Contributed
    to the Biological Sciences, 184
  Common Biological Prefixes and
    Suffixes, 201
  Speakers for Biology Classes and
    Science Clubs, 202
  Sample List of Good Teaching
    Practices, 203
  Suggestions for Instant Laboratories, 204
  Forty Ways To Save Time and Money
    in the Biology Laboratory, 207
  Basic Equipment for the Laboratory, 209
  Field Trips, 210
  Details on Working Out Use of the
    Chi-Square Method, 212
  A Unit of Study, 214
  Lesson Plans, 214
  Building a Teaching File, 215
  Introducing a New Subject in Biology, 217

B. Working With Protists, 220
  Collecting, Culturing, and Using
    Protists, 220
  Tetrahymena pyriformis in the High
    School Biology Course, 226
  Using Simple Stains for *Tetrahymena*
    and Other Protozoa, 228
  Diatoms, 231
  Study of Living Protozoa by the Use of
    Vital Stains, 234

C. Using Plant Materials, 236
  Techniques for Isolating and Culturing
    Fungi, 236
  Staining and Clearing Plant Material, 237
  Coal Ball Peel Technique, 239
  Special Greenhouse Techniques, 240
  Plants for the Classroom, 243
  Liverworts and Mosses (Bryophytes), 245
  Ferns, Club Mosses, and *Equisetum*
    (Pteridophytes), 246

Part 1
TEACHING TODAY'S BIOLOGY, 1

1. Impact of the New Biology, 2
2. Teaching Form and Function
     Together, 20
3. Artistry in the Classroom, 36
4. Personalized Teaching, 50
5. The Biologist in His Community, 64

Part 2
CREATING EFFECTIVE
ENVIRONMENTS, 75

6. Planning a Science Center, 76
7. Planning a Science Laboratory, 96

Part 3
SELECTING USEFUL TOOLS, 117

8. Ways To Use Basic Materials, 118
9. Planning Field Trips, 146
10. Shaping Biology Units, 164

Contents

Culture and Care of Insectivorous
Plants, 246

Things To Do With Insectivorous
Plants, 247

Collecting, Culturing, and Using
Freshwater Algae, 248

D. Techniques Using Animals, 252

Uses of Frogs in Teaching Biology, 252

Collecting Insects, 255

Equipment for Surveying Aquatic
Populations, 259

Developing a Series of Chick Whole
Mounts, 260

Simplified Method of Bone Staining With
Alizarin Red Sulfate, 262

Collecting and Banding Birds, 263

Directions for Making Bird Study
Skins, 263

Preparation of Museum Skins of
Small Mammals, 265

Care of Cold-Blooded Vertebrates, 267

Preservatives and Fixing Materials, 273

Injecting Animals, 275

Collecting and Preserving Small
Invertebrates, 277

E. Books, Magazines, Periodicals,
Journals, 280

Magazines and Periodicals, 280

Preparing a Manuscript for Publication,
281

Suggested Topics Teachers May Write
About, 282

References for the Biology Library, 283

The "How To Know" Series, 285

BSCS Pamphlets, 285

The Current Concepts in Biology
Series, 285

Modern Biology Series, 286

References for Photography, 286

Selected References for Teachers, 287

F. Sources of Materials, 288

Transparencies, 288

School Laboratory Furniture, 288

Greenhouses and Greenhouse
Equipment, 289

Commercial Cages and Traps, 289

Prepared Microscope Slides, 289

Frogs, 289

Chromogenic Bacteria, 290

Microbiological Culture Media, 290

Antibiotics, 290

Commercial Enzymes, 291

Enzymes and Fine Chemicals, 291

Microscopes, 292

Nature Recordings, 292

Film Loops, 292

Bulletin Board Materials, 293

Lettering Materials, 294

Biological Stamps, 294

Closed-Circuit Television Equipment, 294

Educational Films, 294

Materials for Flannelboards, 295

Free or Inexpensive Printed Materials, 295

G. Projects and Class Studies, 298

Laboratory Exercises With Enzymes, 298

Some Basic Tests for Plant Foods, 301

DNA Molecule From *Grocericus storeii*, 302

Exercises on Photosynthesis, 303

Fire and Respiration, 306

Exercises in Plant Respiration, 307

Diffusion, 310

Osmosis, 312

Chromatography of Ink, 314

Chromatography of Amino Acids, 315

Chromatography of Leaf Pigments, 317

A Demonstration of Stomate Action, 317

Measuring With the Microscope, 319

Suggested Projects for Ecology, 319

Biology Projects That Have Been Done
by High School Students, 324

Behavior Studies, 326

Suggested Field Projects in Mammalogy,
328

Regeneration and Replacement Studies,
328

Cell Studies, 329

Uses for the Kymograph, 331

Index, 333

PART 1

TEACHING TODAY'S BIOLOGY

CHAPTER 1

Impact

of the

New Biology

WHAT IS BIOLOGY? Is biology really new? After all, life has been on this earth for several million years. And though there have been gradual changes in form and physiology, biologists generally agree that living things are more alike than they are different.

What has changed, then? Our understanding of life! Reading through old journals and texts we can trace the increasing depth of our understanding of biology. Perhaps some articles from the *Scientific American* magazines of one hundred years ago will help explain it. Though this science journal was devoted largely to inventions and mechanical improvements of machines, it still contained articles on "natural science" such as "The Growth of Vegetation Near Copper Mines," "Malaria and Its Remedies," "New Insects Injurious to Vegetation," "Geology and the King Crab," and "The Merits of the Catawba Grape in American Winemaking."

Compare the titles of those articles with these taken at random from recent issues of the same publication: "The Synapse," "Genes Outside the Chromosomes," "How Cells Make Antibodies," "The Hemoglobin Molecule," "The Genetic Code of a Virus," "The Sarcoplasmic Reticulum," and "The Control of Biochemical Reactions."

This clearly represents a change in subject matter emphasis over the past century. "But," you say, "haven't all fields of knowledge changed in the last one hundred years? Equally great discoveries have certainly been made in history, mathematics, psychology, chemistry, and physics." This is true; but discoveries in these areas, while reflecting changes, do not reveal departures nearly so radical as those in biological science. Spectrographic analysis, nuclear fission, and the laser have modified chemistry and physics, but these developments have not changed the field of physical science to nearly the same degree that some of the recent biochemical discoveries have changed the field of natural science.

The swing from the descriptive, morphological research of earlier biologists to the functional, physiological methods of today's workers is suggested by publications like *Scientific American*. But you can pick up almost *any* newspaper or magazine and read articles on DNA, cytology, and physiology. For example, an issue of *Today's Health* carried articles on "Building Blocks of Life," "Medicines of Tomorrow," and had a photograph of a plastic model of a cell on the cover. An accompanying explanation of the cover picture discussed structure as revealed by the electron microscope. Similar articles on biological science are found in the *Family Circle* magazine, *Look,* and *Life.*

Some biochemicals have become almost as commonplace as the simple-looking equation $E = mc^2$. What we read in the papers and hear on the street reflects only a small percentage of the deeper discovery that is constantly carried on in laboratories the world over.

THE IMPACT OF HEADLINES

Our popular literature is full of bold announcements that indicate the way biology is bending. For example, a new type of microscope using radio frequencies has been developed at Western Reserve University to study atomic and molecular structure in living cells. Ultracentrifuges spin at Oak Ridge's Gaseous Diffusion Plant in the Biophysical Separations Laboratory where viruses are separated for producing new vaccines. At Northwestern University a scientist is investigating a means for determining the effects of space travel on biological rhythms of man here on earth. He feels that phenomena such as gravity and magnetism which the travelers will encounter in outer space may have profound effects on earth-oriented people.

A few decades ago this type of scientific news might have been headline material. Just a little over 20 years ago a handful of scientists working at the Rockefeller Institute in New York announced the results of an experiment they had been working on for three years. Oswald Avery, a man in his 60's, and two younger collaborators, Colin MacLeod and Maclyn McCarty, had completed work on a project dealing with the bacterial species that causes a certain type of pneumonia. The three

Fig. 1.1—Today, much of the printed matter on our newsstands carries articles that report some rather involved scientific research. The fact that such work is coming into closer contact with nonscientists is an indication that biology is understood more and more by the general public.

3

scientists had proved that it is possible to change this particular form into another one simply by introducing a solution of deoxyribonucleic acid.

This news did not rock the world. For one thing, not many people could understand the significance of it. For another, the world was concerned with other problems at that time—problems such as Normandy and Iwo Jima and Hiroshima. The awfulness of World War II eclipsed seemingly impractical and unimportant things like DNA.

Even though this news was not in headline form, it was eventually to play its part in elevating biology to the proper place that it now occupies as a science. In fact, biology seems to have become so interwoven with other sciences, notably chemistry and physics, that it is hard to tell where the borderlines are. Work on the DNA molecule is a good example of the coalescence that has occurred in merging biology with the physical sciences.

Recently, at a meeting of the American Physical Society, Dr. Peter Fong of Cornell University reported the results of his work on the double helical structure of nucleic acids, the molecule's mechanism for unwinding itself to expose its active sites, making possible replication of its structure. The American Physical Society is obviously a group of physical scientists. A few years ago their interest in what a biologist would have to say would have been superficial indeed. But now they and other groups (including the American Chemical Society) take a real interest in the new developments in biology. All of this is evidence that biology has shifted its position—it has changed its image. Since the field of natural science itself has made this change, the teachers of science are faced with a double responsibility—that of understanding the change

themselves, and then of working out the best teaching methods for guiding students through the changes. Certainly, as never before, biology teaching must be creative.

TRANSMUTATION OF A TERM

The term "biology" has undergone a more complete change than almost any other label in education. James L. Slattery, writing in the *Kiwanis Magazine*,[1] sums it up this way:

Developments in biology are occurring at so frantic a pace today that only a computer could really keep up with them. And many of these developments are achievements of the first magnitude. Such, for example, was the spectacular triumph of two research teams, one at the University of Pittsburgh, and another in Germany, who independently achieved the long-awaited synthesis of insulin. The increase in the incidence of diabetes has created fears that the demand for insulin might outrun the supply, if insulin continued to be obtainable only from animal sources. The brilliant biochemical accomplishment of the American and the German research teams makes it likely that synthetic insulin will be available in unlimited quantity in the near future. Such achievements give an attractive added meaning to the definition of biology as "the science of life."

It has long been defined, but without much justification until recent times. To begin with, biology had never really become a science in the way that physics and chemistry had. And the word "life" had become burdened with so many meanings that it amounted to little more than a catchword. If anyone knew what "life" was a hundred years ago, it was certainly not the biologists.

---

[1] "Biology Comes to Life," Part I October 1964, Part II November 1964, and Part III December 1964 and January 1965. The fact that we find a comprehensive article on biology in a magazine for businessmen, such as the *Kiwanis Magazine,* shows to what extent popular news media have influenced the thinking of the man-on-the-street regarding science.

He continues to develop the idea that both physics and chemistry had been true experimental sciences for a long time, while biology just recently gained this distinction.

Even in the 1880's when both physics and chemistry had become spectacularly successful sciences, biology was still wandering in darkness and appalling confusion. The difficulty was that biologists still did not know what it was that they were studying. They were preoccupied with two basic problems—the same two that had fascinated Aristotle, the father of biology. What is the "essential difference" between living and non-living entities? And what is the secret of the "form" of a particular type of living being; and how is this determining form transmitted from a parent creature to its offspring?

Indeed, biology has come a long way. The dramatic breakthroughs of the twentieth century contrast sharply to nineteenth-century biology. Why? It is not that biologists of the past were incapable workers nor that they lacked intelligence.

The answer is related to advances in other sciences. Until they had information from the physical sciences, biologists were not able to understand the secrets of DNA. Yet DNA had been known for almost 100 years.

Though the problems had been identified, earlier biologists could not solve them, so they had to do what *could* be done as they worked in the area of descriptive biology.

THE NATURE OF TRADITIONAL BIOLOGY

From Galen to Schwann "biologists" were really anatomists, morphologists, and taxonomists. Without the technical knowledge and machines that are commonplace in today's laboratories, these scientists had to be content with describing the general appearance and internal structure of plants and animals, what varieties of external forms they had, and where they fitted into the classification system. By the end of the nineteenth century the Linnaean system of taxonomy had been in use almost 150 years, and almost a million plants and animals had been placed into phyla. Just categorizing the specimens was an arduous task. In addition, certain changes had to be made in the original system that Linnaeus had proposed. These modifications, incorporated as the work of the systematists progressed, tended to make "systematic biology," as it was called in those days, a slow and somewhat tedious science. There were discoveries to be made, but the discoveries were not of an experimental nature. No hypotheses were tested and turned into theories in biological laboratories as were the investigations that prevailed in the chemistry laboratories of Lavoisier, Bunsen, and Wöhler. And quite a different "scientific method" was being used by the biologists than by the physicists. The former observed and described what they saw, while the latter manipulated, tested, and applied mathematical proofs to the natural phenomena that they unraveled in their laboratories. Because biology was essentially descriptive until comparatively recent times, while the physical sciences were experimental, it might seem reasonable to consider that biology was late in becoming a science. The physical sciences had their own clear-cut dogma, based on solid experimental evidence. But the dogma of biology was not yet clearly defined. While basic research and experiments were continually contributing evidence, the foundations of biology such as evolution and biochemical mechanisms were yet to come.

THE COMPARTMENT PLAN

As biologists studied the great variety of living forms, they rapidly accumulated a mass of information—information that could not possibly be assimilated by one man. The concept of compartments or subdivisions of biology grew, simply because it was more efficient for one group of scientists to concentrate on classification, another on form, and still another on function. Even these categories were split; plant taxonomy *and* animal taxonomy became separate entities. Soon each had more than it could do, with the mounting mass of information appearing in the journals devoted to taxonomy.

From the standpoint of specialization, biology moved ahead of its sister sciences of chemistry and physics in the late 1800's. While general chemistry split into inorganic and organic chemistry, and physics had its areas of optics, mechanics, and electricity, it was the natural science of biology that became truly departmentalized. Specific areas or fields were brought into existence so that each biologist could pursue his major interest. *Anatomy*, the study of internal structure, gave rise to *histology*, the study of tissues, and *cytology*, the study of cells. The new science of *genetics*, spawned by Mendel, drew from each of these three, and other areas as well. It soon became apparent to taxonomists that each phylum might become an area of science. *Bacteriology, mycology, protozoology, helminthology, entomology, herpetology, ornithology*, and *mammalogy* grew and made their contributions to scientific thought. Biologists working in the relatively recent field of physiology soon became interested in problems of digestion, excretion, respiration, and endocrinology.

Appendix A 184 indicates in chronological order some of the leaders and scientists who have contributed in various ways to the knowledge and development of the biological sciences.

We do not mean to imply that simply because one is a protozoologist he cannot or will not know anything about physiology, taxonomy, histology, and all the other "ologies" of life science. These fields of study are related to each other inseparably. Let us suppose, for example, that an *ecologist* was studying the relationship of the distribution of certain species of diatoms to the degree of water pollution. He would need to know a great deal about *diatom taxonomy* just to identify the species he found. The metabolic processes of diatoms and the relationship of their metabolism to the water chemistry would require an understanding of *biochemistry* and *plant physiology*. Other organisms in the ecosystem, such as aquatic insects, bacteria, and fish, would require additional specialized knowledge. To deal with the problem in depth a number of experts would have to work together. Dr. Ruth Patrick, head of the Department of Limnology at the Philadelphia Academy of Natural Sciences, carries on such work daily in an attempt to determine levels of water pollution. She employs a staff of specialists who help in surveying the "health" of a stream. The staff includes a chemist, a bacteriologist, a zoologist, an ichthyologist, an entomologist, and an algologist. Dr. Patrick herself specializes in diatoms.

No matter what problem you choose in biology, you must deal with a number of interwoven disciplines. Eugene Odum, writing in *Fundamentals of Ecology*,[2] speaks of biology as a "layer cake." He says:

---

[2] W. B. Saunders Co., Philadelphia, 2nd Ed., 1959.

We may divide it horizontally into what are usually called "basic" divisions, because they are concerned with fundamentals common to all life. . . . Morphology, physiology, genetics, ecology, and embryology are examples of such divisions. We may also divide the cake "vertically" into what may be called "taxonomic" divisions, which have to do with the morphology, physiology, and ecology of specific kinds of organisms. Both approaches are profitable.

Are these categories, established many years ago, still valid dividers for the broad area of life science? What impact have the new developments had on these older lines of thought? Let us examine a few of the contributing factors to the quiet revolution that has been gathering momentum recently and see what effect this revolution will have on the biology teacher.

THE HIGH RPM OF BIOLOGY

For the past thirty or forty years there has been almost one revolution per minute in biological science. The beginning is hard to pinpoint; one could say it started in the early 1930's when the electron microscope was developed in Germany and later brought to the United States. One could consider Hans Krebs' 1937 explanation of the basic steps in intermediary metabolism as the true starting point of modern biology. One could point to the work of Watson and Crick and Wilkins on DNA, for which they received the 1962 Nobel Prize, as another notable milestone. Add to these major marks the hundreds of accomplishments in research, most of which never get outside the readership of the technical journals that carry them, and you begin to get some idea of the magnitude of research in biological science that is going on in thousands of laboratories today.

One index to the quantity of research that is changing the face of all of the sciences is money. According to an article in *Time* magazine,[3] the National Science Foundation spent over *360 million dollars* in 1963 for the upgrading of science education and scientific research in the United States. This is one and one-half times as much as private philanthropic institutions, such as the

_____

[3] Copyright Time Inc., May 4, 1962, p. 67.

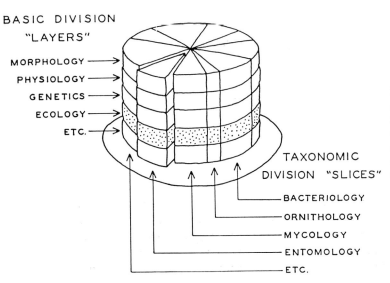

BASIC DIVISION "LAYERS"

MORPHOLOGY →
PHYSIOLOGY →
GENETICS →
ECOLOGY →
ETC. →

TAXONOMIC DIVISION "SLICES"

BACTERIOLOGY
ORNITHOLOGY
MYCOLOGY
ENTOMOLOGY
ETC.

Fig. 1.2—The layer cake of biology relates all fields of study to one another. (From Odum, *Fundamentals of Ecology*, 2nd Ed., W. B. Saunders Co., Philadelphia, 1959.)

8

Ford Foundation, spent. Started on a shoestring at the suggestion of President Roosevelt in 1944, the NSF has become Big Brother to a large number of institutions over the country, contributing a great deal to the revolution in teaching and learning. *Time* reported some of the school reforms this way:

Unforeseen in 1950, the most impressive impact is on education. Though it works through colleges and universities, NSF has played an indispensable part in reforming school math and science. Joining private foundations, it solidly backed university scientists in the mid-1950's when they began helping the schools after decades of disdain. The scholars produced new thematic courses geared to the real structure of subjects, most notably the high school physics course designed by MIT Physicist Jerold Zacharias' Physical Science Study Committee. NSF so far put some $5,000,000 into the project, which now involves 20% of all United States high school students taking physics.

NSF has similarly launched 897 summer and year-round institutes to retrain some 35,939 college and high school teachers. And this summer [1962] at 154 institutions . . . NSF will muster some 7,500 "high ability" tenth- to twelfth-graders for a rich taste of college science.

Another clue to the quantity of research can be found in the indexes of scientific journals. The index of the *Annals of Applied Biology* for the year 1917 indicates 27 entries. The same journal carries 79 entries in the index of the January 1965 issue. This is a 192% increase. Dr. Harvey Diehl, professor of analytical chemistry at Iowa State University, estimates that in chemical literature alone publication of articles has doubled each 10 years since the turn of the century. Extrapolating, he facetiously concludes that, for this rate to remain constant, by the year 2200 every man, woman, *and child* in the United

States will have to be working in the field of chemistry!

You, as a biology teacher, will feel the pressure of new ideas almost before you begin teaching. The number of publications you must read, if you are to keep in touch with current ideas, totals more than a dozen. The new apparatus and the techniques that go with it require new skills and new training. The broadening of biology to include an intimate association with chemistry, physics, and mathematics requires each teacher of biology to have a comprehensive background in the physical sciences. To understand where biology is going it is important that you become familiar with traditional biology, and then the revolutions that have changed it.

With this in mind, let us now take a look at the present status of teachers who are trained in our colleges and universities for a lifetime of biology teaching.

WHERE ARE WE NOW?

Research biologists have caused the vast and complex changes that have created the revolution in biology and helped form a "new biology." But there has been a gap between those who do the active research and those who study biological science in high school and college. We have illustrated the fields of study of biology in Figure 1.2 as a kind of "layer cake." Yet this layer cake is not truly representative of the notion of modern biology as it is viewed by those who work on its living edge. Such an illustration of biology, with its vertical and horizontal layers, might have properly represented it as it existed a decade or two ago. But now such headings as "biophysics," "biochemistry," "cell biology," "biomedical electronics," and "biological fine structure" have invaded the field. Where would these fit into the layers? Actually, they have done more than simply invade biology, for in addition to taking their place with the "oldsters" in the curriculum—areas like morphology and taxonomy—they have changed the study of the nature of life.

However great this change in traditional curricula may seem to be, many college students still are educated to be biology teachers by taking cytology, mycology, parasitology, and other compartmentalized courses. It is only natural that a teacher trained in this way should try to offer his own biology students a once-over-lightly sample of the areas of study he knew as a student in college, emphasizing those areas he himself liked best.

Partly because of departmentalized training of science teachers, there was departmentalized teaching of beginning students by those teachers. High school biology was for years a survey course—a year's look at the diversity of plants, animals, and protists, ending with an emphasis on human anatomy and physiology. In a sense the entire history of biological findings was telescoped into two semesters of high school biology. This systematic study of the phyla of organisms was termed "organismic biology" or the "phylogenetic approach." As each phylum was studied, a representative of the group was chosen as a "typical" specimen for close examination. The method was the same as the traditional biologists had used for many years. External characteristics of shape, size, and color were noted, described, and named. Then internal study was carried out through dissection. Again parts were described and named. A high point in almost every biology course was the dissection of the earthworm, grasshopper, and frog. We do not mean to condemn this type of study. It is valu-

able, and still retains its place as a part of the study of life. The point is that such study is essentially *descriptive* rather than *experimental*. Students are doing *exercises,* not *experiments*. They are confirming the presence (or absence) of structures in each organism. There is no sense of discovery here, no wondering about what the outcome will be, for each student knows that he will find a liver, a heart, and a stomach when he dissects a frog. He merely is confirming a statement in the laboratory manual.

Studying biology phylum by phylum is a little like watching a football game through binoculars. It is possible to focus sharply on each player. The quarterback, halfback, and tackle can all be seen with clarity; the subtle characteristics that make them good teammates may be studied carefully. The disadvantage of watching a game this way is that you miss the interdependence that makes eleven players a *team*. You must see the whole field of action. It is the same with biology. To understand organisms you must see the interrelationships of structure and function—against a background of biochemistry. These are the threads of the new fabric of biology, and are important themes in this text.

THE NEW BIOLOGY

Should we continue to teach phylogeny? Should not a biology student know something about the taxonomic spectrum of plants and animals on the earth? Should not he know something about their morphology, physiology, taxonomy, genetics, and ecology? It is true that the knowledge in each of these fields of biology has widened and deepened in the past few years, but should this change the *teaching* of biology?

A large group of American biologists thinks so. In 1959 the American Institute of Biological Sciences organized a group to discuss what high school students were learning—and *not* learning—in 10th-grade biology. The group, called the Biological Sciences Curriculum Study, looked closely at the content of many high school biology courses and saw that the students were given a great deal of information, but not necessarily understanding. They were not organizing the facts into concepts. The BSCS concluded that high school biology students were learning about the *products* of science but not its *process*. The students were being taught the *words* of biology but not its *ways*. This is what Claude Welch, a supervisor of one of the Biological Science Curriculum Study writing teams, had to say about the kind of biology his team was attempting to write:

> Excitement in science, we think, results from a happy combination of failure and success. Science so far has failed to solve the riddle of life, but it has been just successful enough to whet the appetite. Scientists are excited about the things they understand; but they are also excited about the things they *do not* understand, and this mixture of success and failure urges them on.
>
> This is what you can expect: . . . An approach which assumes an understanding of the means by which scientists solve problems is as important as the problems themselves.[4]

It was the feeling of this committee that biology should be a study of the underlying principles of life—ideas such as photosynthesis, digestion, respiration, protein synthesis, genetics, and evolution. It should *not* be a survey of the plant, animal, and protist kingdoms.

[4] From the preface of the paperbound edition of the *BSCS High School Biology Blue Version Experimental Text*, 1961. Used by permission of the Biological Sciences Curriculum Study.

Furthermore, the biology course as a whole should be built around an inquiry or problem-solving approach. While the problems of the biologist are important, the method used to solve them is even more important. BSCS felt that high school biology should attempt to teach a *method of problem solving*.

The Forest or the Trees?

Which method would create the more meaningful picture of biology for your students? Would you favor the phylogenetic approach, or an approach built on the method of solving problems in biology? Probably you will agree that a balance must be set between the two methods. Sometimes you must look at the forest, and sometimes you must look at the individual trees. Sometimes you must look at the entire range of biology, and include the processes, structures, and behavior of the organisms in order to formulate a certain biological concept. At other times it is necessary to study individual groups of organisms. Form and function must be handled in such a way that they complement one another. It is not possible to create an accurate picture of biology for your students simply by naming parts, using scientific names, and describing shape, size, and location of body organs.

We do not mean that all scientific terminology should be dropped from creative biology teaching. But word lists should not be ends in themselves. Just because a student can name the parts of a flower does not mean that he understands the role flowers play in plant reproduction. Students must learn enough terminology to accurately describe ideas and relate concepts to one another. Mark Twain[5] recognized long ago that

[5] *A Tramp Abroad,* Harper and Brothers, New York, Vol. 2, p. 19.

some terminology was necessary in any field. The following is his description of how they hitch up horses in Europe.

The man stands up the horses on each side of the thing that projects from the front end of the wagon, throws the gear on top of the horses, and passes the thing that goes forward through the other ring and hauls it aft on the other side of the other horse, opposite to the first one, after crossing them and bringing the loose end back, and then buckles the other thing and wraps it around the thing I spoke of before, and puts another thing over each horse's head, and puts the iron thing in his mouth, and brings the ends of these things aft over his back, after buckling another one around under his neck, and hitching another thing on a thing that goes over his shoulder, and then takes the slack of the thing which I mentioned a while ago and fetches it aft and makes it fast to the thing that pulls the wagon, and hands the other things up to the driver.

While the memorizing of terms is far from the learning of biological principles, the same type of useless memorizing can occur in classrooms following the most modern approach. Suppose, for example, you could observe a discussion of elementary biochemistry in a sophomore biology class. It might seem at first glance that this teacher was engaging in up-to-the-minute use of some of the most important aspects of the new biology. But the assignment the teacher makes furnishes a clue to the type of learning his students are actually required to do. As a part of tomorrow's lesson he writes on the board, "Be able to write the chemical formulas for adenosine triphosphate, adenine, guanine, cytosine, thymine, and chlorophyll."

Teachers tend to use memory work because it is the easy way to test a student's knowledge. The much more difficult problem is to test a student's understanding. In Chapter 10 we suggest

Fig. 1.3—Scientific names may sound strange, but they have an important function. This plant has over 100 common names in the United States. Would you rather learn all one hundred, or the single scientific name *Capsella bursa-pastoris*? (One common name, Shepherd's Purse, is based on the shape of the fruit.)

some creative ways of evaluating student achievements.

The correct balance between descriptive biology and fundamentals of biology, between phylogeny and the principles by which living organisms operate, is a delicate balance. At times, in certain places in the course, phylogeny must be emphasized. At other times, general principles of biology must become the theme. The balance you finally achieve will depend on your training and your creative ability. Your formal education is your responsibility. Your creativity, we hope, will be influenced by this text.

Other Implications of the New Biology

The balance between organismic biology and principles of biology is not the only problem to be solved by today's teachers. Here are some other questions of equal importance:

1. How much emphasis will you place on metabolic cycles?
2. How much time will you spend in the laboratory?
3. To what depth will you teach concepts of human reproduction . . . of evolution . . . of biochemical phenomena?
4. How much field work will you include in the course?

In addition to these questions about the subject matter, you must decide *how* you want to teach. Will you be lecturing? Will you be using some form of student-oriented discussion? You must also decide the balance between laboratory and nonlaboratory time. Do you want the students to work in the laboratory on certain days during the week, or should they remain at work in the laboratory for several weeks at a time? Will they work in groups, or do individual work? Will they be doing open-ended *experiments* or confirmatory *exercises?* These, and a number of other questions are discussed in later chapters of the text. However, one recurring question centers on the great emphasis that chemistry has assumed recently, and the im-

pact of that emphasis on high school biology. No matter what type of overall program you operate as a biology teacher, the proper handling of chemistry, wherever it occurs, will be vital to a successful teaching program. Why has chemistry become so important in biology? Perhaps some background will help answer this question.

The Impact of Chemistry

Though nucleic acids were discovered in nuclei of cells as early as the 1860's, their significance did not come into sharp focus until the early 1950's when a new term was coined: deoxyribonucleic acid. The occasion was Wilkins, Watson, and Crick's publication of research that indicated DNA might be the chemical substance composing the genes. They worked on DNA because earlier workers had identified it as important heredity-determining material in the nucleus. If their results were correct, the chemical basis of genetics, and even life itself, might eventually be understood, and the process of protein synthesis might be explained.

Several years earlier another important discovery in chemistry occurred at the University of Chicago. The Nobel laureate Harold Urey and one of his students, Stanley Miller, set up a rather simple piece of apparatus for passing electrical discharges through an atmosphere of water vapor, methane, ammonia, and hydrogen. In one week they chromatographed the residue and found glycine and alanine, the simplest and most abundant amino acids in protein. Though much was written about the experiment, scientists realized they might be causing more problems than they solved if they tried to draw too broad conclusions from the results of the Urey-Miller work. For instance, some of the writing that followed this famous experiment implied that perhaps life on earth had arisen from the gradual change of primeval amino acids, formed under the same conditions that prevailed in that flask in the University of Chicago laboratories.[6]

In such ways chemistry was injected

---

[6] Additional theories on the origin of life in the sea are found in *Metabolites of the Sea* by Ross F. Nigrelli, in BSCS Pamphlet #7, published by the AIBS in March 1963.

so forcefully into biology that neither of the two sciences will be the same again. It is interesting to check indexes of both chemistry and biology texts to find the subject of nucleic acids discussed in each.

How will you handle the explanation of the Urey-Miller experiment in your classes? Will your students have the proper groundwork to understand its significance? What kind of back-

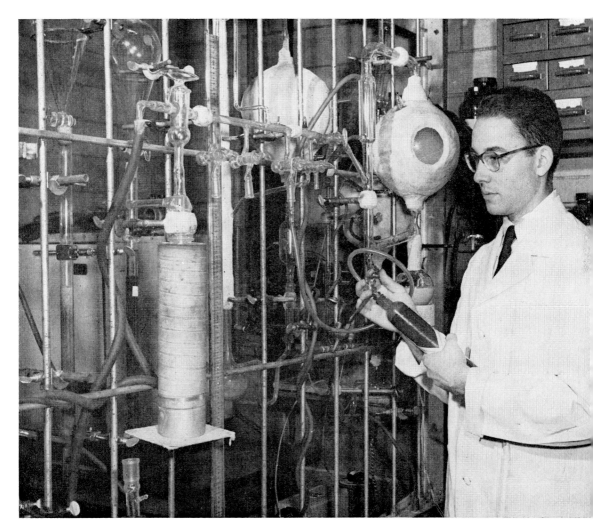

Fig. 1.4—The synthesis of simple amino acids from an atmosphere of gases made an important impact on traditional biology. This is Stanley Miller, who did much of the original work. (Used by permission of United Press International.)

ground in chemistry will you give your students so that they can read their text with understanding? Will they be able to do outside reading in scientific literature and comprehend what they read, based on your teaching?

Chemistry has made its impact on biology as well as on the biology teacher. Your background in chemistry must be strong enough to withstand the impact, for chemistry is now a very real part of biology.

THE LABORATORY: A VEHICLE FOR THOUGHT

We mentioned previously that a biology course should show students how problems are solved, and provide a vehicle for the development of critical thinking. How can this be done most effectively? To answer the question let us turn to an aspect of biology often thought of as one of its lesser parts—laboratory work.

New biology uses the laboratory in a new way. Exercises are still performed, but more often the laboratory is a place for problem solving. Long-range experiments are conducted—experiments without any set answers. There may be blanks to fill in, but the blanks are likely to have many "correct" answers. Many of the experiments never lead to definite answers. They only lead to more questions. These "open-ended" or "unended" experiments show beginning students what science is: It is *doing*, in addition to listening and watching. Proper use of the laboratory, which is a unifying theme of this text, brings students face to face with the ways and means of science.

INNOVATION IN TEACHING

An additional quality, and perhaps the most important one in teaching, is the quiet quest you must make for innovations in teaching. In the early 60's an exciting program was set up at Purdue University under the direction of Samuel Postlethwait, Joseph Novak, and Hal Murray. Together they developed a startlingly new method for teaching freshman botany. In the first chapter of their book they discuss the need for innovation in teaching.[7]

Population explosion, knowledge explosion, updating of high school teaching, and a general increase in the awareness of the importance of education to a successful life have intensified certain problems in education. Specifically, students enrolled in freshman courses in college now represent a great diversity of interests, backgrounds, and capacities. The problem is to provide a learning situation with enough flexibility for each student to make adjustments suited to his individual needs.

In order to accommodate this great range of students, there are three major ingredients necessary in our teaching approach. First, it must include a great variety of teaching and techniques, and must take advantage of all the modern communication media. Each medium must be selected for use on the basis of its effectiveness and efficiency in evoking the appropriate responses. Secondly, these media must be made available in such a way that the students can pace themselves, omitting those things with which they are already familiar and spending extra time where necessary to bring their education up to the level of their colleagues. And thirdly, personal contact with teaching personnel must be made available to provide motivation and interest. Each activity required of the student must be analyzed in terms of the objective and all busy work that does not really contribute to the student's progress must be eliminated. The teaching method must take into account the fact that individuals may perceive ideas and concepts through different channels—some can read perceptually, while others must manipulate and handle materials to obtain an adequate understanding of the concept

----

[7] *An Integrated Experience Approach to Learning*, Burgess Publishing Co., Minneapolis, 1964.

involved. If the goals can be carefully and critically identified, a student arriving at these goals independently along some avenue other than that provided by the teacher, should be accorded the same respect as one who has attended carefully to the details of a program outlined by the instructor. A teacher can only provide guidance, facilities, and motivation. The student must do the learning. All too often, teachers develop pet exercises or programs and every student is required to perform faithfully each item if he is to receive a satisfactory grade. Often these pet approaches make little or no direct contribution to the learning process or the student's needs, and only students who are willing to conform and can properly interpret the teacher's wishes can obtain good grades. Creative students who will not submit to the regimentation may receive low grades regardless of their knowledge of the subject.

The last sentence of the authors' statement is quite an indictment, not only of a great many teachers, but of a great many teaching methods. How can you be certain that your daily teaching remains vital, fresh, and alive? Your training and intuition, properly used, will be the best safeguards you have.

THE IMPACT ON YOU

The creative biology teacher is a hard-working individual. He teaches because he likes the work. He teaches because he likes the subject. He teaches because he likes the students. But he knows that any profession—even biology teaching—will have its share of thorns. Biology teaching is especially difficult because of its dependence on the wide background of the teacher and his ability to keep pace with new developments and current literature. In addition to coping with these problems the teacher must cultivate the student's critical mind and intellectual curiosity. New biology is sure to make its impact on teachers who are recent graduates as well

as teachers who have been in the business 30 or 40 years.

It is too soon to know just what the outcome of this change in biology will be. There are indications that the emphasis on molecular biology represents the view of research scientists rather than the mainstream of biology teachers. It is true that the investigative laboratory approach in high school biology today is a model of the method the research scientist follows.

Perhaps modern biology, like modern physics and modern chemistry, represents a pendulum swing to the far right or the far left. Such swings are needed to call attention to problems and point out roads of reform. Modern "approaches" and "studies" have always left an aftermath of change. But we must remember that there is more than one way to teach biology, and just because a new program exists does not mean that it must be adopted per se. In a recent revision of his excellent *History of Biology,*[8] Eldon J. Gardner puts molecular biology in perspective with respect to the entire field.

Molecular biology is . . . based on the premise that biological mechanisms can be reduced to inanimate chemical and physical processes. In mid-20th century, biology is dominated by chemistry and physics. Difficulties arise, however, when attempts are made to account for the entire organism on the basis of molecular biology. Genetic coding, which represents one of the most important discoveries of the present century, will not account for all embryological and evolutionary changes among living things. If all living processes could be reduced to chemistry and physics, there would eventually be no need for biology. Living organisms, however, are more than bundles of DNA. Since many interactions occur about the molecular level, areas other than molecular biology will continue to have a place in biology. Darwinian biology,

---

[8] Burgess Publishing Co., Minneapolis, 2nd Ed., 1965.

for example, benefits greatly from the contributions of the molecular biologist but it is not likely to be replaced by the current modernized . . . biology. Undeniably, however, the molecular biologists and the many specialists in related fields are now solving problems that have long awaited an explanation. Biology owes them much, but it can't afford to lose sight of the whole just because certain segments are particularly vigorous.

Gardner apparently feels that certain aspects of new biology may be incorporated into older programs. Each teacher must first understand a new program in its entirety and then extract from it those aspects he can feel most comfortable teaching.

CREATING CHAPTER CONCEPTS

1. Make a list of 10 developments in natural science in the nineteenth century that you would consider the most important.
2. Compile a list of the "Ten Most Famous Biologists" from the period of 1850 to 1900, and a similar list for the period from 1900 to 1950. List the accomplishment of each person.
3. We mentioned the American Physical Society and the American Chemical Society as groups of nonbiologists who would be apt to have new developments to report in the area of biology. List 5 other groups outside the area of biology which could receive reports from their members concerning biological research.
4. Experimental excitement was thought to prevail in the laboratories of Lavoisier, Bunsen, and Wöhler. What basic contribution did each make to science?
5. Using a catalog from the college or university you attend, work out the curriculum you would consider most ideal for the person who is to become a biology teacher. How do you think this curriculum would differ from one that might have been suggested for a biology teacher of 50 years ago?

6. Which of the following fields do you think requires the broadest background—ecology, physiology, or genetics? Support your answer by listing the courses you think a college major in each area should have.
7. Devise a biological crossword puzzle that would utilize new concepts and terms found today in biological science. (For clues, refer to the *American Biology Teacher*, April 1964, p. 281.)
8. Prepare a list of the kinds of institutions, foundations, and agencies that provide money for research. Annotate your list so that you will not list agencies that overlap in their function. (This activity will require several hours of library work, and it may be a part of an assignment that is not due until later chapters in the text are discussed.)
9. Prepare an annotated list of classical biology texts, indicating the particular area of biology for which you think each book is valuable.
10. We mentioned that if you are to keep in touch with current biological thought you must read more than a dozen publications. Make a list of the 12 publications you consider most valuable to you, a biology teacher. Include subscription cost and publisher's address so the list can be distributed to other members of the class.
11. What is meant by the *phylogenetic approach* to teaching biology?
12. Do you feel departmentalization of an undergraduate's training in botany, zoology, chemistry, physiology, cytology, and taxonomy weakens or strengthens his preparation for teaching a course in general biology?
13. Devise a technique for insuring that students understand terms and build concepts from them *without* causing rote memory.
14. Write a brief paragraph about the activities of the Biological Sciences Curriculum Study.
15. List 10 areas of biology that you plan to emphasize in your teaching, and write a sentence or two about each of them. Include the time (in days or weeks) that you think would be required to teach each area.

16. Prepare a series of overhead projection transparencies or charts that show the growth of specific areas of biology. A large spiral or inverted pyramid might serve as a geometrical shape on which to base your drawing.
17. Begin a collection of important lectures that have been given by well-known scientists, dealing with the changing emphasis on certain aspects of biology. One important source of reprints of such lectures is the National Science Teacher's Association, 1201 16th Street Northwest, Washington, D.C. 20036.
18. List 5 research projects that could involve the joint efforts of a biologist, chemist, and physicist. Include an explanation of the responsibility each would have in the project.
19. Write a paragraph comparing white light to biology.
20. Take a series of snapshots to illustrate how research and understanding of the life sciences has influenced our daily life. Some suggested pictures include frying a hamburger, shooting a pheasant, pumping gasoline into a car, mowing a lawn, and spraying apple trees for insects.
21. Make a collection of clippings from current newsstand sources that indicates the extent to which the popular press writes about the new developments in biological science.
22. Select one of the following topics and compile a list of references about it, using the *Reader's Guide to Periodical Literature, Biological Abstracts, The Quarterly Review of Biology,* or the *Science Citation Index:* biological clocks, ATP, structure and function of chlorophyll, energy-carrying compounds, or animal behavior.
23. List early discoveries, in addition to the discovery of nucleic acids in the 1860's, that lay unused for scores of years before they took on special significance.
24. The Urey-Miller experiment of the early 1950's was recently repeated by a high school student at a science fair. Work out a procedure that could be used by one of your students for constructing and performing a similar version of this experiment.
25. What is the difference between organismic biology and functional biology?

EXPANDING THE CHAPTER

APPLEBAUM, CARL; QUITTNER, GEORGE; and WEITZ, SHIRLEY. 1963. Twenty-Five Years in Biology. *Journal of Biology,* Bronx High School of Science, pp. 11–13. This article is an excellent review of the discoveries in biology over the last 25 years.

CAMPBELL, L. 1963. National Academy of Sciences: 100th Anniversary Program. *Science* 142(592):561–63. This article contains a summary of the important developments in the life sciences in the past 100 years.

GROBSTEIN, CLIFFORD. 1964. Background Thoughts on Curriculum Planning in Biology. *Bioscience* 14(9):29–33. Some suggestions are made in this article about the course content of a modern biology program.

HEISS, E. D., OBOURN, E. S., and HOFFMAN, C. W. 1950. *Modern Science Teaching.* The Macmillan Co., New York. A book of teaching methods for the physical and natural sciences.

HURD, PAUL D. 1963. Science Teaching for a Changing World. *Scott, Foresman Monograph on Education.* Scott, Foresman and Co., Chicago. This publication presents an interesting discussion of the impact of the new biology.

JEVONS, F. R. 1964. *The Biochemical Approach to Life.* Basic Books, Inc., New York. A book that contains a comprehensive discussion of the present status of biochemistry.

KLINGE, PAUL. 1964. In My Opinion: A Good Teacher. *The American Biology Teacher* 26(8):564. Some criteria that determine the qualities of a good biology teacher.

KORN, JOSEPH. 1964. Research and Responsibility. *Journal of Biology,* Bronx High School of Science, p. 4. An editorial on the use of good judgment by scientists who give the results of their research to the general public.

MIEL, A. 1961. *Creativity in Teaching.* Wadsworth Publishing Co., Belmont, Calif. A collection of essays by outstanding educators on the use of creative techniques in teaching.

OSBORN, HENRY F. 1927. *Creative Education.* Charles Scribner's Sons, New York. An older book that contains some timeless suggestions for good teaching.

PORTMANN, A. 1964. *New Paths in Biology.* Harper & Row, Publishers, New York. A uniquely written explanation of the change some of the new discoveries in biology are making on the world.

ROSEN, ROBERTA. 1962. An Interview With Dr. Alfred Mirsky. *Journal of Biology,* Bronx High School of Science, pp. 12–13. An article presenting a discussion of a well-known biologist's contribution to science.

SANDS, LESTER B. 1949. *An Introduction to Teaching in Secondary Schools.* Harper and Brothers, New York. A book containing many time-tested practices that are still useful in science teaching.

SIMON, H. W. 1938. *Preface to Teaching.* Oxford University Press, New York. This author provided insight into science teaching that is still practical today.

SIU, RALPH G. H. 1964. The Phoenix. *Bioscience* 14(9):34–39. This satire on biology was written from the historical viewpoint.

WILES, KIMBALL. 1952. *Teaching for Better Schools.* Prentice-Hall, Inc., Englewood Cliffs, N.J. In this book emerging trends in science teaching are discussed.

CHAPTER **2**

Teaching Form

and Function

Together

ONE OF THE BASIC yet difficult problems facing the biology teacher is how to mesh the teaching of structure and function.

Anatomy and morphology—those areas dealing with form—can easily become studies in terminology. One high school biology teacher felt that having students memorize the 208 bones of the body was good biology teaching. He made no attempt to correlate a knowledge of human anatomy with the function of bones, bone structure, and joints, nor to compare bones in the human skeleton with the bones of other mammals. No evolutionary significance was given to the study of bones or their relationship to each other.

Anatomy and morphology are not the only subjects susceptible to term-teaching. Physiology and other areas also have their share of special words. To understand why form and function are sometimes taught as "fact-oriented" sciences we will briefly examine their background.

A HERITAGE OF TERMS

Though it is difficult to say who the first biologist was, we do know that biology got an early start. Among its early workers—or at least among its *thinkers*—

Aristotle is frequently mentioned. The questions he asked were not questions about physiology, biochemistry, or taxonomy. He wondered about some of the *structures* he observed. Even without knowing anything about cells he helped clarify the distinction between tissues and body organs. A passage from his translated works[1] illustrates his remarkable insight.

Some parts of animals are simple, and these can be divided into like parts, as flesh into pieces of flesh; others are compound, and cannot be divided into like parts, as the hand cannot be divided into hands, nor the face into faces. All the compound parts also are made up of simple parts— the hand, for example, of flesh and sinew and bone.

While this shows the unusual clarity with which Aristotle understood anatomical relationships, apparently he was far ahead of his time. During the succeeding centuries few others contributed to our understanding of life, although two important treatises were written. Theophrastus described plant structure, and Galen investigated human anatomy. But understanding the form and function of cells had to wait until Robert Hooke began writing about them over 1,800 years later.

Even Hooke did not foresee what tiny functioning units a living cell contained. The functions of organelles like ribosomes, mitochondria, and lipid-protein membrane systems were not clarified until the invention of the electron microscope in the 1930's.

Not much happened during the first 15 centuries A.D. to move biology ahead of the other sciences. In fact, during that time science itself was not clearly

---

[1] *Historia Animalium,* The works of Aristotle translated into English (Vol. 4), Oxford University Press, Inc., New York, 1910.

defined. However, in the period from 1600 to 1900 many of the sciences seemed to bloom at once. Biology was among them. New facts, discovered by dozens of workers, became available to scientific thinkers all over the world. Of course, advances in communication were helpful. With the invention of movable type, the publication of scientific research speeded up a great deal.

It was during this era of rapidly expanding research that the great strides in the structural studies of biology were made. Men like Vesalius, Harvey, Van Leeuwenhoek, Schleiden, and Schwann were doing the original research of their time—on the *structure* of plants and animals. This kind of work was extremely valuable for those who came later—for men like Hans Krebs, James Watson, F. H. C. Crick, Linus Pauling, and Albert Lehninger. Early work laid the foundations, through the study of structure, for later work. And the later work has been largely a study of function.

During the 1800's the great biological anatomists did much of the scientific writing. Their kind of biology, called *descriptive biology,* set the tone for much of the teaching. As an example of the detailed description that grew out of structure-oriented research, look at the following excerpt from a German text written by Richard Hertwig,[2] one of the foremost teachers of zoology in the late 1800's.

Structure of the Nucleus. In its minute structure the nucleus affords a wonderful variety of pictures varying according to the objects chosen, but which are not sufficiently understood to permit of a single description accepted by all. According to their reactions to stains two substances in

particular are distinguished: chromatin or nuclein, which is easily stained by certain staining-fluids (carmine, haematoxylin, saffranin), and the achromatin or linin, which stains not at all or only under special conditions.

The achromatin forms a network or reticulum (according to another view a honeycomb structure) filled with a nuclear fluid, bounded externally by a nuclear membrane, easily isolated in large nuclei. If little nuclear fluid be present, and the reticulum consequently be coarse-meshed, the nucleus seems compact. If the fluid be abundant, the nucleus appears vesicular. This is especially the case when the lines of the framework are separated by considerable amounts of nuclear fluid.

The chromatin enters into close relations with a less stainable substance, the plastin or paranuclein (also sharply distinct from achromatin). In the nuclei of Protozoa plastin and chromatin are usually intimately united, the first forming a substratum in which the latter is embedded. The united substances are most frequently closely and regularly distributed as fine granules on the reticulum, so that the entire nucleus appears uniformly chromatic. More rarely the mixture collects into one or more special bodies, the chromatic nucleoli. The nucleolus is ordinarily a rounded body, more rarely branched.

Until recently the emphasis in biology was on structure or anatomy. This detailed description of structure, in which every visible part was named, left a backlog of terms that prevailed in biology texts well into the 1960's. It has caused teachers of contemporary biology to take a fresh look at the pressures put on students who must learn this kind of science. Emphasis on accurate terms, while essential to the researcher, has caused concern among educators. Studies of structure and the memorization of lists of biological terms have come in for criticism. In a paper presented at the 36th Annual Meeting of the National Association for Research in Science

[2] *A Manual of Zoology,* 2nd American Edition from the 5th German Edition, Henry Holt and Co., 1905.

21

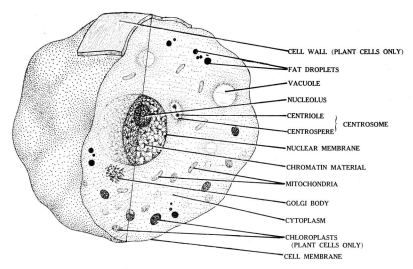

CELL WALL (PLANT CELLS ONLY)
FAT DROPLETS
VACUOLE
NUCLEOLUS
CENTRIOLE
CENTROSPERE } CENTROSOME
NUCLEAR MEMBRANE
CHROMATIN MATERIAL
MITOCHONDRIA
GOLGI BODY
CYTOPLASM
CHLOROPLASTS (PLANT CELLS ONLY)
CELL MEMBRANE

Fig. 2.1—Compare this drawing of a "typical cell" with the present concept of a cell in Figure 2.2. Which structures have been discovered in the past 30 years?

Teaching,[3] David P. Butts of the University of Texas summarized a popular viewpoint among science educators today:

A major goal of science instruction is the development of conceptual understandings. This objective is not a goal of acquisition of correct answers . . . an answer which can be found in a book, but to attain understanding of the relationship which connects the answer to the problem. . . . There is much support in the current literature for the view that direct experience with the raw phenomena is the necessary and probably sufficient condition for data to be perceived and for perceptions to be welded into concepts.

Biology teaching is departing from a descriptive study and from the emphasis on physical shapes, sizes, colors, and kinds of plants and animals.

WHY HAS IT CHANGED?

Part of the pressure has come from teachers and scientists. The recent cur-

riculum revisions in physics, mathematics, biology, and earth science were carried on by teams of high school and college teachers.

Technological advances have contributed to the change. Development of more precise measuring instruments, machines to extend man's senses to the microscopic realm, and means of making accurate analyses of the chemicals of life and life processes have increased our understanding of biology. This increase is beginning to show in teaching. Pictures and diagrams of electron microscopes, molecules, X-ray spectrographs of biochemicals, and quotations from the research work of important scientists are a part of our teaching materials now.

A third factor has been the increased preparation of the students. Much of "high school science" has been put into the junior high and elementary school programs. The results are not always satisfactory, but an overall trend can be recognized. Students are entering high school with some knowledge of atoms, molecules, photosynthesis, respiration, phylogenetic relationships, and

[3] "The Degree to Which Children Conceptualize From Science Experience," *Journal of Research in Science Teaching*, Vol. I, Issue 2, 1963.

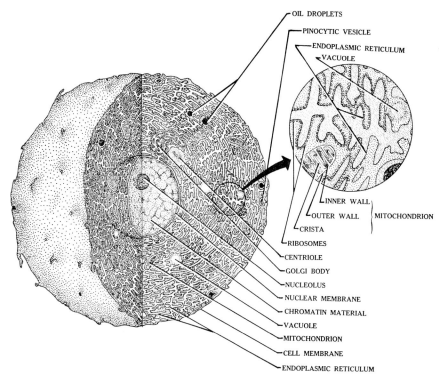

OIL DROPLETS
PINOCYTIC VESICLE
ENDOPLASMIC RETICULUM
VACUOLE
INNER WALL
OUTER WALL — MITOCHONDRION
CRISTA
RIBOSOMES
CENTRIOLE
GOLGI BODY
NUCLEOLUS
NUCLEAR MEMBRANE
CHROMATIN MATERIAL
VACUOLE
MITOCHONDRION
CELL MEMBRANE
ENDOPLASMIC RETICULUM

Fig. 2.2—Compared to the cell in Figure 2.1, this modern picture looks like something entirely different. Much of what this picture shows is the result of work with the electron microscope.

taxonomy. Sometimes their knowledge is incomplete and superficial, but sparks of interest are often created in this way.

FUNCTIONAL BIOLOGY

It is not fair to blame the early biologists for emphasizing a study of structure. After all, our present understanding of the complex biochemical and physiological mechanisms would not have been possible 100 years ago. Apparatus for doing the work had not been invented then. The sister sciences of chemistry and physics, which complement the biological sciences so closely, were still infants. In fact, it was less than 70 years ago that the proton and electron were discovered. The use of radiotracers for elucidating biochemical pathways, and many of the techniques

for tissue culture and experiments *in vitro,* were not to come until the 1920's and 1930's.

In the following paragraph from a modern zoology text,[4] compare the emphasis on function of the nucleus with Hertwig's emphasis on structure, cited earlier.

If the nucleus, by virtue of its genes, is the control center of cellular activities, then the cytoplasm is the executive center. In it the directives of the nucleus are carried out. But it should be emphasized at once that such a functional distinction between nucleus and cytoplasm should not be taken too rigorously. Although the nucleus primarily controls, it also executes many directives of the cytoplasm; and although the cytoplasm primarily executes, it also

---

[4] Paul B. Weisz, *The Science of Zoology,* McGraw-Hill, Inc., New York, 1966.

influences many nuclear processes. A vital reciprocal interdependence binds nucleus and cytoplasm, and experiment has repeatedly shown that one cannot long survive without the other.

Weisz describes each organelle in detail, outlining both its structure as we know it through electron microscope studies, and its function as we know it through modern methods in biochemistry.

As you compare the example from Hertwig's text to the example we have just quoted, you can see how biology teaching has changed. Structures are no longer used exclusively for description. They are used in conjunction with explanations of function. In fact, knowledge of structure clarifies function. Unfortunately, because the function of the nucleus was hardly known 100 years ago, a descriptive, structural approach was about all any biologist could take in either teaching or research.

A QUESTION OF IMPORTANCE

Few viewpoints in teaching are either black or white. Most of them fall into a middle, gray area. The important question of *relative emphasis on structure and function* in teaching is in the gray area.

It is only when we come to look at how an animal works that we begin to understand the meaning of its structure; the long neck of a giraffe is a rather meaningless thing until its function is known, browsing the leaves off tallish trees, say acacias. The tallness of the acacia in turn is also meaningless if one does not understand that the function of the leaves demands that they be at the top to absorb the energy of sunlight.

The structural quality of tallness in acacias has evolved because of its functional advantages, with less shadowing by other trees and hence more hours of sunlight, and less loss of leaves by the browsing of animals. In both the animal and the plant,

function points the meaning to structure. There is thus a basic borderline field between structure or anatomy and function or physiology—usually called functional anatomy, though it could equally well be called structural physiology.[5]

As Dr. Hocking points out in this example, structure and function are not separable. They must be considered together. This summarizes the viewpoint of biologists today. They realize that a knowledge of structure without its functions is largely meaningless. Investigation of the function of an organism or a part of it must be carried out concurrently with an investigation of structure.

Form and function can be interwoven at all levels, from molecule to man.

STRUCTURE AND FUNCTION IN THE HIERARCHY OF SIZE

The Molecule and Man

For many years no one knew the functions of chromosomes and genes. Walther Flemming, who coined the word *chromatin* in the 1880's after he stained the nucleus, was one of the first to observe mitosis. But what puzzled him was this: after division, each daughter cell had as much chromatin as the original cell. He carefully reported the structures in *Cell Substance, Nucleus, and Cell Division* in 1882. He had an inkling of the function of chromatin, but neither he nor anyone else at that time was able to furnish details about its influence over the cell's activities.

Other men in genetics made important, long-standing contributions to our understanding of mitosis—Mendel and his peas, De Vries and his primroses, and Morgan and his fruit flies—all added pieces to the puzzle of genetics. But the questions still remained: What role did

[5] Brian Hocking, *Biology—or Oblivion*, Schenkman Publishing Co., Cambridge, Mass., 1965. ($2.95 cloth; $1.25 paper.)

genes play? What were they made of? How did they work?

The answers are recent history. An understanding of the structure of DNA furnished biochemists and geneticists with the mechanics of its function. For, even on the molecular level, structure and function are inextricably bound together. Trying to separate one from the other makes them both less meaningful.

THE MEANING OF MOLECULAR STRUCTURE

Several features of the DNA molecule make it a good example of the interdependence of structure and function on the molecular level.

First, the doubling of the chromatin that Walther Flemming observed is thought to be accomplished by an unzipping action of the DNA molecule. Looking like a zipper wound around a broomstick, we think this double helix comes apart. The sides of the helix, with their cross links broken, are exposed to attract nucleotides from the nucleoplasm and surrounding cytoplasm. New nucleotides are believed to become attached to each half of the original double helix. Electrostatic forces play a part in the attachment process as well as in the formation of new bonds through the loss of water molecules from participating compounds. But aside from the mechanics of DNA replication, more basic questions remain to be answered.

What determines the sequence of the nucleotides? How do the purines and pyrimidines line up in the correct sequence?

Chemists tell us that a purine cannot bond to another purine; it requires a pyrimidine. And not any pyrimidine will do. The purine *adenine*, for example, will join to the pyrimidine *thymine*, but not to its sister pyrimidine cytosine. The purine-pyrimidine bases sequence themselves as a result of their chemical attraction based on their structure, and the sequence is the same as

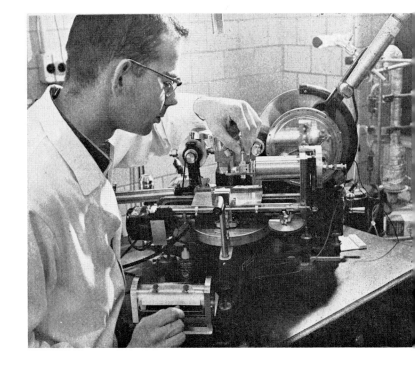

Fig. 2.3—After working out the structure of nucleic acids, scientists were able to predict how the acids functioned. This X-ray diffraction apparatus helped biochemists interpret the structure-function relationships.

that in the original double helix of DNA.

How does the double helix stay twisted together? Again we turn to the chemists for an explanation of function based on form. Chemists tell us that hydrogen bonds—those weak attractive forces of the positive proton nuclei of hydrogen atoms for negative electron clusters—are responsible. Hydrogen bonds form between purines and pyrimidines. Since there is a purine or pyrimidine in each nucleotide, hydrogen bonds keep the nucleotides together, forming "rungs" in the ladder of DNA.

The double helix structure helps explain another function of DNA. This is the change in density of visible chromatin between interphase and prophase. Imagine a helical spring stretched out very long. This represents DNA in interphase. Of course, there is more than one long strand. There are many strands on each chromosome, and a number of chromosomes in each nucleus. (Each string of DNA is thought to be a single gene, but there are thousands of genes on each chromosome.) So the stretched strands of DNA piled on top of each other resemble a network. Now imagine each strand shrinking in length. It gets thicker as it gets shorter; it is more distinct, and easier to see. In this condition chromatin becomes chromosomes.

The extent to which you correlate structure with function on the molecular level is in part dependent on how much chemistry you can meaningfully teach to your class. Unless you can explain how purines differ from pyrimidines it will be difficult to do an effective job in teaching about hydrogen bonding. And unless you show the generalized structure of deoxyribose or a phosphate group, along with one of the bases, students will have difficulty in learning how nucleotides hook together.

We do not mean structural formulas

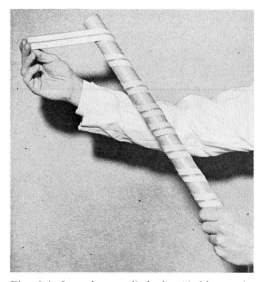

Fig. 2.4—Sometimes called the "ladder molecule," or "zipper molecule," the structure of DNA can be visualized with models. This long zipper wound around a cylinder simulates the structure of a nucleic acid. In what ways does DNA differ from this model?

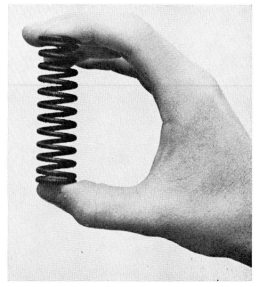

Fig. 2.5—Like this helical spring, DNA is thought to wind around a "core" of constant diameter. Sometimes DNA is described to resemble a "spiral staircase." What is the difference between a helix and a spiral?

of complicated organic molecules should be memorized. Stylized drawings or symbols will serve the purpose. What *is* necessary is the explanation of why DNA functions the way it does—in terms of its structure.

THE STRUCTURE AND FUNCTION OF ENZYMES

Using some of the most complex molecules known to man—the enzymes—you can illustrate another case of unity of form and function.

No one has seen an enzyme molecule or watched it work, but we have been able to contrive a model for its action. The model helps us understand enzyme action and predict results.

We *do* know that enzymes are complex proteins. They are molecules with weights of 100,000 or more. They are twisted and chainlike, with knobby functional groups sticking out from the central "backbone" of the molecule. These facts about enzyme structure help explain function.

The molecules on which enzymes act—the substrate molecules—are also complex. Usually, however, the substrate molecules are smaller than the enzymes, but both substrate and enzyme seem to have shapes that complement each other. That is, where a functional group of the enzyme juts out, the substrate will have a cavity; and where the substrate structure has a bump, there will be a depression in the enzyme. This jigsaw puzzle effect of the enzyme-substrate complex is important in the function of enzymes. It has been compared to a vise holding a chain.

Suppose you wished to cut a chain in two. By clamping the chain in a vise you could hold it firmly for cutting. Enzymes act like vises, holding large starch polymers, lipids, or proteins in fixed po-

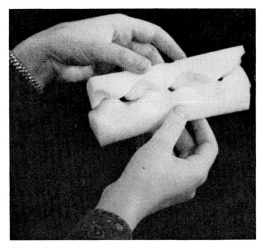

Fig. 2.6—Enzyme molecules are among the most complex structures. To help visualize how their structure is related to function, think of a peak-and-valley surface. The enzyme "peaks" fit into the "valleys" of the substrate molecule.

sitions for hydrolysis. Enzymes also act as vises when they hold small molecules for joining during the process of dehydration synthesis.

THE NEXT LEVEL

Molecules are organized into cells. On the cellular level, structure and function are uniquely bound in processes that make protoplasm more than the "jelly-like substance of life."

Take the chloroplast, for example. Albert L. Lehninger, a noted physiologist, compares a chloroplast to a battery. He bases his comparison on the structure of the stroma and grana—the sandwiched layers that make up chloroplasts. In a *Scientific American* article[6] he emphasizes how a close look at the structure of a chloroplast has helped our understanding of its function.

As the highly organized and sequential nature of the photosynthetic process sug-

[6] "How Cells Transform Energy," Vol. 205, No. 3. September 1961.

gests, the chlorophyll molecules are not randomly situated or merely suspended in solution inside the chloroplasts. On the contrary, the chlorophyll is arranged in orderly structures . . . called grana, and the grana in turn are separated from one another by a network of fibers or membranes. Within the grana the flat chlorophyll molecules are stacked in piles. The chlorophyll molecules can therefore be looked on as the single plates of a battery, several plates being organized as in an electric cell, and several cells in a battery, represented by the chloroplast.

Muscle cells also illustrate the dependence of function on structure. Starting with a description of muscle cell structure, John Pfeiffer[7] says:

There is still another set of cables in the muscle cell. Packed inside each fibril are hundreds of filaments, the smallest component of muscle. The filaments come in two sizes, a thick form and a thin one. When a cross section of a fibril is magnified about 200,000 times by the electron microscope, these superfine filaments are seen to be arranged in a geometric pattern in which thick and thin filaments alternate.

Research thus reveals striated muscle as a system of fibers, each of which is a bundle of fibrils, with each fibril being a bundle of filaments. These structural details are meaningful, however, only insofar as they explain how a muscle works.

Make careful note of Mr. Pfeiffer's last sentence—". . . structural details are meaningful . . . only insofar as they explain how a muscle works." His interlocking of structure and function is typical of the dual emphasis scientists today place on the importance of these two basic aspects of biology.

Then he makes use of another effective teaching device, the analogy. To clarify the appearance of a biological unit, like a cell or tissue or entire organism, analogies are often used. They

[7] Reproduced from *The Cell,* Time-Life Books, Copyright 1964, Time Inc.

are particularly effective when it comes to explaining function in understandable terms.

As a muscle fibril contracts, the filaments do not seem to become shorter. This suggests that the thin filaments may be arranged so that they slide between the thick filaments. This theory might be illustrated by imagining two wooden disks studded with projecting needles. As the disks are brought closer together, the two strands of needles mesh and slide past each other, and the space between the disks decreases without any contraction of the needles. The limit of contraction is reached when the needles studded in each disk touch the opposite disk.

Continuing with an analogy in which nerve-muscle activity is compared to the work that electricity can do, John Pfeiffer explains how energy causes muscle contraction.

Any description of the mechanics of muscle action would be incomplete if it did not account for the power supply which enables a muscle to contract. Muscle is no exception to the rule that doing work requires energy. The generation of muscle energy, like that produced by an automobile engine, calls on electrical and chemical forces.

The "firing" of a muscle fiber—the contraction, or twitch, which occurs when the muscle cell is properly stimulated—begins at a point where the associated motor nerve is joined to the outer membrane of the muscle fiber. The resting muscle cell, which is charged with energy, goes into action when this nerve delivers a signal which "tells" it to contract. The transmission of this message involves two sequences of electrical activity and an intervening chemical event.

The action begins with a series of electrical pulses flashing along the various nerve fibers from the brain to the nerve-and-muscle junction on the muscle-fiber membrane. When the pulses reach this junction they initiate a chemical reaction which releases a squirt of a substance known as a neurohumor. The neurohumor somehow changes the properties of the muscle cell's outer membrane so that the cell releases its pent-up electrical charge. This electrical discharge spreads over the surface of the cell.

With this discharge the muscle fiber is almost ready to contract. The second sequence of electrical activity then occurs. An action message is transmitted from the surface of the muscle fiber to the innermost contracting elements, the fibrils. This takes several thousandths of a second. When this message is received, the fibrils contract. The total time for a single muscle-cell twitch—for stimulation, contraction and subsequent relaxation—is about a tenth of a second.

Structure and Function in Trees

"The stems of trees," writes the biologist Howard J. Dittmer, "are beautiful examples of the amount of growth that can be accomplished by an apical bud and an active cambium." He describes a *Sequoia gigantia,* called the General Sherman, which is 272 feet tall. Although a fire destroyed the stem apex years ago, the trunk, 180 feet above the ground, is large enough in diameter to support an automobile.

Through natural selection, this species—the largest of any living thing on earth—has inherited genetic combinations with survival value. Long roots extend far out into the soil. Thousands of meristems on its branches become active each spring, forming the leaves that furnish food. As the twigs elongate each season, more mileage is added to the tree's plumbing system of xylem and phloem. It is in the xylem that an especially striking example may be found showing complementarity of structure and function.

Xylem vessels, the tiny open-ended pipes stacked end to end, conduct water up. They vary in shape and diameter from one tree to another, but they all have one function in common—they furnish water for the important process of

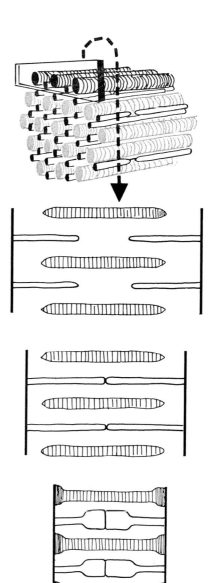

Fig. 2.7—Scientists are not exactly sure how muscle fibrils contract, but the important structures that make up muscle have been identified down to the molecular level. One hypothesis states that the actin and myosin layers slide over one another, causing contraction of the fibril. (From Trump and Fagle, *Design for Life,* Holt, Rinehart and Winston, Inc., New York, 1963.)

30

transpiration, the name given to the loss of water vapor from plants' leaves.

If you study xylem vessels carefully with a microscope, unique details in the design of the vessels show up with remarkable clarity. One of the most striking structural features is in the vessel wall itself. As indicated in Figure 2.8, the wall is not smooth. It is thickened by plates, bands, or patches of crystallized cellulose. These *secondary thickenings* are laid down on the primary cell wall during the differentiation of each xylem vessel.

For a long time the function of

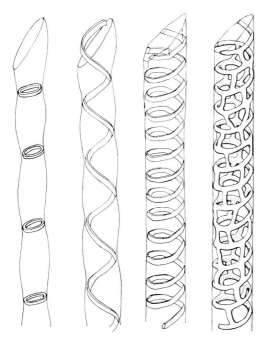

Fig. 2.8—These thickened bands in the walls of xylem vessels withstand the "suction force" of transpiration. They are often compared to steel coils inside vacuum cleaner hoses, which also serve the function of preventing collapse. (From Figure 7–7 from *Biology,* Third Edition, by W. H. Johnson, R. A. Laubengayer, L. E. DeLanney, and T. A. Cole. Copyright © 1956, 1961, 1966 by Holt, Rinehart and Winston, Inc. Reproduced by permission of Holt, Rinehart and Winston, Inc.)

these secondary thickenings in the xylem was not known, but a German botanist, investigating the rise of water in trees, stumbled onto an explanation. Peter Farb[8] describes the experiment this way:

He chopped down a 70-foot oak, immersed it in a vat of acid, to kill the living cells, then dipped the trunk in water—which nevertheless continued to move to the top of the tree.

The explanation would seem to lie in the nature of water itself. Water molecules tend to stick together; they have what scientists call great cohesiveness. If, under certain conditions, water is enclosed in a thin tube, it remains as a solid column, resisting even the immense pull of 5,000 pounds per square inch. Plant sap is not quite so cohesive, yet it has a tensile strength which could—in theory—pull a column of sap as a continuous stream to the top of a tree 6,500 feet high. But what force pulls the sap up? Partly it is the upward pressure exerted by the water absorbed by the roots—but mostly the answer lies in pull from the top by the leaves.

Pull, not push, was the answer. But how could the terrific forces of transpiration pull wirelike strings of water and sap to such heights? Would not the conducting tubes—the xylem vessels—collapse?

The answer to that question lies in a study of xylem structure. Remember the thickenings we mentioned? Like the steel coil in a vacuum cleaner hose, the secondary walls in vessels keep the cells from collapsing. The function of vessels in transpiration depends on their structure. Without reinforcement, the cells would collapse like soda straws through which you try to drink a too-thick milkshake.

The counterparts of xylem vessels are phloem sieve tubes, which usually conduct watery fluids down the plant

[8] Reproduced from *The Forest,* Time-Life Books, Copyright 1961, Time Inc.

stem. Although phloem cells are involved in transport, they have a different structure than xylem vessels. Sieve tubes have no peculiar thickenings in their walls. This suggests that there is little or no suction-force inside sieve tubes. Experiments with phloem cells bear out this hypothesis. Phloem cells apparently must bear no exceptional pressures, either internally or externally.

A Time and Place for Teleology

Interplay of form and function is exceptionally clear in examples like xylem vessels. Now let us say a word about how the vessels acquired their ability to withstand high pressure.

Confronted with data about transpiration, students can be led to draw the correct conclusions about the process. They can understand that without thick walls the xylem vessels probably could not stand the forces involved. But do students understand how the xylem vessels got the thick walls in the first place? This is where teleology comes in. It is the false reasoning process that is sometimes used to explain why plants and animals function the way they do.

For example, why do plants bend toward the light? Scientists tell us that auxins are involved. These growth substances are inhibited or destroyed by light. On the light side of a stem, then, auxins are not as active as they are on the side away from the light. On the darker side auxins function normally and speed the growth of cells. Auxin-influenced cells grow faster on the dark side, causing the stem to bend. It bends toward the light, pushed in that direction by the uneven growth.

Teleology and similar false reasoning, called *anthropomorphism,* both ascribe supernatural causes to some of the activities of plants and animals. Plants

Fig. 2.9—Sometimes models clarify hard-to-visualize processes. This instructor is using an oil can to represent a plant that bends toward the light. Using this example he can explain the role of auxins in plant growth.

do not bend toward the light because they *want* the light, for plants cannot sense their wants and then fulfill them by conscious effort.

If you say that xylem vessels are reinforced because they *need* thickened walls to withstand the pressure, you are offering a teleological explanation. What, then, is the answer?

The answer lies in an understanding of genetics, evolution, and natural selection. Through genetic change, certain cells receive an arrangement of chromosomes different from the ordinary. These new arrangements, called mutations, make the cells look and function differently. Usually, when mutations change a cell, the cell becomes less fit for survival. It dies. In natural selection the less fit generally die, and the better fit survive. Mutations that result in thickened xylem walls equip the cells to withstand their environment.

Genetics, evolution, and natural selection give explanations for the existence of structures like thickened xylem walls. Teleology does not. In biology, and in all science, there is little room for teleological explanations.

Other Levels of Sophistication

We have discussed form and function on the molecular, cellular, and tissue levels. But examples equally as valid could be chosen from the organ, organism, species, or community levels as well. From an insect's compound eyes to the fringed, plankton-straining plates of the baleen whale, form and function are most meaningful when considered together.

TEACHING FORM WITH FUNCTION

Some teachers seem to feel that physiological processes can be described, and experiments and exercises can be performed, without much mention of the structures pertinent to the processes. On the other hand, some teachers feel that straightforward description of the physical properties of molecules, cells, tissues, and organs constitutes sufficient "biology," without correlating them with function.

In a high school biology unit worked out by the Biological Sciences Curriculum Study the inseparability of structure and function is emphasized. Their *Complementarity of Structure and Function*[9] uses subject material

[9] A. Glenn Richards, BSCS Laboratory Block, D. C. Heath and Co., Indianapolis, 1963. (Used by permission of the Biological Sciences Curriculum Study.)

from the physical sciences as well as from the biological sciences. Students study how levers work. They examine muscles, bones, cilia, and the process of cyclosis in cells.

For a correct appreciation of structure we need to think in terms of the function it performs; for a true appreciation of a function we need to consider the structure or structures that are performing the function. Structure complements function. Or function complements structure.

Function is limited by structure. The standard automobile that performs well on a hard, smooth-surfaced road bogs down quickly in a muddy, rutted field. The tractor that is geared to operate successfully over rough terrain cannot compete with an auto on a good road. An earthworm can work its way under the soil with agility, but in traveling over the surface it is outclassed in performance by many other living things.

Function is illuminated by a study of structure. We are going to use our observations of biological design or structure as a basis for understanding the *how* and *why* of performance. We are curious about how things operate, and why. Satisfying this curiosity in the study of biology, the term *functional anatomy* covers the simultaneous consideration of structure and function, which is the study we will undertake in this series of exercises.

Function complements structure. Similar actions can, for instance, be accomplished by highly different structures and by strikingly different forces. With a series of examples with which you will work, we propose to illustrate how and why this is so. Each example and experiment will contribute its part to an understanding of the structure-function relationship. This relationship can become apparent to the unaided eye, as in observing the parts and operation of wings and legs. With a microscope the complementarity becomes apparent in very tiny things such as microscopic organisms and microscopic or even submicroscopic parts of individual cells which are, of course, the building blocks of living matter.

To teach structure-function relationships effectively, choose good examples. Try to do as much laboratory work as possible, using the examples you have chosen or examples that are very similar. And finally, design your tests and evaluations around the concept that structure and function are not at odds with each other, but that they work together.

CREATING CHAPTER CONCEPTS

1. Refer to Hooke's original work and make a list of the descriptive terms he used. How many of these are of value to us today?
2. Consult a text on the history of biology to find out the contributions these men made:
    a. Vesalius
    b. Harvey
    c. Leeuwenhoek
    d. Schleiden
    e. Schwann
3. List some reasons for the change in biology from description of structure to explanation of function.
4. Cite some discoveries in biology that have been made within the last 10 years. Which of these would have been possible without modern technology?
5. Which molecules, in addition to the nucleic acids and enzymes, carry on functions directly related to their structure?
6. Make a model of a myofibril of muscle, using tiny wires. Then make a model of the actin and myosin layers, like those found in striated muscle. Try explaining muscle action to a class (1) without the models, and then (2) with the models. Which presentation is more successful?
7. List some plant structures that perform functions specifically related to their structure.
8. What is the difference between teleology and anthropomorphism?
9. What is the difference between a teleological explanation of phototropism and a scientific explanation of it?
10. Look over the BSCS Laboratory Blocks. Which of them teaches structure-function relationships especially effectively?
11. Prepare a side table demonstration to illustrate adaptive design in bones from various chordates. Use a worksheet of your own design to guide students' thinking when they study the demonstration.
12. Obtain a collection of bone X rays from a hospital. Use them to illustrate relationships of structure and function.
13. Develop a "Photo Essay" display for a bulletin board to show structure-function relationships. Subject matter might include bird nest types, beehives, or types of plant root systems.
14. Write an essay on "The Use of the Micromanipulator as a Tool in Understanding Function and Form on the Microscopic Level."
15. Use straws and spiral notebook binders to explain structure and function in the xylem of plants.

EXPANDING THE CHAPTER

BURNETT, A. L., and EISNER, T. 1964. *Animal Adaptation.* Holt, Rinehart and Winston, Inc., New York. An interesting approach to the study of why animals are able to survive in a variety of places, under a variety of conditions. Peculiar structures are discussed in relation to their functions.

GRIFFIN, DONALD R. 1962. *Animal Structure and Function.* Holt, Rinehart and Winston, Inc., New York. Several of the important body systems of animals are discussed, stressing the relationship between structure and function.

HANSON, EARL D. 1964. *Animal Diversity*. 2nd Ed. Prentice-Hall, Inc., Englewood Cliffs, N.J. A phylogenetic survey of the animal kingdom. This paperback text is a handy reference for an overall understanding of the types of animals there are, and their evolutionary relationships.

KAHN, FRITZ. 1943. *Man in Structure and Function*. 2 Vols. Alfred A. Knopf, Inc., New York. There are some excellent diagrams and photographs in this book, comparing man-made structures to body structures. For example, the cross-bracing of the Eiffel Tower is shown beside a thin section of bone, which is cut to expose the intricate network.

LOEWY, ARIEL G., and SIEKEVITZ, PHILIP. 1963. *Cell Structure and Function*. Holt, Rinehart and Winston, Inc., New York. Structure and function are related on the molecular and cellular level, as this paperback text emphasizes.

MARK, HERMAN F. 1966. *Giant Molecules*. Life Science Library, Time Inc., New York (Silver Burdett Co., Morristown, N.J.). This book emphasizes structure and function on the molecular level, giving an up-to-date account of organic and biochemistry.

MOORE, RUTH. 1964. *Evolution*. Life Nature Library, Time Inc., New York (Silver Burdett Co., Morristown, N.J.). An interesting book about the chemical and biological evolution that has accompanied life on earth. In Chapter 2, "A Theory That Shook the World," the author brings out the role that structure-function relationships played in Darwin's formulation of a theory of evolution.

NOURSE, ALAN E. 1964. *The Body*. Life Science Library, Time Inc., New York (Silver Burdett Co., Morristown, N.J.). A well-illustrated book about the major human systems, their structure, and their functions.

SIMPSON, G. G., and BECK, W. S. 1965. *Life*. 2nd Ed. Harcourt, Brace & World, Inc., New York. This college biology text emphasizes evolution, and the relationship between structures and the functions they perform.

WALLACE, BRUCE; and SRB, ADRIAN. 1964. *Adaptation*. 2nd Ed. Prentice-Hall, Inc., Englewood Cliffs, N.J. Phylogenetic relationships and adaptations of animals are stressed in this book, which is a part of the Foundations of Modern Biology Series.

CHAPTER **3**

Artistry

in the

Classroom

"IN A SENSE," says *Time* magazine,[1] "...computers, films, labs, and TV notwithstanding—nothing much has come along in 2,400 years that essentially improves the Socratic pattern of a learned man plus a group of students, but the pattern can work out in sharply varied and instructive styles." *Time* agrees that good teachers are still the key to successful instruction.

The article describes some hand-picked present-day teachers as "articulate . . . lyrical . . . passionate. . . . The lectures seem to be spontaneous drama." The teacher "seeks 'an electric exchange' with students . . . he conveys a conviction that is exciting, beautiful, profoundly moving. . . ."

Flashing words—words that convey color and excitement and vitality—are used to describe almost all events in our lives *except* those in a classroom. Teachers, it seems, should conduct routine discussions, give the usual, expected lectures, and speak in the erudite, measured English often associated with the teaching profession. Yet, creative teachers are artists. They play roles, invoke feelings, show temperament. They depart from the norm, try new ideas, play hunches, establish unorthodox patterns for the sake of more effective instruction.

Someone has said that creativity begins where know-how leaves off. It is often at this point that one person gives up because he has no answer, while another finds himself improvising, discovering answers, and using skills he did not know he had. So the technician gives way to the artist. Creative experience through teaching may be something we do not even know we want when we choose to try teaching. Most of us, if we have some degree of sensitivity, learn the feel of creativity as we teach. . . . As our insights into the teacher's role deepen we begin to see and feel more satisfactions and pleasures than at one time we had believed possible. If these satisfactions are important enough to us, we begin to make plans—conscious choices—that increase the chances that the satisfactions will occur on other than accidental bases. Being highly sensitive and aware, making value judgments and choices—all are essential ingredients of teaching as an artistic process.[2]

Qualities of successful businessmen are also those of successful teachers. Likeability without condescension, intelligence without domination, friendliness without favoritism—these are a few facets of the personality of a successful person regardless of occupation. Even the small things count. The kind of clothes, the tone of voice, the way of talking—these can add up to a successful or unsuccessful image. Whatever your attitude and behavior, they will have a powerful impact on those you teach.

IMPACT ON PEOPLE

Qualities that make good teachers have become the object of research in social psychology. Professor John D. McNeil reported some of the characteristics of effective teachers in the *Phi Delta Kappan*.[3]

[1] Copyright Time Inc., May 6, 1966, pp. 80–85.

[2] Evelyn Wenzel, in *Creativity in Teaching*, edited by Alice Miel. Copyright 1961 by Wadsworth Publishing Company, Inc., Belmont, California. By permission of the publisher.

[3] March 1958.

The teacher has the complex problem of giving each individual satisfaction as an individual, protecting the group as a whole, and satisfying his own aspirations. To do this, he seeks to evoke the involvement of every student in the activities and in the determination of unit objectives.

This successful teacher tends to be more interested in the students than in subject matter and is not punitive. Generally, he keeps his supervisory role impersonal and allows reproof to be seen as growing from the nature of the situation rather than from himself.

The teacher seeks to encourage and reinforce interpersonal contacts and relations throughout the group structure so as to strengthen it. He seeks to reduce class tension and conflict. In permitting students to formulate ideas and have firsthand experiences, students come to understand and retain the knowledge with which the ideas and the experience deal.

Students say that their ideal teacher "listens carefully to your viewpoint, explains what's wrong and what's right about it, and lets you argue it out instead of flatly telling you you're wrong. The teacher has a fairly good knowledge . . . and is interesting. He has the personality to keep order in the classroom and is not afraid the pupils will dislike him if he does. He accepts a joke but doesn't let it go too far."

It would be helpful for beginning teachers to know what specific qualities will make them successful. Unfortunately it is not easy to pin down the aspects of teaching that make some instructors better than others. Usually it is a combination of several ingredients. When one or two traits are highlighted as "the most important" they lose meaning because of their isolation from others.

Good teaching can also be analyzed from another standpoint. It can be considered as a combination of tangible and intangible components. The intangible aspect—the feeling, attitude, or spirit of good teaching—is very difficult to learn in a methods course. Yet it is this sense of teaching that separates the artist from the nonartist.

By following specific guidelines the new, unpracticed teacher can walk in the footsteps of others who have proved successful. For, while fine teachers have a great deal of talent, a certain amount of talent for teaching exists in everyone. If you make use of key techniques, that talent can be developed and used as one of your most valuable teaching tools.

TOWARD ARTISTIC TEACHING

Whether you choose to lecture to students or conduct classes through question-and-answer discussion, you will have to do a great deal of talking. The talking you do is not the conversational, everyday type. Delivery of lecture material must be organized, arranged in a logical sequence, and presented with the best possible techniques of delivery. The vocabulary you use, your mannerisms, the rise and fall of your voice, and the bearing you present to the audience are aspects of your method of delivery.

Delivery

A great orator, when asked to name the three essentials of public speaking, replied: "First, delivery. Second, delivery. Third, delivery." According to Gilbert Highet,[4] the single most important quality in lecturing is delivery.

Delivery depends on the voice and on gestures. Of these two, the voice is far more important. It must be clear. Not many teachers make the mistake of speaking inaudibly. Yet some talk too fast to follow, while others chew their words, or gobble them, or hiss and splutter them, until the pupils tire of making the effort

---

[4] *The Art of Teaching*, Alfred A. Knopf, Inc., New York, 1950.

37

required to sort out the meaning from the noises, and take their revenge by mocking the noise. If you have a regional accent, or a curiously pitched voice, or a difficulty with any particular letter, watch your pupils' eyes. You will soon learn when they have ceased to understand and are merely listening for you to quack or hoot.

Mannerisms

Some of the most effective, organized, knowledgeable teachers ruin their delivery of subject matter with distracting mannerisms. With some instructors, figures of speech become clichés. Try keeping track of phrases like "In general we find . . . ," "Nevertheless . . . ," "By and large . . . ," or "For the most part. . . ." With students such scorekeeping becomes a game, and the reason they may look attentive is not that they are interested in the biology you are talking about, but in your poor choice of words.

Clichés distract, but so do visual mannerisms. Take the case of the neck-stretcher, for example. This nervous instructor develops the distracting habit of stretching his neck out of his collar every few minutes. For him the response is automatic. For his students it becomes the focal point of attention. Other mannerisms to watch for are eye-blinking, knuckle-cracking, pocket change-rattling, ceiling-watching, and pacing. This last bad habit is characteristic of lecturers who have a fairly large blackboard in the front of their classroom. While talking they make a few notes on the board, then pace to the other end and list some more items. They walk back and forth nervously, transmitting the nervousness to the students.

One form of verbal mannerism lies in the way sentences are punctuated. There are logical pauses in spoken sentences, just the way commas are used to break up printed sentences. These pauses give

emphasis to important phrases, or to words within phrases. In fact, the listener receives valuable clues from the speaker when there are pauses in the speech. Momentary pauses give the listener time to reflect on the significance of the previous statement or to prepare himself for the relevance of one that is forthcoming. Improperly placed pauses can put your audience to sleep, especially when you halt at illogical places:

> "Mr. Chairman. Ladies. And Gentlemen. I have come here. Tonight. At the request. Of my good friend. The member of this Congress. From. The 5th. District. Whom I. Am glad to see. Among us. On this. Occasion."

Hearing delivery like this is like reading under a flashing, intermittent light. It makes you nervous. It distracts. And yet, it is a common fault of teachers. Guard against. It. There is. Only. One fault. Worse than making. Irrational pauses. In your. Sentences.

Punctuating your delivery with "er" and "a" makes pauses even worse. You are marked as an amateur who has not learned the most fundamental principle of public speaking. To avoid this disturbing habit make sure you know your subject well. The use of "er" usually means that the speaker is groping for the proper word to continue his talk. Sometimes he is just plain nervous, but usually he has not prepared himself sufficiently and is confused about the word order and proper choice of language to express his ideas. Plenty of beforehand preparation and a thorough knowledge of subject matter will help forestall this annoying mannerism.

Eloquence and Logic

Thomas Henry Huxley, one of the most profound lecturers of the nineteenth century, was known as a spell-binder. He was eloquent. He could hold the attention of hundreds of students as if he were talking to each one individually. "He gave you in 50 minutes a striking analysis of two or three phenomena in nature which did not seem quite cognate. He glanced at the clock, and in the remaining 10 minutes put them all together, showed their analogies, and left us with a sense that nature was 'not without a plan.' "[5]

Logic can captivate minds and hold attention like no other device, either physical or intellectual. Pure logic, the well-ordered sequence, the finely-honed explanation—these can be delivered with little attention to methods of good speaking. But logic without eloquence is like light without heat. It lacks something. Couple logic with a well-modulated voice, proper inflection, and a gently moving, sincere eye contact, and you have eloquence. These qualities, worked together, make a spellbinder. They will make your students remember your classes and remember what you say. They will help you "sell" biology even to the student who seems uninterested.

Eye Contact

"I once had a teacher," writes a student, "who spent the entire semester teaching chemistry to the top of the lecture table. Now and then he would make a humorous remark to a ring stand, and frequently he would cast a glance toward a friendly Bunsen burner, but he spent most of the time instructing the soapstone surface."

The trouble with this chemistry teacher is that he did not talk *to* his students, he talked *at* them—and the students felt the difference.

[5] Sir W. H. Flower, "Reminiscences of Professor Huxley," *North American Review*, 1895, No. 161, pp. 279–86.

Practice eye contact when you talk, letting your eyes move from one person to another about every ten to fifteen seconds. Eye movement more rapid than this will make you appear nervous; if you maintain eye contact much longer it might make the listener nervous—he may wonder if you are talking exclusively to him.

Voice Modulation and Volume

Voice modulation is the process of making your voice rise and fall in pitch so that it does not become a monotone. This technique of public speaking is like most of the others we have mentioned; it must be practiced to be learned. While it is difficult to describe voice modulation with written words, some excellent examples exist. Turn on your radio and listen to the news. One reason good newscasters can talk for fifteen minutes without sounding monotonous is that they have well-modulated voices. Notice how the voice of a newscaster rises and falls, how the key words are inflected. Listen for the ends of the sentences. They are dropped rather than raised as in asking a question. Learning how to modulate your own voice, then, is like "playing by ear." Try to imitate the speaking style of people who have trained their voices.

One difficulty in practicing these points to become a more effective speaker is that you do not sound the same to others as you sound to yourself. To discover idiosyncrasies in your speaking habits you must hear yourself as others do. This is where a tape recorder comes in.

Make tape recordings frequently, both to practice good speaking habits and to check other aspects of your delivery. When you record your voice "live" in the classroom, conceal the

40

microphone and tape recorder so that students are not distracted by them. Put the microphone behind a few books on your desk, and keep the recorder under the desk or demonstration table. Figure 3.1 illustrates one setup for recording.

When you play back the tape, listen for distracting mannerisms, clichés, voice modulation, inflections of key words in sentences, sentence "punctuation," and voice volume.

This last point—voice volume—often escapes teachers who are otherwise fairly effective speakers. The best way of adjusting your volume is to be able to judge it yourself, and correct it as you talk. But it may be difficult to remember to adjust voice volume while you are trying to remember the proper interval for maintaining eye contact and all of the other details that go into the package called "effective speaking." One sure way, then, is simply to ask the class. If they can hear you clearly in the back row, you apparently are speaking loudly enough. On the other hand, a soft voice is a subtle means of "crowd control." By speaking softly you can make the students strain to hear what you say, and frequently they will become quiet almost without realizing it. The class will become especially quiet if you

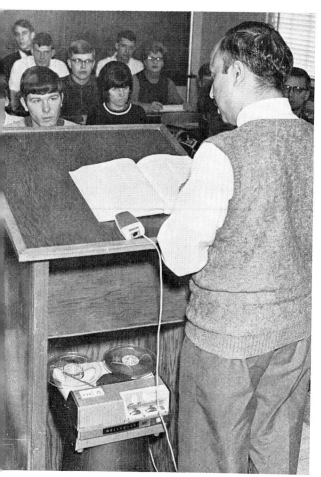

Fig. 3.1—Make a tape recording of your lectures and class discussions. Then play back the tape, analyzing your delivery, voice modulation, and speech mannerisms. Conceal the tape recorder, as this picture shows, so that the students are not aware of it.

begin the hour with questions of a quiz, orally delivered, softly spoken.

THE ART OF CONVERSING

Among other things, Robert Louis Stevenson was somewhat of an expert in human relations. On the ability to draw out the other person and make him feel vital in a conversation Stevenson said, "You start a question and it's like starting a stone from on top of a hill; away the stone goes, starting others."

Nardi Campion[6] writes, "How wonderful when a conversational stone we throw out starts an avalanche of response; when our interest and concern open a wide path to another's personality. How wonderful—and how rare! So often we feel shut out, unable to establish real contact. Yet, by learning the magic power of asking the right question at the right time, we unlock the floodgates of communication."

The art of asking questions is part of a larger art—the art of teaching. People do not like to be told, they like to be asked. Yet as teachers we are constantly telling. When we *do* ask questions they usually call for didactic, factual answers. Rarely do we ask a student to give his impression or his interpretation of biological phenomena. We may ask him to explain something that is in the book or that we may have told him in the first place. Questions like "How do you feel about that?" "Why do you believe that?" or "What reasons can you give to support your answer?" are leading questions. They will draw the student out and require his expression of fresh, original thought. Your analysis of his answers will determine how you structure further questions and how you direct the learning of the class.

_____
[6] "Ask, Don't Tell," *Reader's Digest* (condensed from *Christian Herald*), August 1966, p. 49.

Frequently teachers are faced with the problem of finding a good beginning for class discussion. Or they need a dynamic introduction to a new topic. Here are some suggestions for opening class discussion, or preparing the class for the study of a new topic.

QUOTATIONS: Use quotations from scientists, writers, or famous historical figures. Chapter 8, footnote 9, lists a book of poems in science that might be helpful. Bartlett's book of famous quotations also is an excellent source.

DEMONSTRATIONS: Begin class with an exercise or investigation that had been set up previously. The effect of centrifugal force on plant stems, for example, could be shown to your class. Appendix G 298 lists a number of exercises that could be adapted as demonstrations.

PICTURES: From magazines, books, and newspapers you can get a number of good ideas for starting class discussion. If you collect from these sources during the year and file the material by subject area, you can turn to the picture file for ideas when you want to introduce specific areas of study.

TAPE RECORDINGS: Using recordings from radio and television, or from records, you can build a collection of teaching "openers." Often you can get taped copies of important speeches and interviews by writing the radio station on which you heard the information.

FILMS: You may choose to use only a portion of a film, rather than the whole film. A scene in a motion picture might provide the appropriate introduction to a topic you wish to develop. As a springboard for discussion, portions of motion pictures are often more effective than the entire film.

SLIDES: Color slides are excellent "openers." You can use slides of nature, apparatus, or pictures copied from books and magazines. For example, in a unit on physiology, you might begin with a slide presentation on the kymograph and its uses.

Points on Pronouns

"I want . . ." makes a command of gentle requests. "Would you like to . . ." or "I'd like for you to . . ." softens the effect and is a more tactful way to accomplish what you want. Other pronouns that are often overworked are "you" and "your."

If you and the class will be working through an exercise or demonstration together, you may wish to use "we." But if you are going to grade papers while the students read an assignment in class, you should not make the assignment by saying, "For tomorrow's lesson we will spend the remaining ten minutes studying the text." The students may read, but you will be correcting papers.

Chalk Talk

Of all the audiovisual devices that have been developed over the years, the common blackboard (or modern "chalkboard") is still among the most widely used, and chances are that it will remain high on the list of teaching aids. Good blackboard technique is essential in all teaching. It is especially vital in biology where a number of drawings are made in view of the class, in addition to written material. If you have to make a sketch on the board to clarify a process, or show the flow in apparatus, talk to the students as you draw. Turn around from time to time when you are at the board. When you turn to the class, do not stand with your back to the drawing. Tell the class what you are drawing, then stand aside for them to see it. Many of the students will want to copy the drawing for their notes. The illustrations in Figure 3.2 emphasize important techniques.

The Overhead Projector

If you prefer to face the class as you talk, and like to teach from prepared diagrams and drawings, you will enjoy using the overhead projector. Designed to project images five or six feet square, the "overhead" is being used increasingly as an important instructional tool. Any number of objects can be used on an overhead projector. The common approach is to write or draw on the stage with a felt-tip pen, but you can prepare your own transparencies ahead of time and use them in place of on-the-spot drawings. As the instructor in Figure 3.3 illustrates, it is possible to use plastic embedded specimens on an overhead. In fact, any three-dimensional object that is transparent can be shown to the class on this projector. There are also available good commercially prepared transparencies (see Appendix F 288).

THE OVERHEAD PROJECTOR DO'S AND DON'TS

DO—Put the projector and screen diagonally across the front of the room to provide maximum visibility. Keep the light path perpendicular to the screen. Room lights may be left at normal level.

DON'T—Turn the projector on and leave it on. Use the light only when you want to bring attention to the screen. Even a lighted screen with no image is distracting. Observe how attention goes to the screen when you switch on the light, then back to you when you switch it off.

Fig. 3.2a—This teacher is using poor technique by covering his blackboard work and by talking to the board instead of the class.

Fig. 3.2b—A much more effective board presentation is demonstrated here. The diagram can be seen and the teacher's words heard by all.

DO—Place the transparency squarely on the projector stage before turning the light on.

DON'T—Turn on the light, then fumble around trying to position the transparency.

DON'T—Stand with your hands resting on the lighted stage.

DON'T—Make rapid, distracting gestures on the stage. Be deliberate in pointing out information, then get off. A pencil is a handy pointer for material on the stage, showing up well on the screen.

DO—Use a grease pencil or felt-tip pen to add information or underline key words for emphasis.

DO—If your hand tends to shake when holding the pointing device, touch it to the transparency to steady it.

DO—In general, keep your information toward the top of the stage. That brings it to the top of the screen for optimum visibility.

DO—Use the technique of *revelation*. This is accomplished by covering all but the first point to be discussed. Place a plain sheet of paper over the transparency, then move it down to uncover only what you want to show and discuss at the moment. Many presentations are less effective because the audience reads ahead.

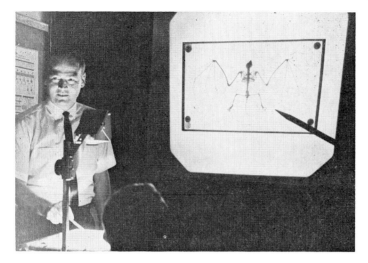

Fig. 3.3—Overhead projectors can be used to show many things, as this picture illustrates. See the list of do's and don'ts that points up important techniques to keep in mind.

DO—Remember that simplicity is important. A transparency can confuse if it is too cluttered.

DON'T—Use color only for the sake of color.

DO—Use color to emphasize concepts or separate ideas. Some simple techniques include: using felt pens for marking the film surface, adding color to lines on the transparency with color grease pencils or color transparent film and tapes, or putting a sheet of colored film over the transparency to give a color background.

DO—Remember that use of the overhead is limited only by your ingenuity and imagination. It is a challenge and a stimulus to put some extra sparkle into a presentation.

DO—Try leaving the overhead in your conference room or classroom and use it whenever possible instead of the blackboard. Then try taking it out after a few weeks and see how much you and the others miss it.

A Word About Words

Dr. Lois DeBakey, a professor of scientific communication at Tulane University, has an unusual job. She is an advisor to writers of technical literature. Part of her task is to help people who write for journals of biology and medicine. Often she follows the advice of Dr.

Cuticle in Herman Melville's *White Jacket*:[7]

A man of true science uses but few hard words, and those only when none other will answer his purpose; whereas the smatterer in science . . . thinks that by mouthing hard words he proves that he understands hard things.

The following list taken in part from Dr. DeBakey's work illustrates how simple substitutes can take the place of words that are unnecessarily "learned."

Hemorrhage	Bleeding
Respire	Breathe
Obese	Fat
Ingest	Eat
Optimal	Best
Utilize	Use
Elucidate	Explain
Substantiate	Prove
Obligate anerobe	Organism that cannot live in free oxygen

We often forget that our students are merely beginners in biology. Many of the words we use have little meaning to the beginning student. The vocabulary we teachers have acquired is the re-

[7] 1849. (Published currently by Holt, Rinehart and Winston, Inc., New York.)

sult of many years of study and reading, while first-year students are getting their first exposure to the technical terms.

In Appendix A 201 are listed some common biological prefixes and suffixes that may be helpful in explaining some biological terms.

Use small words wherever possible. If you *must* choose a scientific term, *use it only if it clarifies* what you are describing. And *use it only if it will be used again* several times. A simple test of the value of a special term is to determine how many times it will appear in future work. If the word is so specialized that it will not be needed at any other point in the biology course, then a substitute

should be found. Your resourcefulness in putting complicated concepts into teachable terms will label you as a creative biology teacher.

DEMONSTRATION DESKS— FROM THE STUDENT'S SIDE

Success or failure of a demonstration is sometimes determined by the planning you put into it, but often even a well-planned one fails as an instructional tool because the "staging" was poorly done. As Figure 3.4 indicates, there are a number of things to keep in mind when you are carrying out a demonstration for your class. The basic

Fig. 3.4—Compare these two pictures. Look for reasons why the upper one illustrates poor demonstration technique. What deficiencies have been corrected in the lower picture?

guideline is: Keep it simple. Keep clutter to a minimum. Arrange the materials in the order you plan to use them so you will not have to search for the proper piece of apparatus when you need it. Remove books, notes, leftover equipment, and all unnecessary apparatus from the demonstration table as you prepare the demonstration.

Keep the student's view unobstructed. Remember, no one can see through a pile of books, flasks and beakers, or a cage.

Talk to the class as you do a demonstration. Tell them what you are doing, explain the function of each piece of glassware and why you are using certain substances. Include the entire class as you speak, remembering the value of eye contact. Avoid talking only to the few students who sit in the front row, or who are closest to the setup.

Stop the demonstration from time to time for questions. Plan places in the process where you can summarize. Pose questions if there are none from the class. Advance a hypothesis that can be proved one way or the other as you continue the demonstration.

Do not be embarrassed by demonstrations that may be interpreted as "wrong." In one sense there are no wrong answers in science. Negative results—those that disprove a hypothesis—are just as valuable as positive results. It is often as valuable to know what is *not* true as it is to know what *is* true.

Students who have difficulty seeing from their assigned seats should be asked to move closer while you perform the demonstration. Not only will they be able to see more clearly what is taking place, but they will feel encouraged to ask questions and enter into class discussion.

THE REALISM OF ARTISTIC TEACHING

When you enter the room of an artistic, successful, creative teacher you can perceive these qualities almost immediately. It is often difficult to put the feeling into words, but you can sense it. There is an air of orderliness in the room, an intangible feeling of well-being. There is an air of anticipation and urgency in the classroom of the effective teacher. Success does not lie in this teacher's youthfulness or maturity, but rather in the knowledge that his students are thinking and learning, and that they will come away from the class better informed than when they entered.

To be completely realistic we must mention another aspect of teaching that often confronts teachers—the ability to deal with unexpected situations. So many "special cases" will arise that it is impossible to list all of them here. Even if we could, you would find it quite meaningless to try to become familiar with all of the exceptions to the rules. Jane Tylor Field[8] puts it this way:

The proper and specific action depends, of course, upon what the incident is; upon the person or persons involved, including you; upon your school administration; and upon your community. But there are certain things it should be helpful to remember.

First, do not let anyone know that you are shook up, either by hysterics, pity, or horror. In the classroom, you are mother, father, counselor, tea and sympathy all in one. You are expected to be a rock, no matter what. So, though you may be blah inside, make like obsidian for your audience.

Second, try to remember any relevant administration policy that may have been

---

[8] "What the Books Don't Tell You," *NEA Journal,* October 1964.

discussed at faculty meetings. If the incident is one of a type for which a formula has been worked out, conform! Do not try to set a precedent with new strategy; for instance, if smoking by students on school property is grounds for suspension, report any offenders immediately. This is no time for a lecture on the possibility of cancer.

Third, if necessary, *ask* for assistance. The nurse, administrators, counselors, and the vast majority of your colleagues are willing to back you up or lend their special skills wherever needed, and teamwork has long been noted for its superiority over lone-wolf efforts.

Along the same line, do not forget to call a parent-teacher conference if it seems wise to do so. After all, parents are people too. Sometimes they are impossible, but for the most part they are like their kids— willing to learn if they think they are hearing the truth.

They may have no idea that their offspring is the class clown or the class black sheep or the class sad sack. Certainly they will seldom be complacent about such revelations. Most parents are only too anxious to help their children be happy and successful, and they know that there is no truer maxim than that success breeds success. Clowning, goofing off, pining away, and like activities use energy which should be put to more constructive use, and parents know it well.

Fourth, do not let a molehill into a mountain grow. Keep things quiet. Don't let everybody get into the act. Don't even let it become an act.

Fifth, be objective. Do not take personally anything that may happen even if it is of an insulting or belittling nature. Remember that no matter how charming or marvelous a person you are, as a teacher you represent authority, and there are times when authority is resented for its own sake.

Sixth, if the event has been one where you were fortunate enough to make some child's life a little happier or more comfortable, do not dwell on that either. Which is to say, don't talk about it. Be discreet. Don't tell everything you know. Just remember the channels by which such results were accomplished, in case you might want to go through them again sometime.

Finally, remember that there will be friction and irritation and pathos and laughter in any job that deals with people, and teaching is certainly one that does. Therefore new little human problems will ever arise and old ones will be repeated. Regard them with humor if possible, profit by past experience, refrain from wasting physical or nervous energy on them, and conserve most of your strength for your biggest job, that of helping students to learn.

Special techniques used by teachers vary, but basically they have the same goal. They wish to stimulate the interest of the student and then develop that interest in a field like biology. Students differ as much in their learning patterns as teachers do in their teaching patterns; instruction must be tailored to individual learning styles.

CREATING CHAPTER CONCEPTS

1. Make a tape recording of a biology teacher in a classroom as he conducts his class. Evaluate his teaching ability and effectiveness as you play back the tape. Which aspects of his teaching are not noticeable when you use the tape alone? Which aspects become *more* noticeable when you can only listen, and not see him?

2. Design a demonstration desk that would provide maximum effectiveness to the students. Consider lighting, water, gas, and air outlets in your design. You may want to design a desk of several levels, or include sections that could be raised and lowered.

3. As a means of self-improvement, design a check list for evaluating a teacher's presentation. Then try it out on your fellow teachers. Include criteria that we have mentioned in this chapter, such as mannerisms, delivery, dress, use of blackboard and instructional media, and methods of handling class discussion.

4. Make a list of "Biological Quickie" statements that can be used to begin a lesson. Indicate the goal of each statement and how you would guide class discussion after reaction to the quickie. Sample quickies: "The earth is only 5,000 years old." (This should lead to a discussion of ways scientists have for determining the age of the earth and life on it.) "The year 2012 will be the end for mankind." (According to Malthus, in that year man will have outgrown his food supply.)

5. Work out some "microteaching" units in biology. These should employ only one concept, or part of a concept, and require 10 minutes or less to teach. List the materials you would need to teach the units, such as single-concept films, overhead projector transparencies, and other instructional materials.

6. List several qualities that you consider a "good teacher" should possess. Do not merely copy these from this chapter, but work out an original list. Arrange the qualities in order of importance, with those that you consider most important first on the list.

EXPANDING THE CHAPTER

BRANDWEIN, PAUL F. 1965. *Substance, Structure, and Style in the Teaching of Science.* Harcourt, Brace & World, Inc., New York. In part this lecture is based on Dr. Brandwein's idea of conceptual schemes in science. He discusses conceptual schemes at all levels, from elementary school through high school.

HASKEW, LAURENCE D., and MC LENDON, JONATHON C. 1962. *This Is Teaching.* Scott, Foresman and Co., Chicago. A general text that answers many of the common questions about teaching. A section of readings, written by leaders in American education, is of special interest.

MARGENAU, HENRY; and BERGAMINI, DAVID. 1964. *The Scientist.* Life Science Library, Time Inc., New York (Silver Burdett Co., Morristown, N.J.). The first chapter, "Hero, and Human Being" contains a collection of anecdotes that depict the scientist as an ordinary human being. Many of the stories are useful to the biology instructor as openings for class discussion, or as introductions for new topics.

MASSEY, HAROLD W., and VINEYARD, EDWIN E. 1961. *The Profession of Teaching.* The Odyssey Press, Inc., New York. The chapters "Effective Teaching Methods," "Classroom Control," and "Common Disciplinary Practices" are most helpful to the beginning biology teacher.

MIEL, ALICE. 1961. *Creativity in Teaching.* Wadsworth Publishing Co., Inc., Belmont, Calif. A number of case studies make this book interesting reading. The creative process and its development by the teacher are important threads of thought upon which the book is based.

NATIONAL EDUCATION ASSOCIATION. 1949. *Toward Better Teaching.* Washington, D.C. While all of the chapters of this book offer suggestions for more effective teaching, Chapters 5 ("Fostering Creativity") and 9 ("Toward Better Teaching") are especially helpful.

CHAPTER 4

Personalized Teaching

ONCE UPON A TIME the animals decided they must do something heroic to meet the problems of the New World. So they organized a school. They adopted a curriculum consisting of Running, Climbing, Swimming, and Flying. To make it easier to administer, all the animals took all the subjects.

The duck was excellent in Swimming; better, in fact, than his instructor. He made passing grades in Flying, but he was very poor in Running. Since he was so slow in Running, he had to stay after school and also drop Swimming to practice Running. This was kept up until his webbed feet were badly worn and he was only average in Swimming. But average was acceptable in school, so no one worried about that—except the duck.

The rabbit started at the top of the class in Running but had a nervous breakdown because of so much make-up work in Swimming.

The squirrel was excellent in Climbing until he developed frustration in the Flying class, where his teacher made him start from the ground up instead of from treetop down. He also developed charley horses from overexertion of certain muscles. So he got a "C" in Climbing and a "D" in Running.

The eagle was a problem child and was disciplined severely. In the Climbing class he beat all the others to the top of the tree, but he insisted on using his own way to get there.

At the end of the year an abnormal eel that could swim exceedingly well, and also run, climb, and fly a little had the highest average. He was declared Valedictorian.

The prairie dogs stayed out of school and fought the tax for public education because the administration would not add digging and burrowing to the curriculum. They apprenticed their children to a badger for individual tutoring. Later they joined the ground hogs and gophers to start a successful private school.

[The above "fable" is an adaptation of a tale attributed to George H. Reavis in *The Cultivation of Idiosyncrasy* by Harold Benjamin, published by Harvard University Press, 1949.]

IDENTIFYING THE NEED

A school for animals would probably fail if it tried to teach everything to all of them. So would schools for people. But people *are* people because they *are able* to enjoy a wide variety of interests. Schools that offer a wide variety of subjects are stimulating. As long as everyone is not required to take all of the subjects, such a comprehensive school provides the best possible means for a well-balanced education. Running is offered in most high schools, but that does not mean everyone has to go out for track. The same is true of swimming. And the same is true of academic subjects.

Take flying, for example. In Atlanta, Georgia, a former Navy pilot started a flying course for high school students. The pilot, Frank Hazelwood, was no professional educator—he simply wanted to do something for the schools his daughters attended. School officials were amazed to find that almost seven times as many students signed up for the flying course as there was room for. One, a former dropout, became so interested in school through the flight instruction that he remained and received his diploma at age 22.

These two viewpoints of curricula—one fiction, the other fact—point up a problem educators have: *planning patterns for individual learning*. Put another way, it simply means that students do not learn as a "class"; they learn as individuals.

PATTERNS OF LEARNING

Dr. Frank Riessman, a psychiatrist at the Albert Einstein College of Medicine, believes no two pupils learn the same things in the same way at the same pace. "Everyone has a distinct style of learning," he says, "as individual as his personality. These styles may be categorized . . . as visual (reading), aural (listening), or physical (doing things). . . ."[1]

There is a need for making a pattern of instruction follow the learning pattern of your students. Only then will they learn efficiently, and only then will you make most progress in teaching them.

Some learning patterns are steady, constant, and consistent. We all know people who attack problems in logical and systematic ways and then work at solving the problems steadily until there is a solution. Some students study like this—a little each day.

On the other hand, some people operate in jumps or spurts. Their activities and work habits are made up of peaks and valleys. They may work with great intensity for a while, then let up and switch to another project. Eventu-

ally they return to the first task and finally finish it.

Both of these learning patterns are common, and both lead to successful mastery of a subject. The point is, each person has his own particular way of learning something. As a teacher, you must recognize this and provide enough variety in learning situations to match the variety of student learning patterns.

INDIVIDUAL BEGINS WITH "I"

Educators feel that the individual needs of students must be of foremost concern. This is a difficult problem because greater numbers of students are entering school each year. Keeping the individual student in mind as high schools get larger is a difficult problem for teachers and administrators. High school curricula have expanded to meet a larger variety of needs, but the problem of individualizing instruction in each curriculum is still not solved despite the increased educational research being done in this field. At a national conference, University of California Professor Richard S. Crutchfield[2] summarized the dilemma: "An educational dilemma facing us today is that we must meet an increasing need for individualized instruction while the continuing growth of our mass educational system makes individualization less and less possible."

Dr. Crutchfield lists three reasons for giving serious attention to the problem of individualizing instruction. First, he believes that there is a sound pedagogical reason based on an individual's background and his particular style of conceptualizing material: "In order that any bit of instructional information—no

---

[1] "Styles of Learnings," *NEA Journal*, March 1966.

[2] Paper delivered at a Conference on Individualizing Instruction at Princeton, N.J., May 11, 1965.

matter how small—be properly understood and mastered by the individual, he must be enabled to assimilate it relevantly to his own preferred and distinctive cognitive style, in such a way as to 'make it his own.' This requires individualized instruction that is geared to the distinctive attributes, needs, and cognitions of the particular person."

Another reason for making instruction individual is related to the learner's need for personal identity: "There is acute need for ways to preserve the student's sense of individual identity in the sea of anonymity flooding the large-scale educational institutions of today."

The third factor that makes personalized teaching so important, says Dr. Crutchfield, is that society will demand leadership from creative thinkers in the future and this type of creativity is best fostered on an individual basis: "It has to do with the changing nature of man's future world. The aims of educational training today must reflect the needs and purposes of tomorrow. I believe that the nature of man's tomorrow is such as to require greater and greater stress on individualized instruction today."

Each of these three factors favoring individualized instruction represents valid and useful guidelines. While each must be considered, the first factor should occupy a prominent place in our teaching. Your ability to design instructive material that allows the student to use it at his own pace, to assimilate it in his own way, in his own style, helps guarantee success. By producing tangible results in his students, the teacher will be more successful; the students will be more successful because they will be able to learn more thoroughly.

The problem of meeting individual differences has been kicked around for years in teachers' meetings, at conventions, and among students preparing to teach. But, while it has received much attention, little has been done to actually meet the needs. Teachers, though sincere, have not had the time to design the unique and imaginative mechanisms for meeting individual differences. And there are other reasons. Scant data exist on the process of learning, our understanding of how people learn, and the pace at which they learn. In addition to this, technology has only begun to make significant contributions in the specialized field of individualizing instruction.

Today many of these factors are changing. Teacher work loads have lessened, especially in secondary school.

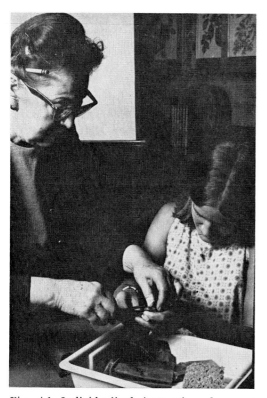

Fig. 4.1—Individualized instruction does not necessarily mean that the teacher and student should be isolated from the rest of the class. This teacher can give individual help during a laboratory period.

The number of "preparations," or different classes that a teacher instructs, has been reduced. Professional associations are putting increased pressure on school administrators to liberate classroom teachers for more preparation time. Instructors of laboratory courses such as biology, physics, and chemistry sometimes have extended contracts that allow them to inventory and order equipment outside school contract days. Wider use of laboratory assistants, "instant laboratory material," and in-service training sessions make teachers' time more available to prepare for specialized instruction.

"OVER-THE-SHOULDER" TEACHING

"Some promising advances have recently been made in the field of learning," says B. F. Skinner[3] of Harvard University. Special techniques that produce "contingencies of reinforcement" have been employed to help us understand how animals learn, for example. The contingencies are simply conditions in the learning situation that require the subject to reinforce what he has already learned, but they attack it from a different point of view, or through a different medium. Reinforcement is most valuable when it occurs a few seconds after the answer is given. The old one-room schools, with the teacher looking over the shoulder of the student as he wrote answers in his notebook, probably satisfied the condition of reinforcement contingency. Unfortunately, it is not the most efficient. "It can be easily demonstrated," Dr. Skinner says, "that unless explicit mediating behavior has been set up, the lapse of only a few seconds between response and reinforcement de-

stroys most of the effect. In a typical classroom . . . long periods of time . . . elapse. The teacher may walk up and down the aisle . . . while the class is working on a sheet of problems, pausing here and there to say right or wrong. Many seconds or minutes intervene between the child's response and the teacher's reinforcement."

This problem of providing reinforcement very soon after a response has been the subject of investigation by the educational researchers who make teaching machines. The solution, which was worked out in part by such scientists, has been possible only because we now have an array of mechanical gadgets that have been adapted to education. These "teaching machines"—and there are many forms of them—let the learner "modify his learning environment" as he sees fit. He can speed up or slow down, call for an evaluation of his response immediately after producing an answer, and backtrack if he wants to strengthen certain areas before going on.

Teaching machines, then, are a compromise between the ideal condition of a 1:1 teacher-pupil ratio and the unideal condition of one teacher trying to instruct a class of hundreds. The machines work only as fast as the "operator" (student) wants them to work, and they have the distinct advantage of providing the right answer at precisely the time the student requests it. No teacher is able to give that kind of individual attention.

The desire for instruction on an individual basis is a strong trait, perhaps even a basic trait of human nature. We all like to feel that the instructor is talking directly to us. Yet many skills, techniques, and abilities can be mastered well without the constant guidance of a teacher. Necessity may require that much of the teaching in the near future

---

[3] In W. I. Smith and J. W. Moore, *Programmed Learning: Theory and Research,* D. Van Nostrand Co., Inc., New York, 1962, p. 24.

be done by programmed instruction. As Eric Bender[4] puts it:

In this ruthlessly automated age, when millions of adults will be forced to master new skills, programmed instruction may prove to be a key teaching technique. Programmed texts for self-instruction are already available in such diverse subjects as logarithms, how to write a memo, second-year German, boiler inspection, and the no-trump bid in bridge.

Most nations today are faced with an "education explosion"— increasingly overcrowded classrooms and only limited funds for more schools and teachers. In addition, in the United States our formidably high level of unemployment indicated the need for a massive effort to educate and retrain illiterates, dropouts, and the jobless. It is clear that our traditional "teacher-in-the-classroom" concept is simply not expansible enough to cope with these new pressures.

Programmed instruction, however, seems admirably suited to fill some of the gaps. Since it is geared to self-instruction, teaching can be done in groups of all sizes; it enables students of wide ranges of ability to learn with remarkable efficiency; and it does not necessarily depend on verbal skills. Although programmed instruction is not designed to replace the teacher in the classroom, it is excellently adapted to take over some of the load. . . .

Just how are some plans of programmed instruction designed to take over part of the teaching load? Are there ways of individualizing instruction in biology *without* using teaching machines? What about fostering creative biology learning through individual student projects, science fairs, and special reports?

FILLING THE NEED

Large numbers of devices, ideas, and programs have already been invented to fill the need for individualizing instruction. We have chosen several to discuss here because of their special significance in teaching biology. They include teaching machines, the audio-tutorial method of instruction, and student projects.

Teaching Machines

In one sense a written test is a teaching machine. So is a worksheet that guides a student by asking questions that must be answered after reading the text. But strictly speaking, neither of these is a complete teaching machine.

Unless a test or examination is handed back to a student with the correct answers written in place of the wrong ones, it does not correct any errors. Even if the test has been corrected with comments, it does not act as a teaching machine to the student. Teaching machines cause learning to take place bit by bit, with each step based on the previous one. Properly programmed, such a learning device will have built-in barriers along the way so that students cannot skip a section. By "feeding" information in small pieces, large concepts gradually form.

The programmer may with some frequency predict error and guard against it. He may know that a certain skill has been studied earlier by his learners, but he suspects that their grasp of it is shaky or even nonexistent. If he wishes to teach something based on knowledge of cube roots, for instance, he sets up a few problems calling for use of such knowledge, warning his learners that this is a special testing situation. At the end of the sequence the program discusses the outcome. Those who do indeed possess the necessary skills are directed to proceed. Those who do not possess the skills are directed to go through a "sidetrack" sequence in which the skills are retaught, after which the learners will return to the main lines of instruction.[5]

[4] "The Other Kind of Teaching." Copyright 1965 by Harper's Magazines, Inc. Reprinted from the January 1965 issue of *Harper's Magazine* by permission of the author.

[5] *Ibid.*

This kind of programming, called linear programming, consists of a series of questions, or "frames." Each frame is essentially a statement with a significant word omitted. The statement is the stimulus and it evokes a response. One such program, used by a Texas insurance company, is called "What Is a Macadamia?"[6] The two frames that begin the sequence are shown below:

Pecans, cashews, almonds, and macadamias are all ———————.
Because macadamias are eaten mostly by hula girls, surfboarders, and beachcombers, we can guess that they are grown in ——————.

The two responses—nuts, and Hawaii—are printed on the same page as

---
[6] Patricia R. Jones, for United Services Automobile Association, San Antonio, Texas.

the frames. The learner covers the responses with a piece of paper while working through the frames. Then he checks himself for correct answers.

The statements are structured so that they assume prior knowledge. If the learner had no knowledge of Hawaii, hula girls, and surfboarders he probably would not respond correctly to the stimulus statements.

It would be far simpler to read the sentence, "Macadamia nuts are grown in Hawaii." But most of us have trouble learning that way. It is far easier to learn if you are forced to relate past information to new facts and concepts. One way of saying it is that we must relate new experiences to old experiences. Linear programming uses this important ground rule.

Fig. 4.2—One way to individualize instruction is through the use of programmed instruction. In the example above, the student writes his response in a blank, then slides the paper down to uncover the correct answer.

The language the programmer uses is another important aspect of "the other kind of teaching."

Cilia and flagella are important locomotor organelles. Flagella are generally long, but cilia are usually ——————.

The grammar of this frame provides the learner with an "either-or" choice. The use of "however" or "on the other hand" would have the same implication. "Similarly," "likewise," "however," "in spite of," and "never" are other grammatical clues. Of course, the stimulus statement above would not make any sense at all if the learner did not know what "locomotor organelles" were. Some prior knowledge is assumed here, which is a basic assumption in all linear programming.

One of the most important considerations the programmer must keep in mind is what he expects the learner to accomplish. In other words, what should the learner be able to *do* after he has worked through the teaching machine? Dr. Robert F. Mager says, "An instructor will function in a fog of his own making until he knows just what he wants his students to be able to do at the end of the instruction." In his book, *Preparing Instructional Objectives,*[7] Dr. Mager says the learner should be able to *construct, compare, write, distinguish, solve,* instead of merely being required to *enjoy, believe, appreciate, understand, know.* In biology such tangible results are not difficult to demonstrate. Interpretation of diagrams, graphs, and charts can be required, and the learner can respond to stimulus statements by making drawings, labeling diagrams, and pointing out errors in setups of laboratory apparatus.

Design of a linear program takes a lot of time and it takes experience, but in the long run it can save you time because it becomes an extension of your classroom teaching. Furthermore, it is a teacher that the student can learn from *at his own pace.*

Pacing is an important consideration in choosing ways of individualizing instruction. One of the most effective means of building learner-controlled pacing into a self-teaching program has been developed recently. It makes use of a specially written laboratory manual and a tape recorder. While many audio-visual teaching aids are a part of the program, the manual and tape are the basic parts. Taking its name from the tape-recorded instructions that "tutor" the student privately it is called *audio-tutorial instruction.*

Audio-tutorial Instruction

"A coed slides into a plastic chair in a soft green three-sided cubicle, consults a mimeographed list, flips a switch, sees a red light blink, dials 1-2-2, pulls on earphones. Into the headset flows the voice of her political science professor, then Adlai Stevenson on the meaning of democracy, finally a discussion of freedom by New York University's Sidney Hook—and thus ends Lecture 1, Second Semester, Political Science 113."[8]

Political science is not the only thing being tape-taught today. In fact, the idea of teaching individual students via tape, with a package of teaching aids included, got its biggest push in the sciences. One popular program, begun at Purdue University in 1961, has been especially structured for the teaching of

[7] *Preparing Instructional Objectives,* Robert F. Mager, Ph.D., revised edition copyright 1963; Fearon Publishers, Inc., 2165 Park Boulevard, Palo Alto, California 94306, paper $1.75; case-bound $3.50.

[8] Copyright Time Inc., February 18, 1966.

botany. Dr. Samuel Postlethwait,[9] who developed the program, calls it an "integrated-experience" approach. He says:

Emphasis on student learning rather than on the mechanisms of teaching is the basis of the integrated experience approach. It involves the teacher identifying as clearly as possible those responses, attitudes, concepts, ideas, and manipulatory skills to be achieved by the student and then designing a multi-faceted, multi-sensory approach which will enable the student to direct his own activity to attain these objectives. The program of learning is organized in such a way that students can proceed at their own pace, filling in gaps in their background information and omitting the portions of the program which they have covered at some previous time. It makes use of every educational device available and attempts to align the exposure to these learning experiences in a sequence which will be most effective and efficient. The kind, number and nature of the devices involved will be dependent on the nature of the subject under consideration.

The term, "integrated-experience" is derived from the fact that a wide variety of teaching-learning experiences are integrated, with provision for individual student differences, and each experience planned to present efficiently some important aspect of the subject. In the audio-tutorial booth, the taped presentation of the program is designed to direct the activity of one student at a time; the senior instructor in a sense, becomes the student's private tutor. It is important to emphasize at this point that the tape represents only a programming device and that the student is involved in many kinds of learning activities. Further, it should be noted that those activities which by their nature cannot be programmed by the audio tape are retained and presented in other ways. For example, guest lecturers and long films are shown in a general assembly session, and small discussion groups are held on a regular basis to provide for those activities which can best be done in a small assembly. Flexibility and independence, accompanied by helpful guidance when necessary, are the key concepts of the approach.

The audio-tutorial method of teaching is similar to the widely used language laboratory. It is one of the most effective means of language instruction. But the sciences, with their endless variety of apparatus, single-concept films, and laboratory procedures, are a "natural" for adaptation to the audio-tutorial method. Some guidelines in building an audio-tutorial program are briefly discussed below.

1. *Scope.* The breadth and depth of each lesson should be carefully analyzed as the lesson is written. For example, you may wish to use an audio-tutorial lesson on "Some Practice With the Microscope." The learning activities that students perform should include a few powerful examples of the capabilities of the microscope. Weed out exercises that have little learning value for the student.

2. *Apparatus.* Long, complex experiments are not satisfactory when adapted to the audio-tutorial method. Try to break up long-range experiments into shorter units. If you *must* use material that requires weeks for completion, remember to refer to the experiment occasionally in the intervening lessons.

3. *Variety.* Use a wide variety of teaching aids and materials. Models, charts, diagrams, pictures, 35mm slides, and living and preserved materials should be available to the student. The single-concept films in cartridge form may also be used frequently to show phenomena that ordinarily would not be observed by your biology students.

---

[9] *An Integrated Experience Approach to Learning,* Burgess Publishing Co., Minneapolis, 1964.

A Look at a Loop

One of the unique applications of teaching films has been developed almost simultaneously with the audio-tutorial method. It is the use of short (3- to 4-minute) film "loops." Encased in a plastic cartridge, the film is spliced head-to-tail. It "plays" over and over, without the need for rewinding. The cartridge is inserted into a shoebox-size projector, and the film threads itself automatically. These films are called "single-concept" loops because they are designed to illustrate only one idea. Sources of single-concept loop films are given in Appendix F 292. When used with audio-tutorial teaching programs the loops augment the lessons with motion pictures of unique phenomena. Such rarely observed events as the release of zoospores in certain algae, the growth and differentiation of the cap of a basidiomycete, or phototropic response in selected angiosperms are available. Using a variety of these loops the teacher can enrich an audio-tutorial lesson with unique scenes of biological events that are often described in the text, but seldom seen. The low cost of the loops and projector makes it possible for most schools to buy them.

Among the numerous advantages of the audio-tutorial method are the following:

1. *The audio-tutorial method is paced and controlled by the student.* He sets the pace. If he wishes to back up, repeat, or speed ahead he can control the tape accordingly.

2. *The lessons can be designed to fit individual student needs at a particular school.* Suppose, for example, that your school has no binocular microscopes, yet you wish to offer basic microscope instruction to

Fig. 4.3—This booth contains a tape playback unit, a single-concept film projector, and other teaching aids that are needed to provide a learning experience for a science student. It represents one approach to individualizing instruction. (Audio-Tutorial Systems from Postlethwait *et al., An Integrated Experience Approach to Learning,* Burgess Publishing Co., Minneapolis, 1964.)

your students through audio-tutorial methods. Unfortunately, the only lesson available includes *both* monocular and binocular microscope techniques. Here is an opportunity for you to design a short teaching program for the particular microscopes *you* have.

3. *More material can be covered.* Recent studies have shown that students do not necessarily learn *better* with audio-tutorial techniques, but they can learn *more.* It is possible to expand certain topics, or treat them with more depth.

Both teaching machines and audio-tutorial instruction are relatively new methods of individualizing instruction. They are designed in part to free the teacher from daily drudgery of repeating basic material—his time can be spent creating new ways to teach. They put the important aspect of pacing into a student's learning pattern. *He* controls his learning rate, not the teacher.

Special Student Projects

While you may not be able to assign a project to every student, it will strengthen your teaching if you can work extra time with several of the students who are especially interested in biology. A student-centered special project, with the instructor as an advisor, generally leads to disciplined inquiry techniques that serve the student throughout the remainder of his school days, and often the rest of his life.

Views vary about student projects in science. Some educators feel that the high school is no place for students to "practice at being scientists." They feel that a true scientist needs a full background in many disciplines—a requirement that no high school student is prepared to fulfill. The high schools' job, they say, is to provide this background. Later, in college, true research projects can be carried on more successfully. In a paper delivered at a regional conference[10] of the National Science Teachers Association, University of Illinois psychologist David P. Ausubel states:

The development of problem-solving ability is, of course, a legitimate and significant educational objective in its own right. Hence, it is highly defensible to utilize a certain proportion of . . . time in developing appreciation of and facility in the use of scientific methods of inquiry and of other empirical, inductive, and deductive problem-solving procedures. But this is a far cry from advocating that the enhancement of problem-solving ability is the major function of the school. The goals of the science student and the goals of the scientist are not identical. Hence, students cannot learn science effectively by enacting the role of junior scientist.

[10] Milwaukee, Wisconsin, October 1, 1965.

On the other hand, John R. Jablonski,[11] Director of Special Projects in Science Education for the National Science Foundation, states:

Emphatically, . . . high school research *can* be done with the proper guidance, selection, and interpretation of information, knowledge, and ideas. The student has the same role as does the working scientist, but to start him on the way, he must receive the necessary guidance. To elaborate again: *Research* is an attitude, a way, an existence. . . .

Can a high school student do research? The answer depends on the total circumstances for each individual.

Mr. Jablonski lists the important elements that make up the attitude a researcher should have. They are:

1. Organization in problem objective
2. Dedication to seeking the truth
3. Open-mindedness
4. Accurate recording of observations
5. Control in the experiment
6. Reporting and possibly publishing the completed work
7. Advancement in knowledge in context or understanding

Probably the most significant point to keep in mind is that *the project must be fitted to the individual.* Obviously, a student with a poor mathematics background would have trouble pursuing a statistical study of population dynamics. Or, if a student has difficulty studying specimens with a microscope and does not like to work with objects that are too small to handle without special tools, he would probably do poorly in a project that required a great deal of microtechnique. Analyze the student. Have him take a self-inventory of qualities that he considers strong, and qualities

[11] *Welch Biology and General Science Digest,* Vol. 14, No. 2, 1965.

60

that he believes are weaknesses in his work habits and abilities, then help him select a topic that could form the basis of an individual project. Several references for kinds of projects and experiments that students have done or could do are listed in Appendix C 236, 237, 239; Appendix D 255, 259, 260, 262, 263, 265; and Appendix G 319, 324, 326, 328, 329, 331.

After the topic is chosen, information should be gathered. Background reading and library research is one of the most essential phases of any investigation. Then, in counsel with the instructor, the student should plan his project, setting target dates for each phase of the work. Finally, when the work is completed, it should be written as a report so that the results can be communicated to others.

A step-by-step plan for structuring a student project has been worked out by the Iowa Junior Academy of Science. We have included basic parts of the plan below.

I. Decide on the nature of the investigation. Will it attempt to:
   Verify physical constants?
   Survey populations of living organisms?
   Construct a model to elucidate a natural phenomenon?
   Study the relationship of natural resources to the living things in the environment?

II. Plan the broad outline. Consider your own background, your desire to pursue the project, the time you can give it, and even your financial resources. Choose a central theme. Start early in the year. If the general topic and nature of the project is well chosen it can be studied for two or three years while you are in high school.

III. Proceed scientifically. Follow these steps:

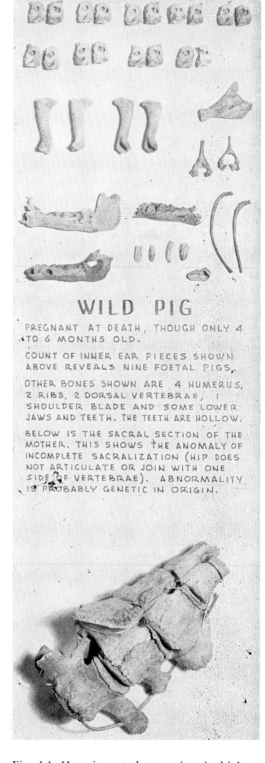

WILD PIG

PREGNANT AT DEATH, THOUGH ONLY 4 TO 6 MONTHS OLD.

COUNT OF INNER EAR PIECES SHOWN ABOVE REVEALS NINE FOETAL PIGS.

OTHER BONES SHOWN ARE 4 HUMERUS, 2 RIBS, 2 DORSAL VERTEBRAE, 1 SHOULDER BLADE AND SOME LOWER JAWS AND TEETH. THE TEETH ARE HOLLOW.

BELOW IS THE SACRAL SECTION OF THE MOTHER. THIS SHOWS THE ANOMALY OF INCOMPLETE SACRALIZATION (HIP DOES NOT ARTICULATE OR JOIN WITH ONE SIDE OF VERTEBRAE). ABNORMALITY IS PROBABLY GENETIC IN ORIGIN.

Fig. 4.4—Here is a student project in biology. The Appendices list more of them. Projects are often more meaningful if the student can draw upon several resources, and use a camera, field techniques, and related laboratory work to complete his project.

Observe the situation you wish to investigate.

Define a problem in the situation.

Make a hypothesis about why or why not certain factors are a part of the problem.

Design an experiment to test the hypothesis. Remember, disproving the hypothesis (negative results) is just as valuable as proving the hypothesis (positive results). It is as helpful to know what the reasons *aren't* as well as what they *are*.

IV. Prepare a report. The write-up of your work is just as important as the rest of the project. The form below can be used when you organize the report.

INTRODUCTION

Statement of the problem. Why did you do the work? List the purpose of the investigation, and discuss how your investigation tried to expand or clarify the knowledge in the general field.

METHOD AND MATERIALS

How was the work done? Describe your experimental apparatus, your way of gathering data, your experimental design, and the type of control you used.

RESULTS

List your data, either in a chart, table, or graph. Report any observations made, and list all facts that were significant to the conclusion you discuss in the next section.

DISCUSSION

What do the results mean? What patterns emerged, or which relationships were meaningful?

CONCLUSION

What is the significance of the results? What contributions have you made through your project to the total problem? What further investigation might be suggested by your project?

BIBLIOGRAPHY

What literature relates to your project? List the entries in the bibliography alphabetically. Each entry should be listed by author, date, title, source.

If you feel that a rigid form for the write-up is "uncreative," you might consider this alternate plan. Arranged as a check list of the ingredients of research and report, it is a series of steps that a student investigator might follow in solving a problem.

My problem and what I learned.

Why I chose this problem.

My hypothesis.

How I tested it.

Some snags along the way.

Help from a scientist . . . or a library, a teacher, someone in industry.

Another hypothesis.

Data.

My interpretation.

References and assistance.

AN INVESTMENT IN TIME

Much of the work we have suggested in this chapter will require a great deal of your time. The design of linear programs to act as teaching machines, the construction of a small audio-tutorial center, and the advising that you provide students working on individual projects will all take many hours. These are hours you could spend otherwise—reading professional literature, developing new units to teach, or for recreation.

The time you spend in developing individualized methods of instruction will pay off in the long run. It will help the slow learner, the late bloomer, the student who works hard but just does not seem to "get it." And the development of teaching aids like these will also help *you*. Once you have a backlog of workable student project themes, several audio-tutorial lessons, and a number of programmed pencil-and-paper "teaching machines" you can begin to use your time for other useful activities.

There is an important use of time for professional work, for writing, for speaking, and for influencing the community outside your classroom. Though these are nonteaching duties, they are a vital part of your image as a teacher. The stature you acquire in the public schools depends on many qualities. Some of these can almost be taken for granted. Others must be consciously developed and enlarged.

CREATING CHAPTER CONCEPTS

1. Make a list of learning devices, types of instructional media, and audio-visual materials that can be used to individualize instruction in biology. Indicate, with a sentence or two, specifically how each should be used.
2. Choose one or two concepts in biology. Then write a short teaching machine on paper, programming the learning of the concepts. Use the technique discussed in the chapter, and follow the form shown on page 55.

3. Collect ideas from professional journals on student projects. Then make a file of these. Organize the file in sequence, in the same way your biology course is organized. This will enable you to suggest individual projects to students throughout the year as you study each topic in class.
4. Write an essay that sets forth your philosophy on student projects, individualized instruction, and science fairs. Do you believe that science fairs, for example, are beneficial to the student, even if he has little aptitude?
5. Plan some capsule, "microteaching" lessons to put on TV tape. Design the lessons so that they develop concepts sequentially and provide individualized instruction to the student who uses them.
6. Write a lesson for audio-tutorial instruction. Choose topics that would be used every year, such as "Use and Care of the Microscope," "Collecting and Pinning Insects," or "Basic Dissection Techniques."
7. Write an article for a professional biology journal, explaining how to individualize instruction for high school biology students. Show specifically in the article how your suggestions will strengthen the learning of the concepts.

EXPANDING THE CHAPTER

BEVERIDGE, W. I. B. 1957. *The Art of Scientific Investigation.* 3rd Ed. A Vintage Book, published by Random House, Inc., New York. In a somewhat philosophical treatment, this book discusses several aspects of the art of investigating a scientific problem in a scientific way.

MAGER, ROBERT F. 1961. *Preparing Objectives for Programmed Instruction.* Fearon Publishers, Inc., San Francisco. This guidebook will help you understand some of the basic criteria in working out programmed instruction.

MILLER, D. F., and BLAYDES, G. W. 1962. *Methods and Materials for Teaching the Biological Sciences.* 2nd Ed. McGraw-Hill, Inc., New York. The latter part of this book contains an appendix full of ideas for individual student projects.

MORHOLT, EVELYN; BRANDWEIN, PAUL; and JOSEPH, ALEXANDER. 1958. *A Sourcebook for the Biological Sciences.* Harcourt, Brace & World, Inc., New York. This excellent sourcebook lists scores of demonstrations, ideas for individual projects, and helpful "how-to-do-it" suggestions for the biology teacher. It is a valuable guide for laboratory work.

PIGEON, ROBERT F. 1964. Programmed Instruction for Biology, *Bioscience* 14(8):21–24. This instructive article defines and illustrates the main features of programmed instruction and gives information on where to find more about the topic.

POND, GORDON G. 1959. *Preparation and Display of Science Materials.* Wm. C. Brown Publishing Co., Dubuque, Iowa. For the student preparing a science fair project, or the teacher arranging materials in the science room, this book contains hints and procedures for making attractive displays.

SMITH, WENDELL I., and MOORE, JOHN W. (eds.). 1962. *Programmed Learning: Theory and Research.* D. Van Nostrand Co., Inc., New York. The chapter by David J. Klaus, "The Art of Auto-Instructional Programming," is especially helpful to the teacher who is preparing individualized instruction.

CHAPTER 5

The Biologist

in His

Community

"AND SO," the speaker concluded, "we teach evolution to students along with the other major concepts of biology."

The speaker was a high school biology teacher. He was winding up a talk to the Kiwanis Club where he had presented some of the important topics that are a part of biology today.

He had begun with the kind of biology most of the men in the audience had studied when they were in school. He remembered with them the frog dissections they had done, and the earthworms and grasshoppers they had carefully cut up. He pointed out the emphasis on human anatomy and physiology.

"Yeah," commented a middle-aged businessman after the meeting, "I never will forget having to memorize the names of all the bones in the body."

"And what about that leaf collection we had to make? That was our kind of biology," said another one of the men.

The teacher spoke for about 30 minutes. Half of the time he reviewed projects the group might remember, but after the audience was "warmed up," he discussed some of the present-day concepts on which contemporary biology teaching is based. He mentioned the need for students to do limited "research" in the high school laboratory

and to develop a critical thinking process through a study of biology.

To show some of the advances made in biological research in the past few years he used as an example the study of genetics. Our present understanding of mutations and "genetic drift" has contributed a great deal to notions about natural selection and evolution. "The gradual immunity of flies to DDT," the teacher stated, "is evidence of natural selection in a relatively short period of time." Citing several other lines of evidence to support his brief mention of some of Darwin's ideas, the teacher concluded his talk by saying that evolution can be made an integral part of the high school biology course.

AN EVOLUTION REVOLUTION

The teacher in the above example was working in a fairly liberal community. Yet there were several people in town who were opposed to teaching "evolution and survival of the fittest," as they called it. Because of their misunderstanding of genetics and the mechanisms by which organisms change over the years, they did not understand how evolution could occur. To make matters worse, they had put the wrong connotation on clichés such as "survival of the fittest" and "man descended from monkeys."

In many communities the citizens seem to become most excited about how many football games are won or lost. They normally do not criticize the academic program, but occasionally parents can cause a revolution in certain high school courses because of one or two ideas that are taught. Evolution can be cause for this kind of revolution. It *can* be, but not necessarily.

Rather than let false notions increase, the teacher in our example decided to meet the challenge head-on. He spoke at several meetings, including service clubs, PTA groups, and some church organizations. Most of what he said concerned the overall biology program at the local high school—the organization of the course, the type of laboratory and field work that was done, and what was expected of the student. But he also informed the citizens of the importance of teaching *total biology* and pointed out the danger of trying to hide certain aspects.

By carrying his program directly to parents and townspeople this teacher met opposition before it had a chance to form. His message was general, and of interest to everyone, whether or not they had children in school at the time. The teacher was shaping an image in the minds of his listeners. He was projecting an image of a professional educator. *Ways and means of projecting a professional image* are important to today's classroom teacher.

CHANGES IN THE TEACHER'S IMAGE

Changes in the public image of teachers have been slow. The road upward has been long and difficult. Tracing American educational patterns in *The Story of Education*, Carroll Atkinson and Eugene Maleska cite some examples that show how society has changed its attitude toward teachers. Remarkably, most of these changes have come about in the past two or three generations.

One teacher, recalling her earlier experiences in a Louisiana school, has told the authors that she had to sign a contract agreeing not to bob her hair or wear slacks. But those local parish board members were quite proud of their liberalism. In a nearby community the situation was even worse; women teachers had to agree not to have dates . . . , to remain in town on weekends, and to retire before 10 P.M.

As a teacher advances in his profession and holds more important positions, he still remains a slave governed by public opinion. Take for example the recent case of the college president in Ohio. An ordinary citizen might divorce his wife, and only close friends and neighbors would pay more than passing attention to the affair. But not so this college president who began courting his dean of women. Someone, presumably a student, gave unwelcome publicity to the courtship by printing on the wall outside his home: "Dottie loves Willie—Willie loves Dottie." Despite this petty persecution the president married his dean of women, and almost immediately was asked by the board of trustees to resign, although they gave him credit for being an exceptionally good administrator. It made no difference that the faculty . . . almost unanimously backed him, that he had served long and honorably as an educator, and had earned a national reputation. The trustees felt that the man's usefulness had ended. Like Caesar's wife, anyone in the teaching profession must be above suspicion.[1]

While the above examples may seem old-fashioned, situations like these still exist. It is almost as if the public expects teachers to be above reproach and beyond criticism. Teachers are often expected to be nearly perfect in every respect.

Are these criteria valid today? Do parents of your students expect you to live differently just because you teach? The answer may not be "no," but it may not be "yes" either. It lies somewhere in between.

While parents serve as catalogs of traits from which their children can copy this habit or that, teachers also fill this important need. As you think back

[1] From *The Story of Education* by Carroll Atkinson and Eugene T. Maleska. Copyright © 1962 by Carroll Atkinson and Eugene T. Maleska. Used with permission of Chilton Book Company, Philadelphia and New York.

65

over your own education you can probably point to one or two teachers whom you never had for a class. But their presence influenced you—their manner, their effect on other students, and their relationships with most of the people with whom they came in contact projected a quality of dedication. Somehow you wanted to be like them. Teachers are in this influential position.

THE TEACHER'S WAY

The way of a teacher is the way of many who deal with the public. Take physicians, for example. They are expected to receive calls at almost any hour of the day or night, to work almost seven days a week, and to offer their leadership in community activities of all kinds.

Ministers are also public employees. They serve as psychologists, marriage counselors, organizers, advisors, custodians, and executives. They are also expected to be the intellectual and spiritual leaders of their congregations.

Like the professions of medicine and the ministry, teaching makes a wide variety of demands on its workers. Many jobs come the way of the teacher. He must serve as Junior Class sponsor, sell candy bars at football games, help organize a homeroom group to build a Homecoming float, take tickets at the school plays, and act as advisor to a club. He is also expected to teach two or three different subjects, grade the papers, and meet the needs of his flock of over 100 students.

Also, teachers are often judged on the basis of isolated events. They are expected to be good examples to youth during their nonteaching hours. They are on public display most of the time. Though this is never mentioned, it is as

much a part of a teacher's contract as his teaching duties.

Often the plumber, shoe salesman, and hardware store manager seem to have more latitude and freedom in their private lives than the teacher. Often the doctor, lawyer, and minister are less restricted by the tacit "laws" that define the responsibilities of teachers in the thousands of communities across our land.

The teacher is subject to the scrutiny and criticism of the parents of children he teaches. But these responsibilities make teaching a profession. Without the well-defined lines of duty it would be just another job. If you are going to contribute to the image of teaching as a profession you should be prepared to follow the prescribed course of action.

Much of the problem of adjustment of a teacher to his community, to his school, and to his students can be solved if he examines his attitudes *before* becoming a teacher. How do *you* look at yourself? Do you enjoy participating in social events? Do you think you would enjoy being an active participant in community affairs? Are you prepared to be an active citizen, helping strengthen the town where it is weak? Much of your image in the classroom will be affected by the way you react outside of class and outside of school. "The trend is to expect reasonable social-moral standards of teachers similar to those observed by all good citizens. A teacher should learn and respect the customs . . . traditions, and expectations of a community while he is there. If they are found to be excessively burdensome or wholly untenable, a new position should be sought for the following year."[2]

---

[2] Harold Massey and Edwin Vineyard, *The Profession of Teaching*, The Odyssey Press, Inc., New York, 1961.

IMPACT ON PEOPLE

The public's concern that a teacher be serious and of strong character is based on the fact that he makes a powerful impact on his students. During an ordinary day students see their teachers as much as they see their parents, but the contact with teachers is not the same as the casual family relationship. Teacher-student contacts are more formal, more specialized, and they usually involve learning of one kind or another. Most of the time that a teacher is talking to a student, some kind of teaching-learning is taking place. Therefore a teacher has an influence over his students that parents do not have. In addition to the teaching of the subject matter, another type of influence occurs between teachers and students. It is the more subtle effect of habits, dress, attitude, appearance, and manner of speaking. During the formative adolescent years especially, youngsters tend to "copy" or "borrow" traits from those people they admire. Most of the aspects of a person's personality are copied and modified from those of other people.

These accessory roles in which a teacher finds himself will make demands on his personality, training, and educational background. Much of the non-teaching activity that we have mentioned above is not discussed in formal college courses in teacher preparation, yet teachers are expected to emerge from college with abilities in several areas outside their subject matter major.

There are several ways to meet the challenge of these outside responsibilities. Certainly common sense offers the most straightforward answer to the questions of how to dress and act. An awareness of surroundings and mores of the community will be your best guide to formulating your habits.

Other methods should be used to keep aware of the professional responsibilities of teaching. They include frequent reading, participation in professional organizations, and in-service training. You have a professional responsibility to yourself to carry on these activities, to remain on the living edge of teaching.

PROFESSIONAL RESPONSIBILITY— TO YOURSELF

As much as we would like to think it, education does not remain modern very long. Within a few years after being graduated from engineering school, for instance, an engineer's college training becomes obsolete. Recent graduates may be more valuable because they have been in contact with the research and development programs of the colleges and universities.

What does the older college graduate do, then? Does he resign himself to a life of fading importance? Does he accept complacently the "planned obsolescence" to which industry relegates him?

While it is true that few educations remain valid over long periods of time, there is no need to assume that a kind of degeneracy sets in the day after graduation. Of course, it *may* occur—but you can prevent it. In fact, you have a professional responsibility to yourself to prevent it and keep "academically alive." A planned program of reading, writing, summer school, and growth through attendance and participation in workshops, institutes, conferences, and conventions will assure you of remaining in touch with vital knowledge.

The Teacher's Reading Plan

Without doing a great deal of extra work in the form of additional college courses, it is possible to upgrade continually your own education by reading. After all, one function of an education is to learn enough about the world of ideas so that it is possible to read with understanding. In science, your educational background makes it possible for you to read with greater comprehension not only your daily newspaper and weekly magazines but also such technical publications as *Scientific American, Science,* and journals in specific areas of biology. We have included a list of professional and nonprofessional journals in Appendix E 280.

The Responsibility of Publishing

You have an additional responsibility to yourself besides reading. That responsibility is writing. While reading keeps you aware of new developments in the classrooms and laboratories of others, writing about your own experiences, ideas, and innovations makes you examine your own work more carefully. You do not need to be in charge of a large research program in order to have material to publish. Many of the solutions you have worked out to your own problems are of interest to other teachers. Topics might include development of a new laboratory technique, an interesting science fair project that a student has done, or the solution of a teaching problem.

Sometimes teachers scornfully say "publish or perish." They feel that writing is an activity apart from teaching. Yet, to keep the feeling of being alive and in touch with the nationwide community of biology teachers it is necessary for the individual teacher to add his opinions and ideas to the hopper. The teacher who has little originality and creative ability will perfect these traits by practicing. Your explanation of some aspect of biology teaching will help you grow professionally because it will give you the valuable experience of putting your ideas into words.

Some guidelines for preparing materials for publication are given in Appendix E 281.

The Value of Belonging

Participation in professional organizations is more than attending meetings and talking with other members. Your participation can grow into a unified voice that has influence on the national level. Certain policies can be formulated by national committees, speaking for the majority of teachers. Pressure by these groups can cause important changes in educational policy.

You will derive a number of benefits from belonging to professional science organizations. Through the journals published by the organizations, the contact with other science teachers, and participation in regional and national conferences you will have a chance to become actively involved in the scientific community. Perhaps the annual conventions are among the most rewarding activities. At gatherings such as these you can meet scores of teachers who have much the same job as yours. You can "trade notes," compare situations and circumstances, and judge your teaching status and capabilities more accurately after finding out how other people do their jobs.

Schools for Teachers

Some statistics indicate that in the average working lifetime of forty years the average person will have to be re-trained to do three different jobs.

Teachers too must be retreaded, but their training cannot wait. We have mentioned three ways of helping stay alive—reading, writing, and participating in professional organizations. But there is a fourth way—in-service training.

With the help of the National Science Foundation in the past few years it has become possible for teachers to go back to college for academic "shots in the arm." Programs to help teachers become familiar with new approaches to teaching biology have been offered. In addition, courses have also been available for teachers who wished to gain background information in specific subject matter areas.

Workshops and short courses help meet the need for keeping teachers updated during the school year. Often outside consultants in specific areas will be called in to talk to teachers in certain academic areas. Sometimes the teachers will travel to neighboring schools for observations and consultations with teachers there. Occasionally the teachers from a given school district will organize themselves into study groups and discuss concepts in modern biology, or they may hold self-help sessions at which they trade notes on teaching techniques, demonstrations, and unique teaching tools.

THE ACTIVE PROFESSIONAL

While it is necessary to be active on the local level, both in your own school and your community, you have a broader responsibility. On the state and national levels there are science-teaching organizations in which you should participate. Many states have science-teaching groups within their education associations. Many of these groups hold annual meetings, publish newsletters, and have regional get-togethers, providing a trading-post for teachers who wish to exchange ideas. Your participation in meetings of this sort keeps you in touch with current activities in all parts of your state. Not only do you get to talk with other teachers in your area of major interest, but you have the opportunity to hear outstanding speakers who can be brought to such meetings. Often there will be panel discussions, short talks, and lecture-demonstration sessions at which classroom teachers present the material. You should participate in activities of this type as another means of remaining in contact with "living" biology teaching.

PROFESSIONAL RESPONSIBILITY— TO THE COMMUNITY

The true professional often puts his responsibilities to others ahead of the responsibility to himself. In fact, one measure of a profession is its degree of service to others. Much has been written and said about teaching as a "service profession."

Who is being served by the teacher? The obvious answer is "students." But the picture is broader than that. The modern teacher has a number of responsibilities outside the classroom—to other teachers and to other members of the community.

Service to Others

"Hardly a day goes by," writes a science teacher, "when one of the elementary school teachers doesn't call me for some test tubes, a piece of rubber tubing, a certain kind of plant, or some type of chemical solution." Teachers in the elementary school often look to the science teachers in the high school for help in getting equipment, setting up

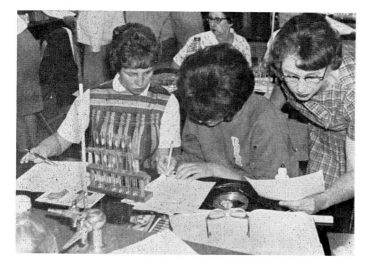

Fig. 5.1—As a high school teacher, you can help others in your school. These elementary teachers are taking a special night-school course in science from a high school instructor.

apparatus, ordering supplies, and providing teaching material. Since the high school teacher is often the only person in town with an education in a specific science, he is looked upon as an expert. "You often have to be prepared to identify someone's rare butterfly, collected during a trip to the mountains last summer; assure a housewife that the insect she caught behind the furnace is *not* a cockroach; or convince a worried parent that the snake his son found in the garden will not poison anyone," says a biology teacher who has answered many such questions.

These requests for help are not unusual; most science teachers consider them part of the job. In fact, they can be used to good advantage by the enterprising teacher. Some taxpayers feel that schools require too many courses that are not practical or down-to-earth. Your contacts with the citizens of your community, in which you have the chance to prove to them that you *do* have a practical knowledge of the world about you, will accomplish an important task. These contacts will convince the townspeople that you indeed have a practical

knowledge of the world—their world. The problems they bring to you may sometimes seem trivial, but they must be taken seriously. Successful public relations can improve your image and stature as a knowledgeable teacher.

Service From Others

Requests for help can work both ways. Consider the teacher who has a Science Club. Students look forward to the monthly meetings, but it is always a problem to get interesting speakers. Here is the place the community can help a school science program. In an article in *Turtox News*[3] E. D. Keiser, Jr., says, ". . . the dilemma of a seeming lack of speaker talent faces most science club sponsors. The individual can avoid such perplexity by devoting time and effort to developing a yearly agenda which utilizes local resources. More often than not, clubs are located in areas having an abundance of potential speakers. It is possible to have an appealing and educational club program by shopping carefully for such talent."

[3] Vol. 43, No. 8, August 1965.

Keiser lists some suggestions for setting up a Speaker's Bureau. He points out that a busy doctor might not have time to analyze the type of audience to whom he is to speak, and then think of a topic specifically for that group. But if the doctor is handed a choice of three or four topics, and a résumé of the type of audience, its size, age level, and background in science, he can more successfully prepare a short talk. In the list that Mr. Keiser compiled (see Appendix A 202) there are over 50 occupations represented with more than 100 suggested titles.

Larger school systems employ science consultants. These experts can be of help to the classroom teacher in several ways. In a National Science Teachers Association publication, written by Kenneth D. George, several specific suggestions are listed for using the services of a science consultant. They include:

1. A conference on ways to improve teaching methods.
2. A planning session on the needs of special groups of students or of individual students.
3. Advice on how and where to do field work.
4. Help in ordering equipment.
5. Upgrading of background through in-service workshops.

The publication also lists some good teaching practices for science teachers. (See Appendix A 203.)

Creating a Good Climate

The image you create depends on your personal attitude, your ability to teach, and your public relations program. The alert teacher who wishes to maintain a healthy relationship with the public, his students, and the administrative officers of his school system must do much more than merely teach. He must stand ready to give help to all who ask for it, and aggressively offer his services as speaker, writer, and participant in many areas of community activities. A list of suggested titles that teachers may wish to write about for local newspapers is given in Appendix E 282.

Through participation in local, state, and national organizations he can exchange ideas and promote some of his own. Through constant awareness of the professional literature he can keep up with new developments in teaching and research. Through in-service workshops, summer school, and science institutes he can continue to build a background in science. These many facets of today's science teacher make his job a difficult one, but the teacher who is active is professional. The active professional teacher with imagination is the creative biology teacher.

CREATING CHAPTER CONCEPTS

1. List some community activities in which you feel the successful teacher should participate.
2. In which of your school's clubs, societies, and organizations do you feel you should be a part?
3. Make a yearly schedule of professional meetings, conferences, and conventions you plan to attend.
4. Write a short paper (3 or 4 pages, double spaced, typed) that could be submitted for publication. Consult the *Style Manual for Biological Journals* (Conference of Biological Editors, 2nd Edition, American Institute of Biological Sciences, Washington, D.C. 20016).
5. Devise a list of subjects for short courses that a biology instructor might offer in adult night school.

6. Compose a list of topics a biology teacher could use in writing half-hour talks for use in his community. The talks could be delivered either "live" or by radio.
7. In what state and national activities should a biology teacher participate?
8. Draw up a plan for a county conference where biologists (both teachers and nonteachers) might exchange information and ideas.
9. Make a list of criteria that can be used to evaluate professional literature. It should enable a student to skim and abstract articles for you so that you

need only refer to the "Abstract File" for information.
10. Keep a daily log or diary of experiences you have as a biology teacher. Use these as topics for newspaper articles.
11. Write a short essay defining the biology teacher's responsibility to his professional societies. Include specific objectives you have in being a member of these groups.
12. Review the Code of Ethics for teachers, published by the National Education Association. How does the Code apply to biology teachers?

EXPANDING THE CHAPTER

GROSS, RONALD; and MURPHY, JUDITH. 1964. *The Revolution in the Schools.* Harcourt, Brace & World, Inc., New York. Some experimental ideas in education are described in this book, including a discussion of the new science curricula.

GRUBER, FREDERICK C. 1957. *Foundations of Education.* University of Pennsylvania Press, Philadelphia. This collection of lectures on education and the educational process provides some background for the teacher who is called upon to speak to civic groups.

JOHNSTON, BERNARD. 1964. *Issues in Education; An Anthology of Controversy.* Houghton Mifflin Co., Boston. As the title implies, this book is based on some of the problems and discontents teachers have. It is interesting reading, and contains much material for talks that teachers might make to civic groups.

———. 1961. *Planning for Excellence in High School Science.* National Science Teacher's Association. A report from the NSTA which expresses the beliefs and hopes of the Association as it aims to assist teachers in analyzing and projecting for science education at the secondary school level.

PLOTZ, HELEN. 1955. *Imagination's Other Place—Poems of Science and Mathematics.* Thomas Y. Crowell Co., New York. As openers and closers for speeches you might make, this excellent collection of scientific poems will be a big help.

———. 1964. *Public Relations Ideas.* National Education Association of the United States. This guidebook to the hows and whys of public relations is especially valuable for the teacher who wishes to project his professional image to his community.

SPAGHT, MONROE E. *The Bright Key.* Appleton-Century-Crofts. New York. The interdependence between business, industry, and education is discussed in this book. The chapter, "What Industry Wants of Education," provides the teacher with the industrialist's view of our educational system.

PART 2

CREATING EFFECTIVE ENVIRONMENTS

CHAPTER 6

Planning a

Science

Center

IT WAS President Garfield who said that a log with Mark Hopkins—a creative teacher and lecturer—on one end and a student on the other was education enough for anyone. President Garfield did not specify where the log should be or from what kind of tree it should come. Perhaps to him it would not have mattered.

What a log this must be now! Schools offer so many different courses that to carry out the example a modern school would have to be a huge collection of logs. Our schools contain few "rooms"; they are made up chiefly of "learning centers." The students do not do "schoolwork"; they "engage in learning experiences." And the activities of students in an English course or in mathematics are quite different from the activities of students in science. The log has become a crowded place and the present teachers and students bear faint resemblance to Mr. Hopkins and his students of 100 years ago. Curriculum change has been only one cause of these differences; our technology has made a large, quiet contribution also.

ENVIRONMENTAL CONTROL

The differences in teaching rooms reflect the wide change of technology in our time. In his book *Teacher and Technology*,[1] William Trow of the University of Michigan describes what the learning environment must be for students today.

For convenience the environment can be divided into three parts—the world of things, of people, and of symbols.

The world of things is the objective world . . . the natural scientists have studied. . . . The world of people is ourselves . . . and the world of symbols, both verbal and numerical, takes over an increasing fraction of the learner's time and interest.

The schools must do an even better job than they have in the past of teaching their students to look and say, to think and do. . . . The task is a continuing one of arranging the environment of the dynamic organism, the growing child, in such ways that he can learn effectively to live in his three worlds of things, people, and symbols.

This three-world life is that of the student when he is at home, or working, or with friends. How can you, the biology teacher, control the student's environment so that this life remains real to him when he enters your classroom? The answer lies in the room itself. The sights, symbols, and even the sounds of science must be a part of the classroom—as much a part of it as the windows, lights, and desks. An empty room that looks like a general-purpose classroom will cause students to react in an ordinary, general way. Create a cubicle of science that will inspire all types of students: those who do no more than occupy space, those who are just average, and those who are eager. It will motivate them to action because they will see biology everywhere they look. For students who have a special curiosity about learning, the creative teacher's biology room will extend their imagination far more than Mark Hopkins' log could possibly have done.

---

[1] Appleton-Century-Crofts, New York, 1963.

Where Do You Start?

A few lucky teachers have the opportunity to design their own rooms. The tremendous building programs of the 1950's and 1960's in which thousands of new schools were built to accommodate the "war babies" born just after World War II are continuing. Not only are new schools being built but old ones are being replaced. Also, there is often opportunity for a teacher to change rooms in a school, moving from an older, general-purpose room into a room that is made over into a biology laboratory. In these cases, though the teacher does not have the chance to plan his new room when the school is being built, he can nevertheless determine how the existing room should be remodeled. Suppose you knew that you would soon be moving your department into a room that was formerly an English classroom. What kinds of things would you have to put into the room to convert it into a place to teach biology?

The *Thirty-First Yearbook* of the National Society for the Study of Education, though it was published in 1932, sets forth a list of timeless guidelines for a comprehensive science room.

A science room is a place where the experience of problem solving is possible.

The plan and design of a science room must provide elements of flexibility.

The design of both classrooms and laboratories should provide facilities for effective teacher demonstrations.

Science rooms should provide facilities for individual laboratory work.

Science rooms should provide certain facilities for objectification by means other than the use of concrete materials (blackboard, bulletin board, display fixtures, visual aids).

These principles can be summarized by saying that the science room must be a workroom where students can "get their hands on science." They must be able to see, smell, and handle science. They must be able to work in small groups, or sometimes individually, on experiments that make biology more real to them. Anything you put into the room must do its share in bringing biology to the student for a close-up inspection. Biology teachers, more than any other group, need tools to teach with. The room you design must have places for these tools.

To Whom Do You Turn

The problems of building or remodeling can best be solved by builders or remodelers. They are problems for the architects, contractors, carpenters, electricians, and plumbers. But, while these craftsmen are experts in their areas, they will need guidance from you since you are the expert in the area of what is needed to make the room "teachable." Here are some suggestions you can follow and some people to turn to for advice in designing a biology room.

1. *Visit other schools.* Drive to schools in your area and look at their biology rooms. Take measurements, list the equipment and furniture they have. Ask for costs. Find out what the teacher likes and dislikes about his room. Ask yourself what *you* like and dislike about the room. Could you teach *your kind* of biology—*creative biology*—in such a place? Why or why not?

2. *Send for catalogs.* Any number of suppliers in the United States specialize in school furniture (see Appendix F 288). Their catalogs include plans and pictures of classrooms the

77

companies have designed for schools all over the country. You can get a great deal of valuable help from these catalogs in choosing furniture and in determining the cost.

3. *Talk to the architect.* If you are involved in helping your principal and superintendent with building plans, you will probably have to talk to the architect sooner or later. Many architects have good ideas about the needs of biology teachers, but some know less than the teacher about the physical layout of a biology classroom. In the latter case, you may wish to make rather firm suggestions. Use your own ideas, based on your experience in the classroom. Try to foresee potential limitations of the room you design and then consult the architect so that the limitations can be corrected.

4. *Write your professional organization.*[2] The National Science Teachers Association will provide members with up-to-date information regarding construction and remodeling of science classrooms. The American Institute of Biological Science also publishes helpful room-planning material.

What Do You Want?

While not all biology teachers will choose identical room arrangements, there are a number of essential items that should be included.

1. Running water to the lecture table, one or two side sinks, and perhaps even to the students' individual laboratory tables.

2. Gas jets at the lecture table and at the students' tables.

3. Compressed air at the lecture table and at the students' tables.

4. Blackboards behind the lecture table.

5. A large lecture-demonstration desk in the front of the room.

6. A preparation room and storeroom, preferably just behind the lecture desk.

7. Bulletin boards on one wall.

8. Display cases on one wall.

9. Storage cabinets under a long side shelf, along one wall.

Armed with the illustrations in this chapter, catalogs, personal talks with other biology teachers, visits with the architect, and information from professional science organizations, the individual biology teacher must make his own room plan.

Face-lifting a Room

Let us suppose you begin teaching in a school that already has an established biology room and the school administration has no desire to remodel it. The room actually may not be too bad the way it is, with long laboratory benches, adequate lecture-demonstration desk in front, blackboards, preparation room off to the side, and numerous display cases on the walls. Would there be a need to make changes in a room of this design? Is there any reason to give it a face-lifting?

We think so. We think you need to fill the room with the sights and sounds and smells of science.

WALL-TO-WALL EVOLUTION

One way of starting the renovation of a biology laboratory involves the so-

---

[2] The October 1961 issue of *The Science Teacher* has an article by Harry K. Wong that suggests many unique ways of remodeling a science room.

Fig. 6.1—Make a list of improvements that could be made in this biology classroom, based on the ideas presented in this chapter.

called "big ideas" of biology—basic biological concepts that give strength to the fabric of the study of life, such as evolution, ecology, the interaction of heterotrophs and autotrophs, energy transfer, and cell replication. Not *all* of these lend themselves to conversion to a wall chart or three-dimensional display, but some of them work out most effectively.

Let us begin with the concept of evolution. What teaching devices can best show the changes of life on earth over the years? The geologic eras of time and the living things present during these periods can be shown in a number of ways. A rope stretched from one end of the room to the other across the top of the blackboard may stand for all known time. Knots tied in the rope signify the arrival of land plants, dinosaurs, birds, and finally man. If the rope is 20 feet long, the last knot, only ¼ inch from the end, represents the short time that man has been on the earth.

If you have adequate space on your walls you might consider having a student paint an entire mural to represent these changes. The steps in working out

such a project are shown in Figure 6.2. Pictures from books and magazines were projected on a long sheet of blank paper bought from the local newspaper. (The last few hundred feet on a roll of newsprint cannot be used in a high-speed press.) By adjusting the distance of the opaque projector from the newsprint, the pictures can be enlarged to the proper size. The entire strip of paper containing the "tracings" of the drawings projected onto it is then taped to the wall where the mural is to be painted. The transferring process, in which the tracings are put on the wall, is accomplished by inserting carbon paper behind the newsprint, and tracing over the drawings a second time. Using moderate pressure from a pencil or ball-point pen, the tracings "print through" the carbon paper, and onto the wall. The entire strip of paper is then removed and the outlines of the figures are painted. Latex paint is easier to work with than oil-base paint, and can be removed easier when it is still wet. A spray coating of clear lacquer is applied for protection from dust and dirt.

An attractive wall mural is not only

a. Select illustration from a book or magazine.

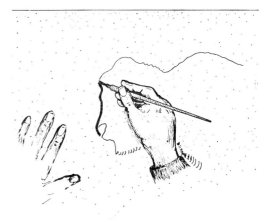

d. Ink figure on wall.

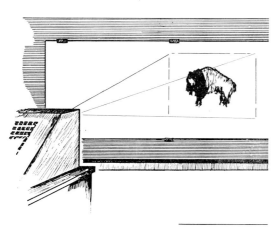

b. Enlarge with opaque projector.

e. Spray lacquer over completed mural.

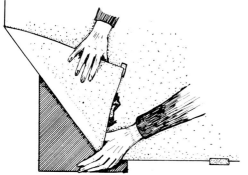

c. Transfer with carbon paper.

Fig. 6.2—These steps can be followed if you wish to make a wall mural for your biology classroom.

Fig. 6.3—The wall mural should tell a story that stimulates students to think and ask questions. Ideal subjects for murals include ecology of a pond, cell replication, protein synthesis, and photosynthetic pathways.

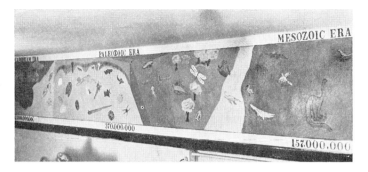

a teaching tool, but provides free advertising. Visitors to your room, including school officials and parents, remember such a mural and begin to associate it with creative teaching. As the image of the biology room improves, your image as a creative biology teacher improves also.

A WALL OF MODELS

Unused wall space—a few square feet here and there in the room—makes an excellent place for a grouping of biological models. The casual visitor to your room—as well as the beginning biology student—will be attracted by models of crayfish, earthworms, flowers, and other specimens. Colorful, three-dimensional enlargements of the human heart, ear, and eye are interest catchers. A larger-than-lifesize jellyfish makes the study of coelenterates more real to the students who have never waded ocean waters infested with *Physalia*. But good models are more than interesting. They are vital teaching aids, as important to creative biology teaching as microscopes, test tubes, and chromatography paper. They teach most effectively the relationship between structure and function, and show vividly the orientation of body organs to one another within an organism.

The expense of commercial models

limits the number a teacher may feel justified in ordering. In Chapter 8 we outline the steps a student can follow in designing and building inexpensive models of biological material, including molecular models. Some student-built, painted plaster models are shown in Figure 6.4, arranged in a wall space four by six feet.

Biology teachers are finding they must teach some chemistry before they can dig into the more sophisticated concepts of present biological thinking. You can help a student understand his chemistry by using styrofoam-sphere models of biological molecules. Of the many arrangements for such a wall of molecules, one grouping, shown in Figure 6.5, puts the models in traditional categories. Carbohydrates, lipids, and proteins are in the first three groups, followed by the energy-carrying molecules of ATP and DPN (NAD), and finally the strikingly similar structures of chlorophyll and hemoglobin—the "pigments of life." Directions for constructing these teaching tools are given in Chapter 8.

A GIANT MOLECULE OF DNA

Besides wall-to-wall murals and models there may be floor-to-ceiling models of molecules.

Look around the classroom. Note

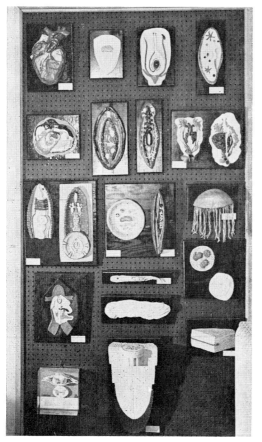

Fig. 6.4—These models, designed and constructed by students, help in visualizing small or complicated structures. Plans for building models like this are given in Chapter 8.

the wasted space. Watch especially for floor space about two feet in diameter, that extends unobstructed to the ceiling. Such a cylinder of space makes an ideal place for a wire and broomstick or garden hose and dowel model of the DNA molecule. Space next to a door, near a cabinet, or in a corner is usually free for such a molecular model.

An example of a teaching model of DNA is given in Figure 6.6. Models of this type are now common in many classrooms. The materials mentioned in the paragraph above are commonly used, but rubber tubing, small sticks, and

plastic pop-beads also make durable, attractive models.

The attention the students give your DNA molecule opens the door for creative teaching in genetics, heredity, protein synthesis, and cellular control. For example, the pairs of bases forming cross-linkages in the molecule might represent the dots and dashes of the Morse code. If so, there would be enough DNA in one human cell to encode 1,000 large textbooks.[3] Stated another way, 20×10^{-12} grams of DNA in the human fertilized egg determine all the characteristics of the mature human being weighing 5×10^{15} grams. On this basis, how much would the adult organism weigh if its DNA were actually as heavy as your floor-to-ceiling model?[4]

Models can misrepresent facts just as easily as they can clarify. DNA may be more easily visualized as a large model, but in actual size "all the DNA determining the hereditary characteristics of the entire population of the earth could be packed onto the head of a pin."[5]

A JUNGLE ON A BUDGET

Biology, the study of life, is often the study of things that are not alive at all. In large numbers of classes it is the smell and feel of something supposed to be *Spirogyra* but which seems more like a mass of black thread soaked in formaldehyde. The student who studies this sad substitute for the glistening, green, gelatinous-covered, filamentous alga not only misses the excitement of seeing

[3] Cricks' original suggestion, according to Loewy and Siekevitz, in *Cell Structure and Function*, Holt, Rinehart and Winston, Inc., New York, 1963.

[4] If your model weighed only one pound, an imaginary organism containing that pound of DNA would weigh 25×10^{25} pounds. Actually, each adult human being contains about *two ten-trillionths of an ounce* of DNA.

[5] Loewy and Siekevitz.

Fig. 6.5—These inexpensive styrofoam models help students visualize the three-dimensional nature of biochemicals. Each of the models in the picture can be removed from the wall for closer study.

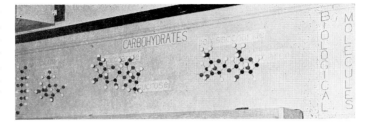

something living, but also gets the wrong idea of what *Spirogyra* really is. This problem of having living material *where* we need it *when* we need it is not new. In the text *Methods and Materials for Teaching the Biological Sciences*, published in 1938,[6] Miller and Blaydes recognized the problem. Their chapter on collecting, culturing, and preserving begins:

> Perhaps no situation offers greater problems to the teacher of biological subjects than the necessity of providing suitable materials for class use.
> Too frequently students know that laboratory or demonstration materials come out of cans or bottles but are unable to trace them further.

Judging from the opening of the same chapter in the second edition of the book by Miller and Blaydes, published in 1962, the problem is still with us:

> There is probably no greater problem for the teacher of biological subjects than the necessity of providing suitable materials for class use.

New approaches to biology make extensive use of living materials, but this does not insure that teachers are going to provide the materials. It is often easier to skip the exercise or experiment that calls for living material rather than order, culture, or collect material. Biology teachers need to include more living material in the biology classroom.

[6] McGraw-Hill, Inc., New York.

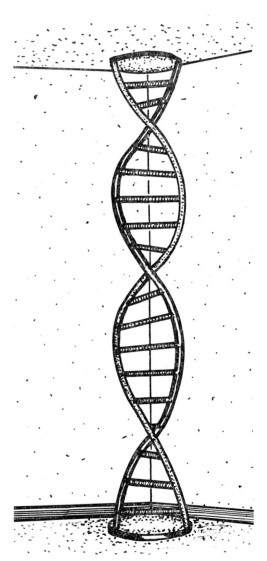

Fig. 6.6—Construction of a giant molecule of DNA is not difficult when it is a group project. This floor-to-ceiling model can be constructed from garden hose and broomsticks. The hose forms the sides of the model; the broomsticks are the crosspieces.

84

Plans for a room jungle make possible the presence of certain vascular plants, easy to maintain, and always available.

Plants are some of the easiest organisms to maintain.[7] By placing them on shelves and window sills a biology room can become a veritable jungle, but a more attractive way to grow a great number of plants for teaching material is to create a planned "jungle" in one corner. Tiers three or four steps high, built as lightweight plywood modules, lend versatility to the basic structure. From the standpoint of teaching, plants that are readily available in one area of the room save hours of before-class preparation. Maintenance of plants is easier when they are placed in a group. If artificial light is required, one group of plants is easier to illuminate than several groups scattered throughout the room. Many schools are provided with greenhouses that may be kept at con-

stant temperature and humidity. This provides a satisfactory environment for growing various kinds of plants and even animals. See Appendix F 289 for sources of supply for greenhouses and greenhouse equipment. Appendix C 240 discusses some good practices in greenhouse techniques.

Not all plants are "house plants"; that is, not all plants will survive well in a dry, temperature-fluctuating winter classroom. And not all house plants contribute effectively to the teaching of biology. In Appendix C 243 we discuss some plants that are satisfactory for class use from the standpoint of both growability and teachability.

LIFE BEHIND BARS

What is a jungle without animals? What is a biology room without something that lives, moves, makes noise, and is interesting to look at? Certainly it is not necessary for you to maintain large stocks of a wide variety of animals, but some livestock is essential, not only to

---

[7] The entire issue of the February 1965 *American Biology Teacher* is devoted to teaching with plants. Helpful suggestions are included for culturing plants that are not only useful in teaching but that grow easily in the classroom.

Fig. 6.7—A corner of the biology room should be devoted to an effective display of plants, not only as an aid in creating the proper environment in which to study biology but also to serve as a source of fresh teaching material.

complete the biology setting, but as a source of teaching material.

Birds. Birds add a bright touch, both in sight and sound. Common birds, including canaries and parakeets, can be kept easily in a variety of cages. Inexpensive cages can be purchased locally, but you may wish to construct a large wire enclosure in a convenient place in your room. Different types of wire from a lumberyard or hardware store can be formed into cages of your own design. A very attractive bird cage is shown in Figure 6.8. It is a unique floor-to-ceiling cage that one of the authors constructed for his room.

While the design of cages may vary, the features that must be built into all of them include a convenient way of cleaning the cage, and feeding and watering the birds.

Frogs. Frogs provide material throughout the year for general class work as well as individual projects. These amphibians are among the easiest of all animals to maintain in the classroom for they have no special food and temperature requirements. Frogs will eat bits of hamburger or earthworms and can stand the temperature changes of a classroom with no apparent ill effects. In fact, we have kept frogs from the time they were collected in the fall until the following spring and they appeared to be as healthy when they were returned to the field as when they were brought in. Frogs need a small, constant dripping of fresh water to keep their skin moist, but that seems to be their only strict requirement. If you have a large, permanent aquarium in your room you can easily convert it to a frog tank. Put some stones in the bottom for the frogs to sit on, and 2 to 3 inches of water in

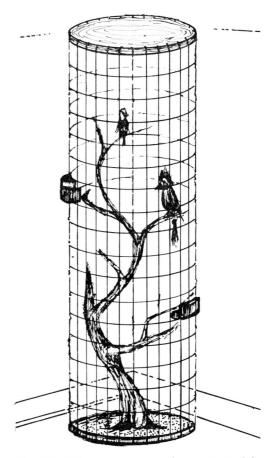

Fig. 6.8—This large cage can be constructed in about 3 hours at a cost of less than $10.

the aquarium. The water should not cover the stones.

If you do not have an aquarium that can be converted to a frog tank you can construct a wire cage, using galvanized wire with a ¼-inch or ½-inch mesh. The cage should be made small enough to fit into the sink in your room so that water from the tap will drip into the cage and keep the frogs moist. The cage shown in Figure 6.9 was put together by a student in about one hour.

The cool, moist atmosphere of a refrigerator makes an ideal environment for frogs. Put the frogs in a plastic con-

86

Fig. 6.9—A small frog cage made from galvanized wire and wood and placed under a dripping water tap in a sink keeps frogs in good condition for several months.

tainer with a loose-fitting lid, and some moist sphagnum moss or paper toweling in the bottom.

Mammals. The dictionary defines menagerie as "a place where animals are kept and trained. . . ." In a sense, your biology room is a menagerie, if it contains birds, rabbits, rats, and mice. While these warm-blooded animals require a little more care than frogs, they are easily maintained in the ordinary classroom without special equipment. Small mammals are important for they not only serve to convey the atmosphere of biology but are important teaching tools.

Sources for commercial cages and traps are listed in Appendix F 289. However, the simplest, least expensive cages can be made from throw-away materials. A one-gallon pickle jar with a wire top and watering bottle makes an ideal mouse house. A homemade jar cage is illustrated in Figure 6.10.

Tanks for Teaching. So much has been written about the "balanced aquarium"[8]

[8] The well-illustrated book *Starting an Aquarium* by Miriam Gilbert is excellent. (C. S. Hammond & Company, Maplewood, N.J., 1961.)

and its use in teaching biology that it seems to have eclipsed much of the other valuable material a biology room can have. While we do not wish to give more emphasis to an aquarium than other teaching equipment, we feel that an attractive aquarium should be a part of the place to teach.

If you are especially interested in raising tropical or local species of fish you may want to devote a rather large area in your room to several aquaria in a lighted display. Lights can be ordinary 25-watt, incandescent bulbs, enclosed in a fixture of your own design. One dealer who raises Angel fish commercially prefers to immerse the bulb about 1 inch so that it heats the water as well as lights. The tanks can be made more attractive by using white silica sand and colored stones in the bottom.

Fig. 6.10—Large-mouth, gallon jars can be obtained at no cost from restaurants or the school kitchen. Screen wire, cut into squares, can then be molded to fit on top of the jars. Sawdust is put in the bottoms of the jars and changed daily.

An attractive aquarium is illustrated in Figure 6.11.

Even if you do not plan to emphasize aquaria, you should have at least one well-kept tank. Students often expect to see an aquarium in a biology classroom, and an attractive one serves the double function of being useful to the teacher, who can cite it as an example of basic biological relationships, and to the student, who may be inspired enough by it to set up an aquarium of his own at home.

Aquaria double as cages for reptiles in both aquatic and terrestrial environments and can serve as terraria for plants. In fact, an ideal arrangement in a well-designed biology room is to have an aquarium and terrarium side by side.

Tanks for the dry-land reptiles, such as chameleons and skinks, must be kept warm and should be held at fairly constant temperatures. In Appendix D

267 we include discussion of the care of these reptiles and also that of other cold-blooded vertebrates, including fish.

A 15- to 25-gallon glass tank with a glass cover makes an especially satisfactory home away from home for exotic reptiles. The wet-land reptiles, such as turtles, do not seem to require conditions within quite so narrow a range.

Mosses and typical terrarium plants, including the sundew and Venus fly trap, require high humidity. We have given instructions for their culture in Appendix C 245, 246 and a list of these plants in Appendix C 243. Some interesting things you may do with them can be found in Appendix C 247.

The Suitable Substrate. We talk about the biotic pyramid, symbiosis, and balance in the world of life, but what do we point to for evidence? We give students generalities about "decomposers"

Fig. 6.11—This lighted aquarium could have a painted mural behind it to add interest. The mural could be designed and painted by students.

and "producers" and mention that bacteria and fungi are the vacuum cleaners of nature, returning partially oxidized nutrients to their more usable form as gases and salts. Will your students have to be satisfied with word pictures of symbionts, or can you point to evidence in the biology setting itself?

The balanced aquarium is one kind of evidence, but an old log covered with lichens, fungus, and moss conveys a more vivid picture of the role these important organisms play in recycling the chemicals of life. The physical shape of the log or branch you use, and the organisms on it will determine the type of display you make.

A suggestion for the use of soybean plants is shown in Figure 6.12 in which a wall space of 1 × 3 feet was used. The plants were held in place with wire, and the poster was attached to the wall with masking tape, rolled into small pieces, and placed at the corners on the back side.

ACCESSORIES FOR TEACHING

In addition to the wall murals, room jungle, giant DNA model, living animals, aquaria, terraria, and a substrate to show symbiosis, several other important items should be a part of the place to teach biology. Some of the most important equipment the biology teacher uses is *not* science apparatus, limited to use in a science room. Items such as an overhead projector, wall-mounted screen, slide projector, tape recorder, and flannelgraph easel all play a vital part in biology instruction. Every creative teacher should have this equipment and make use of it. While the overhead, opaque, slide, filmstrip, and film projectors have become commonplace in today's schools, there are novel ways to use them, especially in

Fig. 6.12—The organisms on these roots provide a basis for an explanation of some of the symbiotic relationships that exist in nature.

biology instruction. These, and their less used partners, including the tape recorder, flannelgraph, and magnetic chalkboard, have a vital place in dynamic instruction. The specific roles of educational media are outlined in Chapters 3 and 10. Instructional resources should be an integral part of your place to teach biology; they should become as useful to you as microscopes, test tubes, and petri dishes.

The room furniture might be overlooked, or taken for granted, since it often seems to be a permanent part of the room. (Sources for school laboratory furniture are listed in Appendix F 288.) Placement of display cabinets, laboratory tables, and the lecture-demonstration desk is an important factor. A large display cabinet is a necessity for the science setting *if* it is actually used for display. Too often room cabinets with glass doors become *storage cabinets* rather than *display cabinets*. Well-kept shelves, free of miscellaneous unsorted bottles, are interest-getters. Jars of uniform size, containing fresh, clear formaldehyde and specimens in good condition—accompanied by appropriate neatly lettered labels—create a display students *want* to look at. Figure 6.13 illustrates what we mean.

By enlisting the aid of some students and providing them with the appropriate resource material, the busy teacher needs to do little more than plan the overall display layout. Students can carry out the basic plan and still be left with enough latitude so they feel they have a creative part in the project. Sometimes it is necessary to "window dress" a cabinet with hangings of colored cloth. Printed signs add a completed look. In his excellent manual on preparation and exhibition of science materials for the classroom, Gordon Pond[9] writes: "In the future, greater efforts will be made to bring science to young people on

terms that will promote understanding and interest. One of the best mediums through which this can be accomplished is exhibition."

Mr. Pond includes a number of practical suggestions to the nonartist on making labels, arranging preserved and plastic-embedded specimens, and creating attractive and informative displays in the classroom, using inexpensive materials.

The refrigerator, autoclave, transfer chamber, growth chambers with controlled temperature and humidity, and the constant temperature oven and incubator should be given their place in the biology room. A number of individual investigations, as well as preparation and culture of material for class use, will require specialized apparatus.

Finally, in any space left in this biology room containing the "things of science" so necessary for today's teaching, a room library should be established. Biology is not only discussion, lecture, and laboratory work. A great deal of information is offered in specific texts, reference books, journals, and magazines, and *it must be available in the biology room* at the time it is to be used. A room library can fill this important need. Limited space and funds may make your reference collection a small one at first, but as you continue to show a need for the materials, the school administration will cooperate. A suggested nucleus for establishing a room library may be found in Appendix E 283 (reference books), Appendix E 285 ("How To Know" series), Appendix E 285 (the BSCS pamphlets), Appendix E 285 (the Current Concepts in Biology series), Appendix E 286 (the Modern Biology series of paperback books), and Appendix E 280 (magazines and periodicals).

---

[9] *Science Materials: Preparation and Exhibition for the Classroom*, Wm. C. Brown Publishing Co., Dubuque, Iowa, 1959.

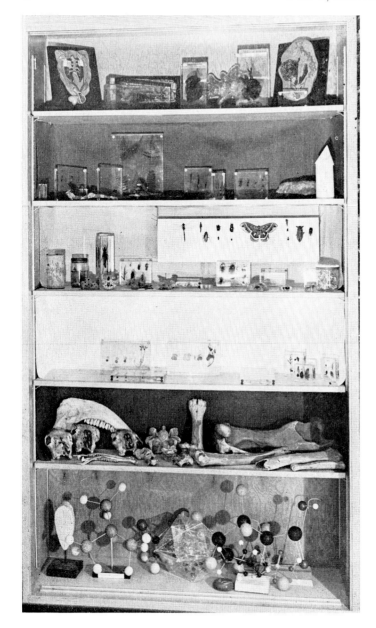

Fig. 6.13—This display cabinet shows what can be done with creative ideas and some inexpensive materials.

REFLECTING PROFESSIONALISM

There was a time when people went to a blacksmith shop to have their teeth fixed. The poor victim would be held down, screaming with pain, as the "dentist" extracted a tooth.

Blacksmiths were not the only people having two occupations in those days.

Barbers doubled as surgeons. For six centuries, starting with a papal decree in 1163, barbers of Europe practiced surgery, performing such tasks as blood letting, lancing of abscesses, and treatment of wounds. The red stripe of the barber's pole today symbolizes the blood of the barber's ancient craft, and the white stripe is symbolic of the bandage.

Fig. 6.14—Controlled environmental conditions make possible a number of long-range experiments with both plants and animals.

Today, as you recline in a dentist's chair, old-fashioned dentistry seems barbaric. Anesthetics, vibrationless drills whirring at 300,000 rpm, and a score of unique tools make the dentist's job far different from what it once was. The skilled physician, in a clean, well-lighted, efficient office with the necessary tools for carrying out the practice of medicine, provides a striking contrast to the early days when barbers were doctors.

Like the early "dentist's" office in a blacksmith shop and the "doctor's" office in a barber shop, early science teachers did their work in uninspiring, nonprofessional settings. Biology lectures were given in drab lecture halls, and class laboratory work was done in dimly lighted rooms that often doubled as chemistry laboratories or storerooms. If someone entered such a room it might be difficult for him to determine what subject was taught there. Sometimes the only clues would be rows of jars filled with preserved plants and animals, or a skeleton hanging by its skull in the corner. Such surroundings, often supplemented by dull teaching, must have indeed made biology a tedious and uninspiring subject.

What will your biology room mean to the student? Will it reflect professionalism? Will it reflect the type of teaching—the *involvement with biology*—that you want to convey? The biology room must be a continuing, quiet source of excitement to the student. It must give him the feeling that he is *in biology,* and that he can get inspiration by a kind of osmosis, from his surroundings.

REFLECTING CREATIVITY

Remember—a creatively designed biology room does not necessarily insure creative biology teaching. Any number

Fig. 6.15—A room library cannot contain the comprehensive list of references that are found in the main library. However, it can become an information center for special topics studied throughout the year.

of people are excellent craftsmen, but they are poor speakers, or organizers, or disciplinarians. To transform the room itself into an appropriate place to teach creative biology you must become, in a sense, an interior decorator. You must have the imagination to see where to put the charts, the grouping of plaster models, the styrofoam models, the animal cages. It is entirely possible you *do* have the imagination and artistic flair required to bring about these changes in your room. *But*—it is also quite possible that you would never be able to teach with the same artistic flair. Your teaching techniques must take on the same new feeling of creativity that your room reflects. Even teachers who have a small amount of imagination find it easier to

teach in the kind of biology classroom we have described in this chapter. The words of an uninteresting instructor seem less drab, the chalkboard drawings seem less sketchy. You can rise to fine teaching in the properly designed room. As you teach in such a classroom your words glisten with new brilliance.

A properly engineered biology setting means as much to students as a well-designed dentist's or doctor's office does to patients. Surely no one would want to return to the days of the blacksmith shop for his dental work. And who would want to have surgery performed in a barber shop? The days of biology teaching in drab surroundings were with us only yesterday. Let us hope they are gone.

Fig. 6.16—The modern biology classroom, properly designed and decorated, is a controlled environment that creates an optimum learning situation. (From E. H. Sheldon Equipment Company, Homewood, Ill.)

CREATING CHAPTER CONCEPTS

1. What technological changes have influenced the construction of biology classrooms and the equipment that is used to teach biology?
2. Design a scale model of a biology classroom, using styrofoam or cardboard replicas for the furniture in the room. A scale of 1 inch equals 1 foot is convenient to use. Build your model on a piece of corregated cardboard and include the four walls for the room, indicating what kinds of displays you would have on the walls.
3. Using catalogs of laboratory furniture, make up an order for desks, benches, and cabinets that you would like to have in the room where you teach biology. Assume you want to accommodate 25 or 30 students and that they are to have both class work and laboratory work in the same room.
4. Assume you have a biology classroom that is complete in every respect except for one of the following. Indicate how you would overcome each particular deficiency.
   a. No compressed air.
   b. No gas jets.
   c. No electrical outlets at the laboratory tables.
   d. No large display cases.
   e. No preparation room.
   f. No storage cabinets in the room.
5. As a group project with several other students in your class, design and carry out one of the following projects. If it is possible, install your projects in the classroom where you meet.
   a. Design and paint a wall-to-wall evolution chart on a long strip of paper.
   b. Arrange a group of painted plaster models to illustrate a particular biological theme.

c. Build a giant DNA model that extends from floor to ceiling.

d. Construct a plant jungle in the corner of your classroom. If it is impractical to bring in potted plants, make cardboard models of the plants.

e. Make a large wire birdcage for the classroom.

f. Compile an annotated list of suitable texts for the room library.

6. Design and construct one of the following:

a. A wall chart that illustrates the relationship between animal phyla and geologic eras.

b. A series of clay models to illustrate the evolution of the horse.

c. A wall-mounted display of *Lepidoptera* common in your area.

(For help in carrying out this project, consult Pond's book on the preparation of biological materials for display.)

7. Make a list of 20 plaster models that would be the most desirable to have in the classroom.

8. What other charts besides the wall-to-wall evolution chart would you consider valuable in teaching biology?

9. Take a field trip to find a small log or branch that has lichens, moss, and fungi on it. Work out a display using the material you find. Include appropriate labels in your display.

10. Work out a way of representing your classroom as a cell and devise ways to represent the organelles. Balloons and other three-dimensional structures can be used to show some of these.

11. Create a mobile, using cardboard and wire (from metal coat hangers), that represents biology. The mobile should be made so that it can hang from the ceiling of the classroom and move freely.

12. Write brief descriptions of some of the superior teachers you have had and include ways they made their classroom seem more closely associated with the subject being taught.

13. Discuss practical ways in which you could remodel your present classroom.

14. Make a list of community resources that could be used as you begin the task of converting your room into a more meaningful place to teach biology.

15. Work out a set of instructions for keeping the following items the year around in your classroom:

a. An ant farm.

b. A demonstration beehive.

c. Living snakes.

d. Tropical fish.

e. Tropical birds.

f. Exotic plants.

16. List 5 items or techniques you plan to use to influence your students to "see, hear, and feel biology."

17. Interview an architect, using a tape recorder, to gather information on designing a biology classroom and laboratory.

18. Visit several schools in your area and take snapshots of the biology classrooms. Have the pictures enlarged and then make a key for each one, indicating the good and the undesirable features of each room.

19. Create a before and after pair of cartoons that depicts the change in biology classrooms over the past 50-year period.

EXPANDING THE CHAPTER

DAWSON, JAMES R. 1964. The Impact of New Curricula on Facilities for Biology. *The American Biology Teacher* 26(8):601–4. In this article the efficient use of space in the laboratory and the influence of BSCS on present biology teaching are discussed.

HEISS, E. D., OBOURN, E. S., and HOFFMAN, C. W. 1950. *Modern Science Teaching*. The Macmillan Co., New York. Some excellent floor plans in Chapter 11 of this book are helpful in remodeling or designing a biology classroom.

KEEFE, REV. ANSELUM M. 1965. Plants for a Window Garden. *The American Biology Teacher* 27(2):118–23. The article contains a list of plants that can be grown in any room with windows.

POND, GORDON G. 1964. *Science Materials: Preparation and Exhibition for the Classroom.* Wm. C. Brown Publishing Co., Dubuque, Iowa. A well-illustrated manual with detailed instructions on making charts, printing signs, arranging displays, and making effective use of space, color, and materials in displaying things of science.

WONG, HARRY K. 1964. Modernizing a Science Laboratory. *The Science Teacher* 31(6):39–41. One of the first articles that should be read by anyone contemplating designing a biology classroom.

CHAPTER 7

Planning a

Science

Laboratory

IN THE MIDDLE OF AUGUST, 1771, an Englishman was busy at work in a corner of his wife's kitchen. Though he was trained as a minister, his hobby was science and he often worked in the kitchen; it was the closest thing he had to a laboratory. He kept a few jars and bottles in the cupboard, along with some chemicals, and in the kitchen he could use the water and heat that were available.

The clergyman-scientist did not work for a university, nor did he have a grant to support his research. In fact, the materials he used and the surroundings in which he worked seem quite unscientific by present-day standards. On this particular August afternoon he was working with a sprig of mint, a candle, and a jar of water. In his notebook he wrote:

Finding that candles burn very well in air in which plants had grown a long time, and having had some reason to think there was something attending vegetation which restored air that had been injured by respiration, I thought it was possible that the same process might also restore the air that had been injured by the burning of candles.

Accordingly, on the 17th of August, 1771, I put a sprig of mint into a quantity of air, in which a wax candle had burned out. . . .

In ten days this man was able to see the result of his simple experiment, for on August 27 he recorded another observation in his notebook:

. . . On the 27th of the same month, another candle burned perfectly well in it. This experiment I repeated, without the least variation in the event, not less than eight or ten times in the remainder of the summer. Several times I divided the quantity of air in which the candle had burned out, into two parts, and putting the plant into one of them, left the other in the same exposure, contained, also, in a glass vessel, immersed in water, but without any plant, and never failed to find, that a candle would burn in the former, but not in the latter.[1]

From these quiet, simple surroundings one of the most basic biological discoveries was made. The discovery: the production of oxygen by plants. The discoverer: Joseph Priestley.

The 1700's were times of great activity in other fields of science. Long-standing theories in the areas of taxonomy, medicine, and comparative anatomy were overturned by a new approach to research. Theories, such as those of Galen's description of circulation, that had stood unchallenged for 14 centuries were crumbling under the first clean-cut scientific application of methods of investigation. For the first time, natural laws that had been handed down from Hippocrates and Aristotle were being examined from a new point of view. Instead of viewing science from the philosopher's bench, biologists were working at science from the laboratory bench. They were beginning to prove (and disprove) hypotheses by experiments rather

---

[1] Mordecai L. Gabriel & Seymour Fogel, *Great Experiments in Biology* © 1955. Reprinted by permission of Prentice-Hall, Inc., Englewood Cliffs, N.J.

than by engaging in debates and philosophical discussions.

Perhaps the problem of biogenesis, or spontaneous generation, illustrates the parting from tradition that biological research was beginning to take.

The origin of life has been a basic problem to biologists over the years, and of course is not completely solved today. But important inroads to the solution of the problem have been made. One of the early attempts to explain it was the theory of spontaneous generation, which held that life could arise from a favorable combination of nonliving materials, involving no parent organisms. An old biological belief, it came into being during the time of the Ionian school of philosophers, who thought that flies and other small creatures were formed spontaneously in the mud on the bottom of streams. The formula for making mice, it was thought, consisted of moist, soiled rags, a dark attic, and some kernels of wheat or pieces of cheese.

For centuries almost no one tried to experiment with the theory of spontaneous generation. In fact, the designation of spontaneous generation as a "theory" was false in itself, for it implied that an observed problem had given rise to a hypothesis that then was tested in a series of experiments before it became a theory. Why no one took the time to perform an *experiment* to prove the hypothesis is a puzzling mystery.

Curiosity about life's origins was the driving force that motivated Francesco Redi to design a set of experiments to try to settle the problem. No one told this Italian physician that he was using the "scientific method." He was simply following a procedure that seemed to him the only logical approach to use in trying to prove whether a statement was true or false. His careful work, reported by Eldon Gardner in *History of Biology*[2] illustrates the clear, incisive thinking of a true scientist.

Redi set out to determine by experimentation whether flies could be produced spontaneously. His procedure followed a precise pattern and it represents an example of applied experimental science. He began by killing three snakes and placing them in an open box where they were allowed to decay. Maggots appeared on the decaying flesh, thrived on the meat and grew rapidly. The maggots were observed critically day after day and . . . several varieties were identified by size, shape, and color. . . .

Redi continued his experiments, testing different kinds of flesh, both raw and cooked. He used meat of oxen, deer, buffalo, lion, tiger, duck, lamb, kid, rabbit, goose, chicken and swallow, and . . . several kinds of fish. Maggots developed on the meat . . . and adult flies of the same kinds that emerged from the maggots were observed to hover over the decaying meat and Redi noticed that the flies dropped tiny objects on the meat. Redi theorized that maggots might be developing from the objects dropped by the adult flies on the putrefying meat.

An experiment was carefully designed to test this hypothesis. Portions of fish and eel were placed in flasks. The openings of some flasks were completely sealed off and the meat was observed through the glass as it underwent decay; comparable flasks prepared in the same way were left uncovered as controls. Flies were soon attracted to the opened flasks and in a few days maggots appeared on the meat. Similar flies were also observed inside. Occasionally maggots appeared on the tops of the sealed flasks. They would wriggle on the surface and appear to be trying to get through the glass to the putrefying meat inside. This indicated that the maggots were developed from the elements dropped from the adult flies and not derived spontaneously on the decaying meat.

[2] Burgess Publishing Co., Minneapolis, 2nd Ed., 1965.

Redi continued his investigation to see if different seasons had an effect on the work, or if various kinds of vessels might have some effect. He also buried some meat underground as a means of preventing its exposure directly to the air and the flies. While Redi did not completely solve the problem, he did bring into focus *the method of solving a problem in science.* Though "scientific method" has become a cliché in popular literature today, its use as a means of attacking and following through a problem to its solution has been strengthened over the years. The scientific method and the laboratory in which it is practiced are the ways and means of science.

THE WAYS AND MEANS OF SCIENCE

The work of Priestley and Redi was indeed significant, for the results of their work helped lay foundations of modern biology. But though the *product* was important, the *process* was just as meaningful. The scientific way of working—the *method of setting up the problem and organizing the research*—is just as important as the problem itself.

The early natural philosophers thought they could solve life's riddles by talking about them. It was rare when someone thought of turning to the laboratory to study natural phenomena. Yet it is the laboratory where much of the real work is done.

Because the laboratory and laboratory investigations play such a basic role in biology and the other sciences, teachers feel it is important to use the laboratory work for teaching. Several objectives are accomplished when students learn in a laboratory: they get firsthand experience with the apparatus and living material that biologists use; they realize that a laboratory is a workshop for testing hypotheses; and they get practice in using the methods of science.

THE MEANING OF LABORATORY WORK

A laboratory is more than a room with worktables, glassware, bottles of reagents, and specimens. It is even more than a place where students can handle things of science and come into direct contact with the organisms and equipment that are discussed in the text. It is the place to do the *labor of science.* In fact, someone has remarked that a laboratory is a place for "more labor and less oratory."

Though laboratories have been the centers of biological discoveries for many years, they have not always been used as places to teach biology. It was long the custom to lecture about biology, and students learned it purely as an academic subject—from books and notes taken during a professor's discussion. No attempt was made to make it a practical experience with plants, animals, and protists. This nonlaboratory treatment of a laboratory science was not confined to biology however; even a subject as laboratory-centered as chemistry was taught in the United States strictly by lecture until 1858 when Josiah P. Cooke began teaching the first student laboratory at Harvard.

What were these teaching laboratories like? Were they similar to the research laboratories of today? Were they

Fig. 7.1—Redi's experiments, now famous, were not accepted widely when he did them 100 years ago. What other significant work on this theory was done at that time? (From *Biological Science: An Inquiry Into Life,* Harcourt, Brace & World, Inc., 1963. Used by permission of the Biological Sciences Curriculum Study.)

● Redi prepared four flasks of animal flesh, and left each flask open to flies.

● Four other flasks were prepared identically, but these were covered to exclude flies.

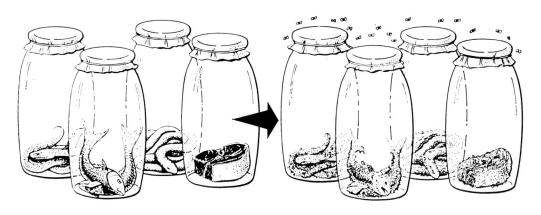

● The final experiment again excluded flies.....

.....but admitted air. Why was this modification made?

places for students to design and conduct true experiments and investigations, or were they places to do exercises that simply substantiated scientific principles described in the textbook? The use of the laboratory in biology teaching today is an outgrowth of its introduction into classroom teaching in the 1800's and its gradual growth since then.

Today, through creative use of the laboratory as a teaching device, it is possible to build an entire course in biology around laboratory work. Before we turn to some suggestions on its creative use, let us examine the mechanisms by which the scientific method can be learned in the laboratory. Let us look at the merits of the *exercise,* the *experiment,* and the *investigation.*

The Value of an Exercise

Like physical exercise, or exercises on the piano, laboratory exercises perfect skill. They train the fingers and sharpen the mind in the techniques of *doing.* Since so much of the work in biology involves techniques that are used over and over, exercises that allow practice of the techniques are valuable. The early biology laboratories were largely places where techniques were practiced by the students. The laboratory manual consisted of a series of "recipes" much the same as a cookbook contains recipes for preparing food. The students followed a prescribed set of directions and usually came out with a predictable set of answers. Some of the exercises were so unimaginative, however, that they could be completed without going into the laboratory at all. Such "dry-labbing" was possible because the exercises usually contained a number of blanks to be filled in with one-word answers, and drawings that could be labeled by copying similar drawings in other texts.

Laboratory exercises continue to serve a useful function and are still widely used in biology teaching. They have the advantage of reinforcing learning by repeating in the laboratory selected classic experiments of biology. The student has the opportunity to retrace the footsteps of biologists who made basic scientific contributions. Even though a student has memorized the fact that mercuric oxide will decompose with heat to yield mercury and oxygen, there is a particular kind of excitement awaiting him who *confirms* this fact. A certain amount of confirmatory laboratory work is essential, especially for beginning students, for it not only shows the student how a particular discovery was made, but it points up places where incorrect conclusions and experimental error may have occurred.

Laboratory manuals based entirely on confirmatory exercises unfortunately can become fill-in-the-blank workbooks. The spirit of original investigation seems to get lost, or at least covered up, in the process of searching for the correct term to write in the proper blank. After all, if a laboratory exercise has been designed to show a certain principle of science, there is not much latitude for answering original questions related to it. Knowing that teachers are busy people with lots of papers to correct, writers of laboratory manuals attempt to reduce the method of reporting conclusions to a series of blanks to be filled in. While this makes laboratory work easier for the teacher to evaluate, it leads the student to begin thinking in terms of a series of fixed, factual answers. Quite often the facts can be located much easier in a textbook than from the laboratory work, and students are quick to realize there is such a convenient shortcut.

The Nature of an Experiment

In contrast to the exercise, which may encourage recall of memorized facts, the experiment requires the student to formulate conclusions on the basis of experimental evidence. Often there is not only one correct conclusion, but several. Sometimes the conclusion proves something that is *not* true, and this can be just as important as knowing what *is* true. A student who works on a well-designed experiment in biology feels that he is solving a problem, and for this reason the use of experiments in biology teaching is sometimes termed the *problem-solving approach.*

Some teachers feel that it is important to perfect technique before doing experiments. Others feel that beginning students should have the opportunity to confirm a large number of known phenomena through exercises before they begin to work on "research problems" of their own design. Of course, both opinions have merit, and the task of the creative biology teacher is thus not in choosing *which* type of laboratory to operate, but *how much* of the exercise, experiment, and investigation to blend together to create the most effective learning situation for his students. Sometimes it is possible to design a given problem either as an exercise or an experiment. Let us consider the following laboratory problem and examine its merits when it is designed either as an exercise or an experiment. The problem deals with the formation of a white powder that appears when solutions of an acid and base are mixed, and then evaporated.

THE WHITE POWDER PROBLEM

Suppose you have just finished a discussion about salts and their importance in biological systems. As a means of becoming more familiar with salts and their formation when acids and bases react, you plan some laboratory work. The chemical reaction between hydrochloric acid and sodium hydroxide will form the basis of the laboratory work in which solutions of the acid and base are mixed on a watch glass and then evaporated over a water bath.

As a confirmatory exercise, the laboratory directions can simply instruct the students to mix the acid and base, evaporate the mixture, and note the formation of a white powder, which is sodium chloride. The fact that the laboratory directions state a white powder will form, and that the powder is sodium chloride, removes all elements of problem-solving from the exercise, and it becomes simply a confirmation of the instructions. There is some suspense as students watch the mixture evaporate, but almost all of the students know before they perform the exercise what the outcome will be. The value in doing laboratory work of this kind, as we pointed out before, is simply in seeing the salt form. There is also some value in practice with the metric system, as the prescribed milliliters of acid and base are measured.

Designed as an experiment, the white powder problem would include the same basic laboratory instructions, with one important change. It would *not* state that a white powder would form after evaporation of the acid-base mixture. Instead, it would ask the question or pose the problem: "Describe the results of evaporation of the acid-base mixture." Then it would ask for an explanation of the results, based on a balanced chemical equation. A properly designed experiment would also include parallel lines of investigation of the problem to rule out certain sources of error. There would need to be provisions for experimental "controls," or experiments identical to the original, but in which one factor was changed. Thus, the first question that asked for an explanation of the results would not lead to a single answer, but instead would require the design of secondary experiments before a complete, satisfactory answer could be obtained. For example, would it be correct to assume that the powder was the result of a reaction between the acid and base? Perhaps either the solution of acid or the base, when evaporated by itself, would yield a white powder. Here, then, is one hypothesis to test by experiment before a complete answer can be given to the original question. Suppose the base NaOH and acid HCl were mixed, then evaporated to form white crystals. It is *not* correct to assume these crystals were NaCl, because further experiment shows that a solution of NaOH will *also yield white crystals*. And what about the water that was used to make up the original acid and base solutions? Could it contain a white powder too? Evaporation of both tap and distilled water helps answer that

question. To accurately pinpoint the source of the white powder in this experiment it finally becomes necessary for the student to use some knowledge of chemistry outside the boundaries of the original problem. After localizing the source of the white powder, and showing it to be formed from the neutralization reaction of the acid and base, he can test it with pH paper. NaCl, a neutral salt, will form an aqueous solution that has a pH of 7. If, on the other hand, the powder was NaOH, the pH of such a solution would be close to 10—characteristic of a strong base.

The white powder problem is an example of a seemingly simple exercise that could be turned into a complicated experiment. Perhaps it seems so complicated that the original objective—that of showing the student the nature of an experiment—was lost. However, the complicated one we have outlined proves a point that is too often overlooked in the discussion of experiments for students of biology. Experiments are never simple; they often ask more questions than they can answer. Experiments need controls as checks on their accu-

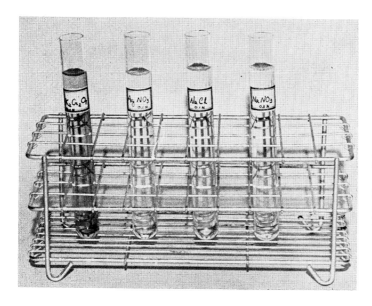

Fig. 7.2—Describe the results if the contents of any two of these test tubes are mixed. On the basis of the possible reactions that can occur, write the procedure as an exercise, as an experiment, and as an investigation. Each write-up should be about 1 paragraph in length. The solutions are (l. to r.) $K_2Cr_2O_7$, $AgNO_3$, NaCl, $NaNO_3$.

racy. They frequently lead into uncharted waters of thought, depending on knowledge and techniques from the complete spectrum of science. A single experiment, properly carried out, can spawn dozens of others. The chain of events triggered by experiments makes up a long-range problem termed an investigation.

The Art of an Investigation

Actually, an investigation is nothing more than an experiment that has no end, or rather, has an open end. Open-ended experiments have several advantages over exercises, but perhaps their greatest value lies in their ability to require the student *to develop a specific problem-solving thought process.* He is required to observe, hypothesize, experiment, and finally theorize. Telling him that these are the steps of the scientific method is one thing; having him follow these steps as he designs an experiment to test a hypothesis is quite another. Anyone can memorize the steps in the scientific method, but only the student who has been trained to think with the method, and who has practiced using it in the laboratory, can arrive at a correct solution.

The Biological Sciences Curriculum Study Committee on Innovation in Laboratory Instruction considers the open-ended, long-term investigation to be of great value in high school biology teaching. In the preface to their laboratory block on *The Complementarity of Structure and Function*,[3] they state:

During the next six weeks we expect you to work and to think like a student or research worker. Your actions in the laboratory or while doing homework should be directed by logical thinking. Each step of your investigations should be governed by the basic questions: *how, what, where,* and *why?* The answers you get will be precisely as correct as your combination of background knowledge, logical thinking, and accurate experimentation let them be.

The scientist does not know all the answers. He is seeking to learn the answers just as you are.

The first experiments could end in failure for either you or the scientist. The scientist often learns more about things that will not work than about things that will work.

The procedure given in this Laboratory Block is one that a scientist might follow if he were studying the same problem.

While this example of a six-week block of laboratory investigation is a recent addition to the excellent material available for teaching biology today, the art of investigation that this laboratory block cultivates is certainly by no means an innovation in science. It is not a recently added trimming put on as an afterthought. Investigations have formed the groundwork of biological research. Scientific investigations are not confined to run a few days, weeks, or months. The story of W. J. Beal and his investigation of seed dormancy is an example of a classic open-ended problem that is scheduled for completion in the year 2040.

Described by Trump and Fagle in *Design for Life*,[4] Beal's work remains an example of a long-term, open-ended investigation.

In 1887 Dr. W. J. Beal of Michigan Agricultural College (now Michigan State University) started an experiment to help answer the question of seed dormancy. The experiment is scheduled for completion in the year 2040. Beal was mainly

[3] A. Glenn Richards, BSCS Laboratory Block, D.C. Heath and Co., Indianapolis, 1963. Used by permission of the Biological Sciences Curriculum Study.

[4] Holt, Rinehart and Winston, Inc., New York, 1963.

concerned with how long seeds can remain alive in the soil, when buried too deep for germination. He dug a narrow trench about 18 inches deep "on a sandy knoll . . . fifteen paces northwest from the west end of the big stone set by the class of 1873." In the trench he buried 20 bottles of seeds that were mixed with damp sand. The bottles were without lids, tilted downward to avoid flooding. They contained identical collections—50 seeds of each species.

Beal had intended that the experiment should run for 100 years. But after the 40th year, when eight kinds of seeds germinated, Dr. Beal's successors decided to extend the planting intervals to ten years.

The exercise, experiment, and investigation represent three different roads, but they all lead to the same objective: the understanding of science as *knowledge* and as a *method*.

Fig. 7.3—Dr. W. J. Beal began an investigation of seed viability that will not be completed until the next century. He set up the investigation in 1887. (From Trump and Fagle, *Design for Life,* Holt, Rinehart and Winston, Inc., New York, 1963.)

CREATIVE WAYS OF OPERATING A DYNAMIC BIOLOGY LABORATORY

Instant Laboratory

Sometimes laboratory work is left out because it takes so much time to prepare for a laboratory session. With the many duties that high school teachers have in addition to their actual teaching assignments, they find little time to devote to planning laboratory work, gathering materials, assembling the apparatus, and then cleaning everything and putting it away after the laboratory class is finished. Preparation for stimulating and challenging laboratory work often is slighted.

To minimize laboratory preparation and the hurried rush for the few final pieces of equipment before class, an *instant laboratory* should be constructed. Simple and inexpensive, the instant laboratory center uses shoeboxes, or other suitable containers, located in a section of shelving. Each box contains the equipment necessary to conduct a specific experiment or demonstration. If the apparatus is small enough to fit into the box, a class-quantity may be kept. For those experiments that require larger pieces of apparatus and more individual pieces, a card listing the missing items and their location in the room (or source of supply) may be put in the box. Labels for the boxes may be handprinted with a felt-tipped pen, using different colors for different types of demonstrations and experiments. For example, those subjects dealing with plant material might be labeled in green ink, those with animals in red, and those with protists in black. The color coding helps identify the contents of the instant laboratory at a glance. Or you might organize the boxes by units. For example, all of the experiments used for a unit on physiology might be in one cluster of boxes, all of those for ecology in another cluster, and so on. You could paint the fronts of the boxes different colors, keyed to the units. Appendix A 204 suggests materials you may wish to include as part of the instant laboratory and storage center.

In addition to serving as a storage center for laboratory material, the instant laboratory can be a central store for all of the miscellaneous material

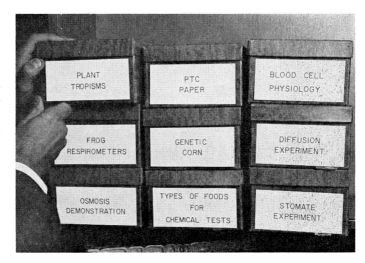

Fig. 7.4—A pile of empty shoeboxes and a little labor are needed to make this time-saving storage center we call the *instant lab.*

106

that must be on hand in a biology room. Include items such as cotton, small hand tools, clean microscope slides and cover slips, string, test tubes, insect pins, masking tape, aluminum foil, corks, droppers, felt-tipped pens, stirring rods, razor blades, dissecting equipment, and lens tissue.

A specific experiment that can be adapted to storage in the instant laboratory deals with diffusion. Designed to show the importance of particle size and its effect on diffusion rate, it makes use of eight-inch lengths of 6mm glass tubing, tiny cotton plugs, and dropping bot-

Fig. 7.6—These students are determining the rate of diffusion of HCl and NH₃. A drop of concentrated HCl is put on the cotton at one end, while a drop of concentrated NH_4OH is put on the cotton at the other end. What is the product of the reaction? Where in the tube will it occur?

Fig. 7.5—Commercial scientific suppliers manufacture instant labs that are designed to provide the teacher with all equipment necessary to perform a specific demonstration. (From The Welch Scientific Company, Skokie, Ill.)

tles of concentrated HCl and NH_4OH. Working in pairs, students place a cotton plug in each end of the tube, then put a drop of HCl on one plug, and a drop of NH_4OH at the same time on the plug at the other end. In a few minutes a white ring of NH_4Cl forms about one-third the distance from the HCl (two-thirds the distance from the NH₄-OH).[5] Class quantities of the lengths of glass tubing, cotton, and dropping bottles are stored in one of the boxes in the instant laboratory, to be pulled out and used with a minimum of preparation.

Side Table Demonstrations

To broaden the scope of laboratory exercise, or to provide supplementary materials for experiments, it is often

[5] Conclusion: Molecules twice as heavy (HCl) travel only half as far as lighter molecules (NH₃). Do you think results would be different if the tube were held vertically instead of laying it horizontally on the table?

helpful to make use of side table demonstrations. Such demonstrations consist of individual specimens which are not available in class quantity, setups of experiments that may be studied by two or three students at a time, clusters of library materials and references dealing with the specific topic on which the laboratory work is centered, or displays of articles and clippings that relate to the laboratory work during that particular period. When preserved specimens of plants and animals are set out as demonstrations, a 3 x 5 card beside them is helpful in guiding students as they study. The explanatory card does not have to carry a complete explanation, for it is sometimes more instructive to have it list a series of questions about the specimen than to have it contain mere statements of information. Fragile and rare organisms embedded in clear plastic are excellent items for display, and when not used during laboratory work they can form a part of the permanent room display.

Side table demonstrations are not meant for casual browsing, but for study. To aid in integrating them with the individual laboratory work that students do, a worksheet should be designed. While it requires considerable time and thought to design these worksheets, a few can be constructed each year until you have a comprehensive file. Of course, questions on the worksheets should be constructed so that they require students to come to conclusions rather than merely recall facts. A worksheet designed to help students learn about annelids might be constructed in the following way:

Function and Structure in Some Annelids

— Leeches —

General description. These annelids lack setae and parapodia. They have anterior and posterior suckers. The coelom is much reduced and filled with a large amount of tissue. The flattened body has a wrinkled appearance because it is segmented, or divided into many short sections.

In the following chart, write a short de-

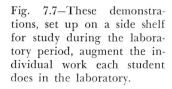

Fig. 7.7—These demonstrations, set up on a side shelf for study during the laboratory period, augment the individual work each student does in the laboratory.

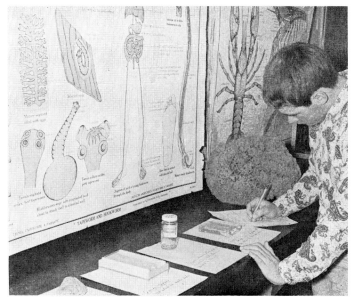

108

scription of the adaptations each structure has so that it may perform its function. Study a living leech, and a preserved, dissected leech for aid in completing the chart.

Structure	Function	Adaptations of the Structure to the Function
pharynx		
esophagus		
crop		
lateral caeca		
stomach		
intestine		
anus		

Observe living leeches.

1. Describe their movement, and list the types of muscles that are apparently present that would cause such movement.
2. How are leeches able to hold fast to a substrate?
3. What structures associated with the head of a sandworm are not found on the leech?
4. What functions can you list for these structures?

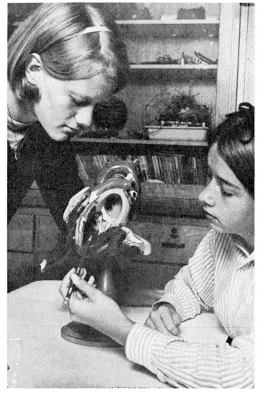

Fig. 7.8—A plaster model of certain animal or plant structures can help students clarify their understanding of the anatomy of living things.

Although plaster models of biological material are always present in the room, either on the wall or in a display case, they should be brought out for close study by the class during appropriate times. If you have a model of a leech, for example, you should include questions on a worksheet that call attention to the leech model and the structures it shows. The questions should cause the students to compare the design of the model with the design of the living or preserved animals themselves.

Occasionally worksheets should contain questions that can be answered only by reference to the room library. Having books and periodicals available in the room itself makes laboratory sessions especially dynamic, for nothing is quite so frustrating as having the desire for a certain bit of information yet being unable to turn immediately to a source that contains it. The room library must become a center for individual study during laboratory periods.

A Teaching Museum

In addition to side table demonstrations and the instant laboratory, a teaching museum can be used to emphasize relationships between plant, animal, and protist groups, or to stress certain relationships in biochemistry, physiology, or cytology. Ideally, such a museum would be in the classroom, but it could be located in the hall outside the classroom, or elsewhere in the building. The museum need not be large; a glass cabinet on the wall will serve for display of specimens, models, pictures, or flannelgraph overlays. If more room is available a floor case can be converted into a museum, and the wall space behind the case can be used to hang a chart that correlates the information on display. An ideal size for such a case is 4 feet long, 3 feet high, and 2 feet from front to back. The wall chart may be made of white oilcloth or paper. Drawn with a felt-tipped pen, in color, the chart can be made from a design projected on the background with an opaque projector, in a process similar to that described for making the evolution mural in Chapter 6. For maximum effectiveness the teaching museum should be changed about every 6 weeks. Though the topics should be of general interest to all biology students, they will be determined in part by the materials you have available for display.

Some suggested topics are:

> Photosynthesis
> Plant Reproduction
> Alternation of Generations
> Ladders inside YOU: The DNA Story

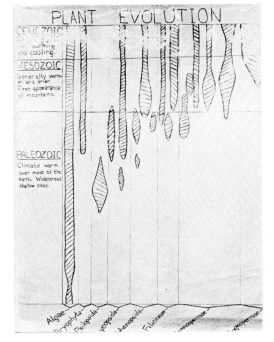

Fig. 7.9—This simple yet effective "museum" was planned by the students and the teacher. A display has been arranged in the case. A chart (above) has been prepared and hung over the case to further explain its contents.

Evolution (This topic should be narrowed, since it is too large by itself. For example, evolution of the horse, or a selected type of insect, might make a more satisfactory topic.)
Respiration
Digestion in plants

For your students to get the most out of each museum, make an out-of-class assignment in which they follow a prepared worksheet, emphasizing ideas brought out in the display. One such worksheet on Plant and Animal Evolution might take the following form:

THE EVOLUTION MUSEUM

Before going to the Evolution Museum to complete this assignment read pp. 16–24 in the text and then answer questions 1–4 below.

1. What criteria do geologists use to divide time into categories?

2. What methods are used today in determining the age of both living and fossilized plants and animals?

3. What evidence is there that plants and animals have changed in form and structure over the years?

4. Why do we think that life on earth probably began with simple forms and evolved to the more complex forms?

—— ANSWER THE FOLLOWING QUESTIONS AT THE EVOLUTION MUSEUM——

5. Which group(s) of animals is (are) present in greater numbers today than ever before?

6. Which major group(s) of organisms appears (appear) to be the most successful in the world today? Why? (Cite two reasons as they appear in the museum.)

7. If a group of animals is present in large numbers, does this usually indicate that there is also a wide variety of species within the group? (Give reason for answer based on what you see in the Museum.)

8. What major morphological characteristics do all groups of animals have in common?

The Laboratory Test

Laboratory tests, sometimes called "laboratory practicals" or "spot tests" should be given at least once a semester during a course in biology, and more often if you have the time to construct them. A laboratory test emphasizes the material studied in the laboratory and includes dissections, demonstration specimens, models, charts, and prepared microscope slides. The test is administered in the laboratory room, with individual "stations" established at each student's place. A station consists of the object over which the questions are asked, and a 3 x 5 card that contains the questions. The answers, a short phrase or a single word, are written on a separate answer sheet. As the students move from place to place they answer the questions in numerical order. A total of fifty questions, covering twenty-five stations, makes a satisfactory laboratory test because it can be finished within a fifty-minute class period. Of course, the physical arrangement of the room, the scope of the material being covered, and the number of students taking such a test are additional factors that must be considered when making up such an evaluation.

If you feel that a fifty-question laboratory test is not practical for your class, you might try a shorter version. Use four or five stations, set up on the window ledge or demonstration desk. During a longer written test students go to the stations in a prearranged sequence, interrupting the written test temporarily.

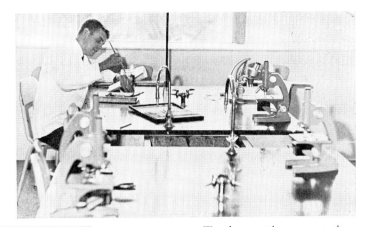

Teacher setting up stations in a lab test.

Fig. 7.10—The laboratory test should evaluate the student's understanding of what he did in the laboratory. Numbered pins may be used to locate specific structures in some of the specimens. The questions are on 3 x 5 cards near each microscope or specimen.

A station showing card with question.

The laboratory with students taking the lab test.

A Room Map

The many pieces of equipment, both large and small, have a way of getting piled into corners, empty shelves, and drawers in the biology room until the teacher himself is not quite sure what he has. Even if the room is used by only one teacher there is a problem of keeping the inventory in order. In many schools two or more teachers migrate into certain classrooms for one or two periods each day. For these teachers who work in your room occasionally, but do not have their office there, a room map is necessary. A room map not only makes possible the immediate location of equipment, but it has the added advantage of serving as a cumulative inventory. Slides, models, preserved materials, chemicals, catalogs of scientific equipment, cultures of protists, and many other incidental items that are used in biology should be listed on the room map.

The map itself should consist of two parts: a drawing of the room, and a list. The drawing can be a floor plan. The location of cabinets, bookcases, desks, and all storage spaces should be shown in the drawing, along with a number or code letter for each. The list is most helpful if it is arranged alphabetically for equipment, and phylogenetically for biological specimens. The use of a typical room map is illustrated in Figure 7.11.

WHAT KIND OF LABORATORY FOR YOU?

Now that we have discussed the place to teach, the tools with which to teach, and the operation of a dynamic

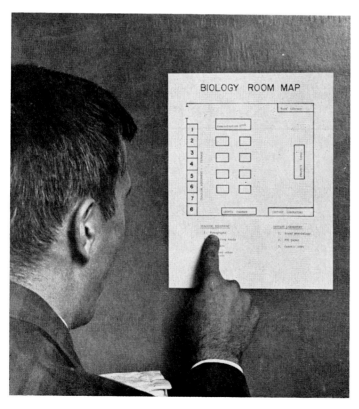

Fig. 7.11—The room map helps students locate materials, and serves as an instant inventory of the equipment in the biology department.

biology laboratory, what kind of laboratory program will you institute in your teaching?

You may not have a definite answer to this question. In fact, many educators are not sure what the relative emphases of laboratory and lecture should be. In a study for his Ph.D. degree, Jacob E. Mayman investigated the effects of the "book method, lecture method, and experimental method" on 500 New York City pupils. He concluded that "on the basis of efficiency as measured by percentile attainment, by lasting impression on the minds of . . . school pupils, and by persistence in memory . . . the three methods rank *experimental method, lecture method,* and *book method.*" In other words, the experimental approach, or problem-solving method of learning, was the most significant in Mr. Mayman's study. In another study, I. M. Allen concluded that the "informal problem method" was more effective in securing learning than the ordinary "text-book method."

While learning through the laboratory apparently insures the students' grasping and retaining more information, it is a slower teaching method than the lecture or lecture-discussion method. Usually the emphasis on laboratory work means that certain blocks of information in the text must be completely omitted from the biology course. Sometimes the problem is not what to teach, but what *not* to teach. As one worker said, "To develop power to think takes time. High school courses should be intensive rather than extensive."

The choice you make for your own biology teaching depends on the types of students you have; the equipment with which you work; the text and laboratory manual you use; your training, background, and interests; and your own attitude toward laboratory work. Properly designed laboratory work in biology can help students develop a problem-solving approach that transcends biology itself and extends into all fields of endeavor. The training to solve problems creatively should be a part of every student's education, for it creates better citizens as well as makes better scientists. Appendix A 207 points out forty ways that you can save time and money in your biology laboratory.

CREATING CHAPTER CONCEPTS

1. List 3 important scientific discoveries that were made in the areas of physical or biological science that came about by the application of the same scientific method that was used by Priestley and Redi.

2. Home-baked bread becomes moldy much faster than commercially baked bread. Outline a method of investigating the cause of the difference in the rate of fungus growth.

3. List 5 objectives you would expect your students to achieve in performing an exercise in the biology laboratory.

4. We mentioned the heating of mercuric oxide as a typical exercise. How could you redesign the exercise to make it an investigation?

5. In the white powder problem, what parallel experiments would be necessary to prove conclusively what powder appeared on the watch glass after the acid and base mixture was evaporated?

6. How does the concept of an *experimental control* apply to the white powder problem?

7. List 5 open-ended investigations that a student could pursue. Include the approximate time each would take, and include sources of information that relate to the projects.

8. Make a complete list of items you would consider essential for the instant laboratory.

9. Make a plan for a teaching museum, including a drawing of a wall chart to hang behind the case and the items necessary for display inside the case. List the objectives you hope to accomplish as a student views the completed museum.

10. Construct a laboratory test of 25 items, covering 12 stations. Write your questions on one side of a 3 x 5 card, and indicate on the other side what item, specimen, or piece of apparatus the questions are pertinent to.

11. What laboratory techniques could be most effectively taught through the use of student exercises?

12. List several advantages of the dry lab. For example, what learning is accomplished in a dry lab on chemical formulas, structures, and balanced equations?

13. Discuss the pros and cons of a laboratory test.

14. "New biology uses the laboratory in a new way!" Discuss the significance of this statement, citing specific examples from current laboratory materials for high school biology.

15. One approach in a high school biology course is to offer a survey of the protists, plants, and animals on the earth today. Another approach employs a study of the functions of systems within organisms, with emphasis on the problem-solving method. Which approach do you consider most valuable for the student who is science-shy?

16. Consult the *Biology Teacher's Handbook* by Joseph Schwab and summarize in two or three paragraphs the scientific method of problem solving that BSCS considers so important.

17. Using Gabriel and Fogel's *Great Experiments in Biology,* choose a specific experiment described in the book and point out where the observation, hypothesis, experiment, and theory steps are in the scientist's procedure.

18. Cite several reasons for the relatively slow understanding and application of the scientific method of problem solving from the time of Aristotle to Pasteur.

EXPANDING THE CHAPTER

AYLESWORTH, THOMAS G. 1962. Four Kinds of Thinking in the Biology Classroom. *The American Biology Teacher* 24(8):597–99. The author provides a discussion of four types of scientific processes that can be used in solving problems.

BERGER, MELVIN. 1963. Using History in Teaching Science. *The Science Teacher* 30(7):24–26. The article discusses the use of history in teaching physical science, but some of the same methods are applicable to teaching biology.

BEVERIDGE, W. I. B. 1957. *The Art of Scientific Investigation.* 3rd Ed. A Vintage Book, published by Random House, Inc., New York. A book that gives a teacher background regarding scientific investigations.

BURKMAN, ERNEST. 1963. A Science Teaching Center for the Southeastern States. *The Science Teacher* 30(8):23–25. The contact of science teachers with industrial laboratories as a means of improving instruction is explored in this article.

DAMPIER, WILLIAM C. 1949. *A History of Science*. Cambridge University Press, New York. This well-organized book provides a history of all branches of science, including biology.

GABRIEL, M. L., and FOGEL, S. 1955. *Great Experiments in Biology*. Prentice-Hall, Inc., Englewood Cliffs, N.J. A collection of original notes and essays written by the scientists who made important contributions to biology.

KELLY, G. 1965. Robert Hooke Plus 300 Years. *Bioscience* 15(6):408–11. In this article the author discusses the contributions of this important scientist.

KLINGE, PAUL. 1964. In My Opinion: Teaching Tools Are Important. *The American Biology Teacher* 26(3):164. In this editorial a well-known biology teacher makes a comprehensive review of the tools necessary for the teaching of a laboratory-oriented course in biology.

———. 1963. The Laboratory: Use and Design. *The American Biology Teacher* 25(5):324–99. The entire issue of the magazine is devoted to the use and function of the laboratory in biology teaching.

MILLER, D. F., and BLAYDES, G. W. 1962. *Methods and Materials for Teaching the Biological Sciences*. McGraw-Hill, Inc., New York. A combined text-source book for biology teachers that combines a large number of useful exercises and experiments.

MORHOLT, EVELYN; BRANDWEIN, PAUL; and JOSEPH, ALEXANDER. 1966. *A Sourcebook for the Biological Sciences*. 2nd Ed. Harcourt, Brace & World, Inc., New York. The authors of this book have organized an excellent collection of exercises, experiments, and techniques for biology teaching.

SCHWAB, JOSEPH J. 1964. *Biology Teachers' Handbook*. John Wiley & Sons, Inc., New York. This handbook provides theory as well as much practical information pertaining to conceptual approaches to modern biology teaching.

SVROCKI, JOHN S., and THOMAS, CHARLES S. 1960. A Picture File of Student Projects in Biology. *The American Biology Teacher* 22(9):544–46. A useful teaching aid for cataloging student projects is suggested in the article.

PART 3

SELECTING USEFUL TOOLS

CHAPTER **8**

Ways To Use

Basic

Materials

EVERY PROFESSION has its own box of tools. The doctor, diemaker, engineer, watchmaker, lawyer—each requires a particular set of instruments. So do biologists.

Biology teachers are craftsmen who mold the thinking process of their students, and to do so they must use a large number of teaching aids. Some of the materials themselves are the basis of study, but more often the living things, visual aids, prepared slides, and microscopes are means to an end, and not ends in themselves. See Appendix F 289 for places to get prepared slides.

SHOW OR TELL?

Warren Strickland, writing in the October 1964 issue of *The Science Teacher,* says that teaching is summed up by this old Chinese proverb: "If I hear, I forget; if I see, I remember; if I do, I know." Almost every present-day text that deals with the teaching of science has a large section devoted to a discussion of equipment. As indicated by the Chinese proverb, however, the trend to use apparatus, living and preserved materials, and audiovisual aids is not really new. Some of the oldest methods books show the early development of the theme that science teaching must be doing, feeling, and touching and not just seeing, reading, and talking.

It is increasingly recognized in the teaching of . . . science that no school now puts its students to the study of text books . . . before they have acquired its primary concepts and facts by the study of objects and phenomena.

This is the meaning of the modern laboratory.

The attempts to teach science from books is still a common error in high schools. Pupils are studying . . . the verbal descriptions of plants when they should be studying the plants themselves. It is in the teaching of the elements of science that the laboratory has its highest educational value.

The "modern laboratory," mentioned in the above quotation, and the apparatus it contained would be quite different from the modern teaching tools we now know and use. The quotation is from *Elements of Pedagogy* written by Emerson E. White *in 1886.* Though the physical arrangement of his laboratory and equipment might be old-fashioned by our present standards, his thinking was certainly up-to-date. The following statement from a 1958 book[1] proves that the value of teaching tools has not diminished over the years but has become even greater.

Science is not "chalk-talk"; it is experience in search of meaning.

There are many kinds of props or aids. Where possible use living materials— the real props. These may be brought to class or youngsters may go to the living materials through field trips.

TYPES OF TEACHING TOOLS

Living Tools

Too often living things are not even considered as teaching tools—a concept

---

[1] Morholt, Brandwein, and Joseph, *A Sourcebook for the Biological Sciences,* Harcourt, Brace & World, Inc., New York, 1958.

that seems especially odd when you consider that biology is "the study of life." Among others, Dr. Paul Dehart Hurd of Stanford University believes that living teaching tools are the most important of all. In a prepared address that he has delivered to scores of science teachers in the United States and Canada he says:

The biologist has had the problem of bringing his laboratory work to life. In a biology class, students are told that biology is the study of life. And then teachers spend nine months trying to prove that statement with a parade of dead, preserved, embalmed, pickled, pressed, embedded, and otherwise immobilized and distorted specimens.

There is seldom the use of frogs that jump, fish that swim, flowers that smell, worms that wiggle, birds that fly, or humans who think.

Learning is truly accurate only insofar as students have opportunities for a true experience with the phenomena or materials under study.

Sometimes the scarcest things in biology classrooms are organisms themselves. There may be tables, chairs, books, microscopes, models, blackboards, pictures, and charts, but no plants, animals, or protists.

FROGS

While frogs have been used in biology laboratories for many years, most of the work done with frogs centered around dissection of preserved specimens. We believe that teachers can use frogs for teaching other aspects of biology than anatomy, even though for years students from Maine to California have cut up thousands of frogs to study anatomical relationships and compare certain systems of the frog with similar systems in man. The study of frog anatomy is important, but living frogs can be used for a large number of important exercises in general biology as well. We list a few of these in this chapter and include more of them in Appendix D 252.

A healthy living frog harbors more than 200 parasites. At least five of these can be investigated without any special apparatus or techniques by beginning students in the high school laboratory. The frog lung fluke *Hematolechus* is easily removed from a pithed frog by slitting each lung with a razor blade. The flat, spotted flukes will begin to wiggle and can be picked up with forceps. Often gravid, the flukes shed their eggs if placed in a petri dish of distilled water. The process can be watched with a binocular microscope.

The large, ciliated protist *Opalina* grows in the rectum of the frog. To remove it for study, slit open the rectum and gently scrape out the contents, transferring them to a drop of water on a microscope slide. With a magnification of 100× you can see *Opalina* (and other protists) moving slowly in the water, using cilia. Additional protists in the digestive tract of the frog include *Nyctotherus, Hexamita,* and *Trichomonas.*

A drop of living blood from a frog can do more than serve as a means for studying the structure of blood cells. The erythrocytes of the frog act as tiny osmometers, indicating whether a solution is isotonic, hypotonic, or hypertonic to frog blood. Put a drop of the solution on a slide, add a drop of frog blood, and make the observations with high power. Plasmolysis of the cells indicates they are in a solution hypertonic to the blood plasma; if they swell and burst, the solution surrounding them is hypotonic to the plasma.

The study of ciliary action comes alive when living cilia from the mouth and esophagus of a frog are observed

119

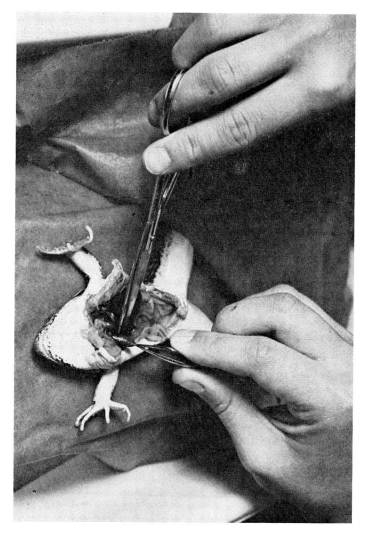

Fig. 8.1—Tiny, spotted lung flukes are found in almost every frog that might be collected in the United States. The flukes can be removed from a frog's lung, as shown, and studied with a microscope.

with the microscope. A piece of ciliated epithelial tissue, mounted in Ringer's solution, can be studied for over an hour before the cilia stop beating.

Corpuscles squeezing through tiny capillaries in the frog's foot provide first-hand experience with the circulatory system. A helpful part of the exercise, described in Appendix G 319, is to require students to compute how large the capillaries and red blood corpuscles are. This might be used to introduce students to the techniques of measuring

with the microscope, and the use of the metric system.

Living frogs make ideal subjects to use for experiments involving a student-constructed "respirometer." This simple device, shown in Figure 8.2, will give moderately accurate results during a class period of forty or fifty minutes, and the quantitative data it provides allow the student to construct a graph that relates oxygen consumption to the temperature of the environment. For a dramatic illustration of the response of a cold-blooded vertebrate to temperature changes, let the students watch a frog jump on the laboratory table. Then immerse the frog in ice water for a minute. This time the frog's responses are fairly sluggish. Recovery is rapid.

So much information exists on the use of frog heart and leg muscle in physiology experiments that we will not explore the details of the process here. (Some experiments are given in Appendix D 252.)

If you are unable to collect living frogs in the area in which you teach, they can be ordered from a number of suppliers and shipped to you. Frogs apparently are excellent travelers, and they usually arrive in fine shape. The price for a dozen frogs, mailed to your school, runs between $1 and $3. Appendix F 289 lists places which supply frogs.

PROTISTS

A great deal of material on the use of the protists—especially the protozoa—is available for teaching general biology. Some of the unique uses are recommended in the examples here, utilizing three different protists: a protozoan, an alga, and a bacterium. They will serve only to introduce you to the use of pro-

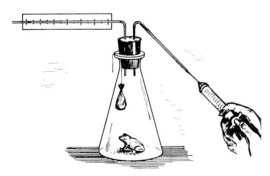

Fig. 8.2—This respirometer measures O_2 uptake of the frog. What provision is made for collecting the expired CO_2?

tists as teaching tools. Ways of collecting, culturing, and using other protists are given in Appendix B 220. Included also are ways of using *Tetrahymena pyriformis* in Appendix B 226, and some simple stains for *Tetrahymena* and other protists in Appendix B 228. In Appendix B 231 reference is made to diatoms and in Appendix B 234 to the study of protozoans by using vital stains.

Paramecium is widely used because it is conveniently cultured and requires a minimum of care. When kept at a temperature of 65° to 75° F, a culture of about 200 ml will keep indefinitely if water is added about once a month, along with 3 or 4 grains of boiled rice. *Paramecium* is often used in behavioral studies to show response to external stimuli such as electricity, light, chemicals, mechanical stimuli, and food. An interesting examination of internal mechanisms can be made when the organism ingests tiny particles of yeast that have been soaked in the indicator Congo red. Changes in pH can be observed as the particles slowly circulate in the cell, until portions of them are finally egested. Starting out at a pH of about 7, the food particles suddenly become acidic; then their acidity slowly returns to a pH of 7 again. Students can watch this occur during one laboratory period, but the process does require extremely careful observation and precise control of the light entering the microscope.

Oscillatoria, one of the blue-green algae, is another interesting organism. Even though it is classed with other algae, it is thought to be more primitive, and probably is a closer relative of the bacteria. Students can confirm this by looking for definite nuclei (they are absent) and thick, cellulose cell walls (also absent). One of the peculiar character-

122

PART 3/SELECTING USEFUL TOOLS

istics of *Oscillatoria* involves its movement: oscillation. If light and temperature on the microscope slide are just right this blue-green alga will slowly sway back and forth. With guidance from you, your students may use the following questions for further investigation. See Appendix C 248 for further information on algae.

1. If *Oscillatoria* has no nuclei, where is the DNA?
2. How can an organism reproduce if it has no nuclei in its cells?
3. Does *Oscillatoria* have chromosomes?
4. What part does the thin gelatinous sheath around the filaments play in the swaying movement? Clue: add a drop of India ink to the wet-mounted slide. Carbon particles in the ink will cling to the sheath, making it easier to see.

The bacterium *Serratia marcessens* is easy to work with. This species, similar to a number of others, forms pigments of different intensity as the temperature varies. Other bacteria behave in a similar way, and can be used in parallel studies. Appendix F 290 lists some chromogenic bacteria and sources other than biological supply houses for obtaining these organisms. Appendix F 290 names some sources for obtaining microbiological culture media. In addition, Appendix F 290 and 291 list some sources for obtaining antibiotics, enzymes, and refined chemicals that might be needed in doing some experiments and demonstrations associated with bacteriological studies.

PLANTS

Discussion in Appendix C includes plant collection, culture, and use in the laboratory (see Appendix C 236 to 251). Of these plants we selected *Coleus* and *Zebrina* for mention here. These two are especially easy to maintain; both can be propagated from cuttings; they represent "typical" dicots and monocots; and they can be used to show internal and external structure of these two subclasses of Angiosperms. In addition, they are appropriate for use in exercises and experiments in basic plant physiology. The cells of *Zebrina* (wandering Jew) secrete oxalic acid, which builds up in such quantity in the cells that the pH of the cytoplasm is lowered. The change in pH apparently activates certain enzyme systems that bring about formation of large, needle-shaped

Fig. 8.3—The blue-green alga *Oscillatoria* is an excellent specimen for study with the microscope, and is available in almost every classroom aquarium. The dark greenish-black scum on the side of the glass is actually a mass of filaments of *Oscillatoria*.

crystals of calcium oxalate. Each student can make a free-hand section of a stem or leaf from *Zebrina* and observe the crystals with a microscope.

Zebrina flowers have accessory filaments near the stamens that can be used to show cyclosis. These "stamen hairs," which are actually a series of cells growing end to end, may be removed, put into a drop of water on a slide, and observed with high power to see the protoplasmic streaming in each cell.

For the observation of osmosis and diffusion in living tissue, the anthocyanin-filled cells of *Zebrina* leaves can be mounted in hypotonic or hypertonic solutions. Because the anthocyanin provides a visual clue to the degree of osmosis, the process is easier to watch than when conventional cells that have colorless cytoplasm, such as *Elodea*, are used.

Coleus, the well-known square-stemmed houseplant, grows easily in the classroom and is useful in exercises and experiments on photosynthesis. The multicolored leaves can be used in chromatography of plant pigments to show that purple and red leaves have almost the same complement of plant pigments as green leaves. In exercises designed to show the areas of starch production in leaves, *Coleus* is excellent material. Cross sections of the leaves reveal the two characteristic cell types in the mesophyll of most dicots: the spongy and palisade parenchyma. When students use fresh material and make sections with a hand microtome or a razor blade they feel a closer contact between theory and practice. When a student can pick a leaf from a plant, section it, and observe firsthand the structural

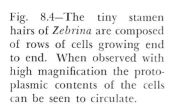

Fig. 8.4—The tiny stamen hairs of *Zebrina* are composed of rows of cells growing end to end. When observed with high magnification the protoplasmic contents of the cells can be seen to circulate.

characteristics that a text discusses, he reinforces his learning.

Preserved Tools

The oldest method of preserving biological specimens is still the most widely used today. It makes use of solutions of formaldehyde, alcohol, and acetic acid, occasionally modified by the addition of phenol or glycerin. One combination of formaldehyde, acetic acid, alcohol, and water (50:2:10:40) is termed FAA. It is a fixative and preservative that can be used for almost all small plants and animals.

An aqueous, dilute solution of formalin is a satisfactory general-purpose preservative that can be used without the addition of acetic acid and alcohol, but some shrinkage and hardening of the tissue will occur. To prepare such a solution, dilute the commercial solution of formaldehyde (which is a saturated solution of formaldehyde gas, HCOH, dissolved in water) with distilled water to make a 5–8% solution of "formalin." Consider the saturated solution of formaldehyde as 100%, for the purposes of diluting.

Ethyl alcohol, diluted to make a 70% solution (70 ml alcohol, 30 ml water) is another useful general-purpose preservative.

The three preservatives listed above are entirely satisfactory for preserving the majority of the specimens you will encounter in general biology. However, there are other preservatives that we mention in Appendix D 273 under "Preservatives and Fixing Materials." Green plants and protists, for example, require a modified FAA solution that will "fix" the chlorophyll and other pigments and prevent them from breaking down.

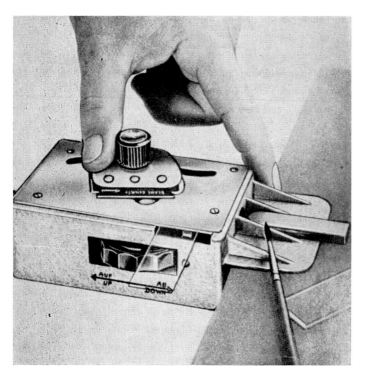

Fig. 8.5—An inexpensive microtome makes possible the sectioning and study of fresh material. Firsthand observation of material from the field makes biology more meaningful and lays solid groundwork for more detailed study of commercially prepared materials. (From Faust Scientific Supply Company, Madison, Wis.)

Some of the freshly collected animals (pigeons, fetal pigs, frogs, and small mammals) can be injected with liquid latex before preservation. Colored latex, forced into the arteries with a hypodermic syringe, will harden overnight. Animals injected in this manner mean more to the beginning biology student because hard-to-find arteries are easily recognized when filled with red latex, and certain veins are almost impossible to find unless they contain hardened blue latex. The rubbery material also makes the blood vessels less apt to break during dissection. Figure 8.6 shows stepwise the process of injecting latex into the arterial system of a fetal pig via the carotid artery in the neck. Appendix D 275 lists other common animals with suggestions for injecting them.

Frozen Tools

Freezing is a newer and more satisfactory method for keeping specimens

Fig. 8.6—Injection of a specimen with latex. The injection process requires only a few minutes, yet it makes the specimens much easier to work with when dissecting. As a check on the extent of injection, the femoral artery in the leg can be observed as it fills with latex.

b. Using scalpel for cutting into neck to expose carotid artery.

a. Fetal pig, ventral side up.

c. Injecting latex into artery.

over long periods of time. Frozen materials retain their original colors (formaldehyde bleaches), remain pliable and soft when thawed (formaldehyde hardens), and have no disagreeable odor (formaldehyde nauseates[2]). Somehow, it seems more natural to dissect a frog that bleeds a little. It seems more exciting to work with butterflies, bees, and beetles that are soft, pliable, and *dry*—just as they are in life. As a storage place for furred animals waiting for dissection or mounting, the freezer is unexcelled.[3]

A great deal of local material can be collected in the warmer months and frozen for future use. Frogs, earthworms, clams, snails, and birds are all obtained easily from the field, but be sure to obtain a collecting license from local authorities. Although most of these can be kept alive as easily as they can be preserved, it is not difficult to wrap them for the freezer if the menagerie is full. Insects should be put into small bottles in class quantity so that it is necessary to thaw out only one small container for each class. Frogs and small mammals should be placed in plastic bags because there is some dehydration of the tissue if they are not wrapped. Algae, such as *Spirogyra*, *Oscillatoria*, *Rhizoclonium,* and *Cladophora,* can be bottled in small collecting bottles or baby food jars and put into the freezer. Some of the algae become dormant when frozen. Certain species of *Oscillatoria*, for example, begin to move gently back and forth when brought up to room temperature on a microscope slide. Fetal pigs from a packing plant can be put in large plastic bags

in class quantity and frozen. Certain body organs, used as teaching tools year after year, keep well in a freezer. One of the authors has a frozen ten-year-old heart from a cow. Each year the heart is thawed, examined by the biology class, and then refrozen until needed again. Kidneys, brains, lungs, and livers from cattle, sheep, or pigs may be kept in the freezer too, to be brought out for demonstration at appropriate times each year. Flowers and fungi can be preserved by freezing, but they collapse soon after thawing and thus cannot be refrozen like animal tissue. For one-time use however, frozen plant materials are excellent.

Mechanical Tools

The kymograph and microtome are two important mechanical tools that should be a part of every modern biology room. They need not be the most expensive; in fact, a kymograph can be made from two peanut butter cans and a dime store display motor for a total cost of less than $2. The nut-and-bolt microtome in Figure 8.7 can be made by a student at home. The use of these teaching tools, whether in formal laboratory work or individual investigation, is important and should be discussed in the biology class you teach. Specific uses of the kymograph and sources of supply are listed in Appendix G 331.

Of all the tools for extending the biologist's senses, the microscope is one of the most useful in many areas of laboratory work. (Appendix F 292 lists supply houses for microscopes.) One important use is helping measure small objects. The diameter of the field of view, calibrated with graph paper, serves as a kind of micron-stick on which to line up cells, protists, crystals, or tissues for an

---

[2] An effective solution for removing the odor of formaldehyde from preserved specimens can be made from these proportions (by weight) of tap water, ammonium phosphate, and urea: 95:5:1. Soak specimen in this mixture a few hours.

[3] Sarah Brown, "Frozen Mice," *The American Biology Teacher,* Oct. 1964.

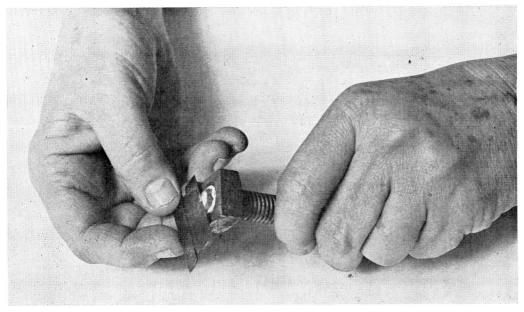

Fig. 8.7—This microtome is made from a 3-inch machine bolt, ⅝ inch in diameter, with an SAE 28 thread. The end of the bolt is filed flat and the face of the nut is filed smooth to provide a uniform cutting surface.

estimation of size.[4] (Appendix G 319 includes instructions for measuring with the microscope.)

In physiological exercises the microscope aids in observing cyclosis, osmosis, muscle contraction, and transport in plants and animals. Imagine the surprise, for example, of the student who discovers that substances circulate in plants just as substances circulate in animals—and that it is possible to actually watch the process. A low-power magnification of the midrib of an *Elodea* leaf clearly reveals xylem and phloem transport.

As a means for comparing relative effects of chemical and physical factors on a given organism use a series of mi-

croscopes, or a "microscope train." With such a train observations can be made continuously along the line. For example, suppose a student wished to investigate the effect of different salt concentrations on the hatching time of brine shrimp. Rather than set up large culture dishes he could carry out the work on the semi-micro level by keeping the salt solutions and shrimp in concave center slides. For convenience and quickness in following the progress he could put each slide on a different microscope. Observations could be made by continuously checking from one microscope to another as the process developed.

The binocular dissection stereomicroscope is more useful in some ways than the more common monocular microscope. Your biology laboratory should have at least one monocular microscope for each student, and one dissecting microscope for every two students. With

---

[4] You can work out a laboratory exercise for your students on measuring with the microscope. Some clues to use are found in an exercise, "How Big Are Cells," on p. 83 of *Laboratory and Field Studies in Biology* by Lawson and Paulson. Holt, Rinehart and Winston, Inc., New York, 1958.

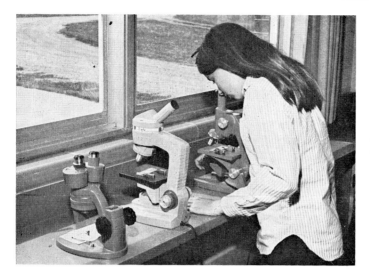

Fig. 8.8—A series of microscopes can be used to observe changes in material that would be difficult to study with only one microscope.

the proper combination of objectives and eyepieces on both types of microscopes it will be possible to view objects through magnifications of $1\times$ to $1000\times$.

Some exciting investigations can be carried on by students who use the technique of microdissection. By drawing glass rods out into fine threads of different diameters and with differently shaped hooks on the ends, a student can study many protists by microdissection. Some of the classic work in microphysiology can also serve as the basis for extended investigation with the microscope. For instance, many years ago German biologists determined that the helical chloroplast in *Spirogyra,* and *not* the cytoplasm, was actually the site of the oxygen-producing reactions of photosynthesis. They adjusted a filament of the alga with glass needles so that a tiny shaft of light could fall on the chloroplast. Then, by introducing oxygen-seeking bacteria they observed the bacteria gathering at the illuminated spot on the chloroplast, while none of the bacteria gathered at a similarly illuminated spot on the cytoplasm. Perhaps you will have a student who would wish to expand this project with the micro-

scope. The investigation could include a correlation between the numbers of bacterial cells that gather and the amount of oxygen evolved. Some genera and species of bacteria are more effective indicators than others. And what about other protists as oxygen indicators? Using the train of microscopes a student could investigate simultaneously the oxygen-sensing capabilities of several organisms.

Besides the microscope there are other visual and audiovisual tools that are vital parts of creative biology teaching. They are tools that aid our two most important senses: hearing and seeing.

The use of instructional media in teaching is becoming so widespread that many books have appeared to explain the place in education of tape recorders, motion picture projectors, overhead and slide projectors, television equipment, teaching machines, flannelgraphs, chalkboards, models, globes, maps, charts, and graphs.

Audio Tools

In Chapter 6 we referred to the sights and sounds of biology. With the tape recorder you can bring the sounds of biology into the classroom or, in a sense, take the class on a vicarious field trip. You may wish to record sounds of local animals or purchase recordings of animal sounds. Used during the study of a unit on behavior, such sounds help students learn about animal communication and territorial defense.

Among the large number of recordings available, bird and amphibian sounds are some of the most interesting. (See Appendix F 292 for places to buy recordings.) In a record made by Cornell University[5] the engineers have recorded the song of a sparrow at normal speed, then reduced the speed by one-half and then by one-fourth. This lengthens the song and allows the listener to hear the pattern of the notes more accurately. Cutting the speed also drops the pitch. Since the song of the song sparrow is made up of some extremely high-pitched notes, it is possible to distinguish individual tones more clearly when they are played at a lower pitch. Using this technique, biologists have determined that some of the notes in the sparrow's song were not single notes at all, but were composed of three notes, forming a triad. Perhaps an interested student in your classes, who enjoys electronics as well as biology, could record the song of a robin or the chirp of a cricket, slow the speed, and investigate the peculiarities of some of the sounds of nature.

The tape recorder can also form the basis of another student project in biology—that of creating an "audio time scale" that represents the evolution of the earth. Students working in small groups can compress the evolutionary time scale into the space of ten or fifteen minutes by recording on tape the major events that occurred over the past few billion years. Instructions for completing such a project, listed in Lawson and Paulson's *Laboratory and Field Studies*,[6] are given below:

The object of this part of the study is to make a tape recording that will give an audible presentation of the relative duration of time spans.

The length of time that the tape will run will probably depend on the length of

Fig. 8.9—Sounds of animals near your school can be brought to the classroom through the use of a portable tape recorder. The collection of sounds of local animals might be the basis of a student project.

[5] "Music and Bird Songs," a 10-inch LP recording with commentary, available from General Biological Supply House, Inc., 8200 So. Hoyne Ave., Chicago, Ill. 60620.
[6] Holt, Rinehart and Winston, Inc., New York, 1959.

your class period. The duration of time covered by the recording should go back to and include the Azoic Era. This recording will probably be most effective if no commentary is used; the Azoic then will be presented by a period of silence . . . since no living things were present.

The evidences . . . should be placed along the tape in their proper sequence and at the proper intervals. To find the interval divide the number of minutes the tape will run into two billion years. This will give the number of years covered by one minute of running time on the tape. The events should be so spaced that the class will become aware of a gradual acceleration in the occurrence of events. The last minute or two of the tape will include so many references that it will be difficult even to mention the very recent occurrence of man on earth.

Radio and television carry a number of programs that can be tape recorded and played in the classroom. Special features on television include interviews with scientists in which they talk about their work, and documentary programs that summarize the events leading up to the development of outstanding milestones in science.

Some of the most interesting sounds of biology are not the serious sounds, but the humorous ones. A number of our American folk songs have been based on biological themes. Use these recordings to inject humor into the biology course when the subject seems to be bogging down. One recording that makes a special hit with the students just after studying a particularly difficult unit in biological chemistry is simply the names of all the chemical elements set to a Gilbert and Sullivan tune.[7] The recording never fails to get laughs from the students, but more important, it breaks the tension that results from

studying pure theory for a length of time.

Visual Tools

THE CAMERA

It seems odd, but the camera is not often considered a visual tool in teaching biology. Yet pictures are sometimes worth more than words in creating a concept or proving a point. Students can use cameras to illustrate reports, summarize projects, and collect data. The teacher-operated camera can add valuable visual teaching material to a cumulative teaching file over the years.

Projects in ecology (see Appendix G 319) mean much more if pictures of the study area are included. Even before going to the field to begin the project it is helpful to consult a picture of the general area and discuss the aspects that will be studied.

Studies in comparative anatomy are hardly meaningful without illustrations, and some of the easiest illustrations can be made with a camera. Suppose, for example, that a student wished to compare the digestive systems of a pigeon, frog, and fetal pig. Dissection time for each specimen would be about two hours, and in that time the students would get little more done than opening the body cavity, locating organs, and tracing the digestive system and its accessory parts. He would have no time for comparison of the digestive systems before the tissues started drying, shrinking, and changing color. He would be unable to study the three systems side by side to work out some of the adaptations each would show for the type of diet and life of the organism. However, with a number of photos of each digestive system he could make comparative measurements and draw some meaning-

---

[7] From the $33\frac{1}{3}$ LP record entitled "An Evening Wasted With Tom Lehrer," available from Tom Lehrer, P. O. Box 121, Cambridge, Mass. 02138.

ful conclusions. The resulting write-up would not only mean more to the reader, but would cause the student to do more meaningful, quantitative work on the problem.

Prepreparation for a field trip can be accomplished by showing a set of colored slides that describes the site to the class.

The step-by-step explanation of a certain laboratory procedure, one which uses rather complex apparatus, can be augmented with 35mm slides the teacher has taken.

End products of long-range experiments can be photographed with the still or movie camera. Imagine yourself before a class, explaining the phenom-

enon of geotropism. During the explanation show the class a picture of a phonograph turntable, revolving at slow speed, with a potted plant on it. Then using a series of pictures you have taken day after day (calendar in the background of the pictures to show successive days) show the changes in plant growth. Will the plant be vertical, slanted toward the center, or tipped toward the outer edge of the turntable? Why?

With a photomicrographic attachment a camera can be used to take pictures through a microscope. With such a device you can build a file of valuable teaching photomicrographs. Students can take photomicrographs that can be used in conjunction with their projects.

With a camera you can borrow ideas from museums, libraries, and the classrooms of fellow teachers. With a 35mm camera loaded with high-speed film you can photograph an attractive museum case or display, enlarge the resulting picture, and then use it as a guide when you want to construct a similar exhibit for yourself. In Washington, D.C., the Smithsonian Institution employs scores of expert craftsmen who build lifelike plastic and fiber models of hundreds of organisms. If you had a series of pictures taken in museums of the Smithsonian, the American Museum of Natural History in New York, and the Field Museum and Museum of Science and Industry in Chicago, the pictures could form the basis for a student project in model building. To copy book titles without writing them down, photograph scientific literature in the stacks of a university library. Once enlarged, the picture becomes a ready reference to important collections of reference works.

The camera you choose for the job can be any one of a dozen types from over 20 different manufacturers. As a

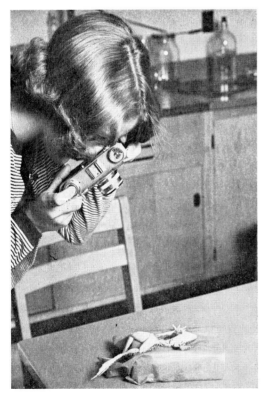

Fig. 8.10—With photos of the digestive, circulatory, or nervous systems of several animals a student can take measurements and make quantitative comparisons.

science teacher you want the one that can do the largest number of different jobs. In each of our examples of ways to use the camera we had in mind one basic camera body: a 35mm single-lens reflex. This type uses film 35mm wide and has only one lens, which serves both as viewfinder and picture-taking lens. This camera has two advantages over all others: it prevents "cutting the heads off" by assuring that you can put on film *exactly what you see* through the viewfinder, and film (both black and white, and color) is more reasonably priced because of its small size. Our list of suggested exercises at the end of this chapter includes the problem of choosing a camera and accessories you think you would find most valuable in teaching. We cannot give you the background in this chapter to enable you to solve the problem entirely. We want to show you how to use a camera as a teaching tool. Then you can read books and pamphlets on photography and visit a store to decide which camera and set of accessories is best for the type of biology you are going to teach. (See references for photography, Appendix E 286.)

THE SINGLE-CONCEPT FILM

In a basement laboratory on the campus of a leading midwestern university a technician sat huddled over a microscope. As he looked through the two eyepieces he adjusted a 16mm movie camera over the third tube. On the stage of the microscope there was a slide holding a few strands of the green alga *Basicladia*. Because this particular alga grows only on the back of a living turtle, the filaments were removed minutes before they were mounted on the slide. The microscopist made a few minor adjustments in the light source illuminating the mirror of the microscope, and then changed from the $10\times$ objective to

the $43\times$. The equipment was finally ready. Now, if the organism would only cooperate. For several days strands of *Basicladia* had been mounted in water on a slide and put on the microscope stage for observation, but nothing happened. Maybe today would be different.

Suddenly the technician saw what he wanted. He pressed an electric foot switch. The camera started whirring. In 2 minutes he had photographed the first 16mm film of active zoospores of *Basicladia*. One year later a plastic cartridge containing an endless "loop" of the film was on the market for $15.

We have just described the process of making one of the most exciting teaching tools available today. It is the single-concept film. Used for several years by salesmen to show clients specialized equipment, and in industry for training new employees, the single-concept film only recently invaded the educational field. Its function: to bring to the students those rarely seen processes of well-known phenomena. Take zoospore formation and discharge in the algae, for example. Teachers for 50 years have been *talking* about zoospores[8] and their role in asexual reproduction and dispersal of the algae. But here the description stopped because no teacher had the time and usually did not know the technique for inducing the tiny, haploid, flagellated cells to break out of their zoosporangia and swim away while students watched through the microscope. The film loop has helped solve this problem.

In several universities across the country a number of research groups have made short, one-idea films to show processes that formerly had been only verbalizations. Films on pithing a frog, pinning an insect, determining CO_2 up-

---

[8] *Basicladia* was not included in their talks for the entire 50 years, however. It was first reported by Hoffmann and Tilden in the *Botanical Gazette* of 1930, Vol. 89.

take in a leaf, and splashing gametes of the bryophytes can now be a part of a teacher's room library. They can be inserted into a shoebox-sized projector and viewed as easily as a book can be pulled off the shelf and read.

The single-concept film (see Appendix F 292) is valuable for many reasons, but one of its biggest advantages is that it does not remove the students from the classroom and take them somewhere for a capsule lesson. The single-concept film is introduced by the teacher, and as such it becomes a real part of his subject matter presentation. Many of the films are silent and must be narrated by the teacher. Thy are teaching tools in the truest sense, for like a wrench or hammer they lie inert unless they are picked up and used by a craftsman. You are that craftsman. Single-concept films enable you to build powerful unit concepts in biology (see Chapter 10).

BIOLOGICAL BLOWUPS

From time to time room decorations should be changed to reflect the unit or concept that is being studied in the text or the laboratory. An interesting way to accomplish this is to arrange a group of 8 x 10 or 11 x 14 enlargements on the bulletin board. They serve the double function of adding atmosphere to the room and being instructional tools. The newest work with the electron microscope is especially meaningful when electron micrographs are put up in the biology room. A key or legend beside each, pointing out the important structures, makes it possible for students to interpret the pictures without much extra help. Such pictures on display become study aids, reinforcing the knowledge acquired in class. You may wish to choose some of the pictures you have taken as display pictures, or use selected pictures from books or magazines. If the latter are used, it is wise to seal them with a heated press and thin plastic film. Figure 8.12 illustrates the process of preparing pictures, newspaper and magazine articles, and student-made drawings for display either on a bulletin board or as pass-around material in the classroom.

A picture series of the stages in the life cycle of an insect common in your area could form an interesting and informative display on a bulletin board. (See Appendix F 293 for some sources of bulletin board materials, and Appendix F 294 for lettering materials.) Several large pictures showing the metamorphosis of a grasshopper, fly, or mosquito can be the result of a student project in biological photography.

Fig. 8.11—This film-loop projector, sometimes called a single-concept film projector, can show up to four minutes of film. The film is a continuous loop, automatically threaded. Thousands of subjects are now available on loops, bringing phenomena into the classroom that most students would never see otherwise. (From Technicolor, Costa Mesa, Calif.)

BIOLOGICAL THEMES

The beauty of nature is often portrayed in postage stamps. There is much evidence that every form of life is depicted in stamps in today's postal art. The main theme of the November 30, 1959, issue of *Life* magazine was a "Rainbow of Stamps." Biological topics much

134

Fig. 8.12—Inexpensive plastic film, bonded to an illustration with this heat press, preserves the picture for years. Instructions for laboratory work, and the room library list of titles can also be protected in this way.

more profound than just the representation of biological forms have found their way into philatelic form. Topics such as malaria, conservation, and even the genetic code have been subjects of stamp production. Abbott Laboratory in the 1962 winter issue of its journal, *What's New*, devoted an entire article to "Stamps Against Malaria." One of the major biological supply companies had as the cover of its August 1962 issue of *Turtox News* a front and back spread of stamps involving biological topics. Appendix F 294 lists sources where biological stamps may be purchased.

Some biological topics found in stamps are:

Insects	Butterflies
Birds	Game Birds
Dogs	Fungi
Diseases	Flowers
Wild Animals	Trees
Lizards	Conservation
Turtles	Fruits
Snakes	Crops
Fish	Invertebrates

Biological themes are also shown in poetry[9] and in art.[10] These can be used successfully as part of the biology teaching program.

EDUCATIONAL TELEVISION

Even Jules Verne would have been startled in 1961 if he could have looked into the sky over Montpelier, Indiana. What might have looked like an ordinary plane was actually a *flying* television studio. Lessons prepared on the ground and recorded on video tape were being beamed from an airplane at an elevation of 23,000 feet to 5 million students in 13,000 schools. The Midwest Program in Airborne Television Instruction (MPATI) was one of the first large-scale programs to get educational television "off the ground," although station WOI-TV at Iowa State University began broadcasting educational television programs in the early 1950's.

When we think of educational television we think of a planned series of programs being transmitted by some kind of Big Brother station in a distant city. But the image of ETV as impersonal and all-encompassing is chang-

---

[9] Helen Plotz, *Imagination's Other Place— Poems of Science and Mathematics*, Thomas Y. Crowell Co., New York, 1955.

[10] Albany Institute of History and Art, *Art in Science*, September 1965.

ing fast. According to Professor W. C. Trow, writing in *Teacher and Technology*,[11] it is because of the several hundred closed-circuit installations in all kinds of schools today. He feels that the medium is still in infant stages. Another authority in the field of educational television, Phillip Lewis,[12] lists these unique possibilities for the future use of ETV.

1. Centralized electronic distribution of motion picture films and other visuals direct to where they are wanted.
2. Remote monitoring of corridors, play, and other areas.
3. Central storage of personnel records or other data that are immediately available for remote visual examination or direct copying from the TV screen.
4. Electronic hookup of widely separated school buildings or locations for conference purposes.
5. Ease of pupil viewing of demonstration procedures in shop labs.
6. Extension and expansion of audience and spectator facilities.
7. Kinescope of video-tape recordings of presentations and programs for repeated use later the same day or in the future.

Since Mr. Lewis was writing about the application of ETV to general education, we may list a few additional suggestions on the use of ETV in teaching biological science.

1. Video tape of rare or unusual biological events such as sperm discharge from antheridia of bryophytes, the fusion of two gametes to form a zygote, conjugation in earthworms, or the expulsion of a contractile vacuole in a protist.
2. Video tape of twig, leaf, or flower characteristics, to be used either as preparation for a field trip or a review of a field trip.
3. Closed-circuit projection, in the biology room, of microscopic objects. This technique works as well with living materials as it does with prepared, stained slides, and assures that all members of the class see exactly what you want them to see.
4. Close-up views of models such as the dissectible human torso. If you are teaching a large group, it is often impossible to pass large models around the class.
5. Video tape of a series of homologous structures in animals to illustrate an evolutionary sequence.

The choice of equipment and its integration into the type of biology program you will operate can be made only after you study the available literature in the field and talk to consultants. We have listed some of the important ETV literature in Appendix F 294.

Other visual teaching tools that are indispensable items in a well-equipped biology classroom include the overhead projector, opaque projector, slide and filmstrip projector, 8mm projector, 16mm projector (see Appendix F 294 for educational films), blackboard, and flannelgraph. The variety of use you make of these visual aids is directly proportional to your imagination. For example, in addition to using the overhead projector to show the traditional transparencies, you can use it to show the details of bones and skeletons of tiny animals. Chemical reactions can be performed in test tubes on the stage of an overhead projector. Bioelectricity can be amplified and put through a projection-type galvanometer on the projector. Lectures and tests can be prerecorded on the roll of crank-through plastic for overhead projectors. One of

---

[11] Appleton-Century-Crofts, New York, 1963.
[12] *Educational Television Guidebook*, McGraw-Hill, Inc., 1961. Used by permission of McGraw-Hill Book Company.

136

PART 3/SELECTING USEFUL TOOLS

the most fascinating applications involves the use of polarized plastic to
simulate xylem and phloem transport in
plants, or circulation in animals. Strips
of polarized plastic are pasted to the
transparency itself while a revolving disc
of polarized plastic is mounted under
the projector's lens. When the disc is set
in motion, dark bands that simulate
moving materials seem to "travel" in the
transparency as the students view it on
the screen.

The flannelgraph or flannelboard
(see Appendix F 295 for sources of materials) can be an especially effective
visual aid in presenting some of the com-
plexities of metabolic pathways such as
glycolysis, the Kreb's cycle, and the
photosynthetic schemes. A flannelgraph
in use is illustrated in Figure 8.13.

Tools You Can Make

When asked which faculty member
was his best friend, a biology teacher
said it was the shop instructor. It
seemed that every spare minute of the
science teacher's time was spent in the
school shop, making tools to teach with.

It is often easier to make teaching
aids than it is to buy them. Some of
them cannot be bought, and some are
too expensive or of the wrong design.
For one reason or another they may not
quite tell the story you want them to
tell.

A good example of what we mean is
illustrated in Figure 8.14 in which
biological molecules are shown. Models
similar to these are available commercially, but none of the commercial
models did exactly what the biology
teacher wanted. These styrofoam models
were made from spheres, toothpicks, and
glue for a total cost of $5.36. The time
it took to design and make them totaled
about 12 hours, but *it was all student
work*. The teacher utilized the artistic
abilities of students who expressed an
interest in doing the work.

Another inexpensive tool you can
make is the insect log, shown in Figure
8.15. An old birch log was used for the
one in the illustration, but any type of
log would serve the purpose. In addition to providing a quick reference to
representative insects in each of the ma-

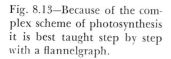

Fig. 8.13—Because of the complex scheme of photosynthesis
it is best taught step by step
with a flannelgraph.

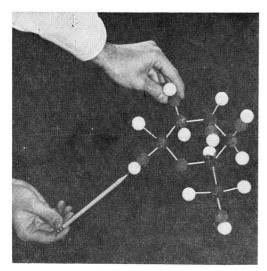

Fig. 8.14—These models of molecules made by students are excellent teaching aids. They are lightweight, unbreakable, and can be easily passed around the class for closer study.

jor orders, it serves as a room decoration that adds organisms to the biology setting. A log of this type can also be used to show various species of the same genus, variation in leaf shapes, adaptations suggested by variations in birds' feet, and types of feathers.

A transfer chamber, such as the one pictured in Figure 8.16 costs about $4. It can be put together easily, using plywood and glass.

Perhaps the most interesting and best looking tools you can make are the painted plaster models. The rubber mold will last indefinitely, so you can replace a model if it is ever broken, or trade molds with other teachers so that you have "copies" of their plaster models. The wall of models we described in Chapter 6 is the outcome of student work, using the technique shown in Figure 8.17.

Making the models is one thing. Using them is quite another. They are not meant to be works of art; they are *designed for use.* The model of the clam,

Fig. 8.15—This insect log, made from an 18-inch length of birch, helps students review characteristics of some of the common orders of insects.

Fig. 8.16—This small transfer chamber was designed and built by a student in 3 hours for a cost of about $4. (From Figure 23–11 from *Design for Life* by Richard Trump and David Fagle. Copyright 1963, by Holt, Rinehart and Winston, Inc. Reproduction by permission of Holt, Rinehart and Winston.)

for example, shows much more clearly than a dissection just how the intestine passes through the heart. The model is not intended to replace dissection as a means of firsthand observation, but a dissection of a clam can be more meaningful if such a model is used for reference in the same way that the laboratory manual and text are used for reference. The model also shows the digestive gland that surrounds the stomach, and it provides a clear picture of the relationship of the brain to other body organs. Beginning students, working on clam anatomy, rarely find the brain. During the dissection they can study both the model and the specimen.

Tools You Should Buy

There are some pieces of equipment that you cannot collect, improvise,

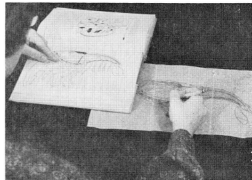

Fig. 8.17—The design, modeling, and casting of a plaster model are preliminary steps to the final painting the model receives. Rubber base (latex) paints are used because they are water soluble and can be mixed easily or thinned with water.

  a. Sketching a crayfish from a picture. (It is preferable to use a real crayfish.)

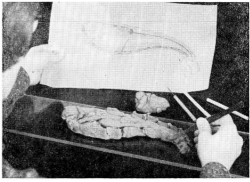

  b. Modeling with clay, following the sketch.

  c. Covering the clay model with latex. Put on in thin layers and let each layer dry before applying the next. Ten or 12 layers may be necessary for a firm mold.

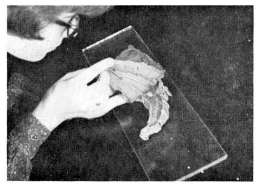

d. Removing the latex mold from the clay model.

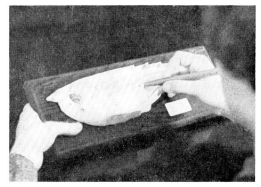

g. Painting and lettering the model. Use any type of paint (water base, oil base, other).

e. Pouring molding plaster mixed with water into the latex mold.

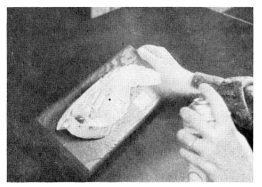

h. Spraying the model after the paint is dry to make it more chip resistant.

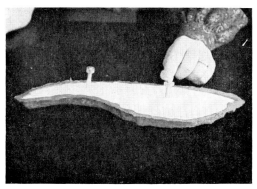

f. Pressing bolts into the soft molding plaster so this model may later be fastened to a board and hung on the wall.

i. Painted model mounted on a board, placed on a pegboard for display.

borrow, or make. They must be bought. We consider the following list of *non-expendable equipment* to be basic for a biology laboratory:

refrigerator
microscopes
greenhouse or climatarium
incubating oven
chromatography supplies
electrophoresis chamber
collecting nets
tree trimmer
torso model
microprojector
assorted glassware

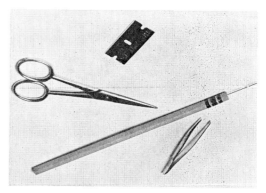

Fig. 8.18—This "dissecting kit" can be purchased locally for less than $1. Replacements made from time to time cost only a few cents.

It is not necessary to send away for every item that you need. On the main street or in the shopping center of towns across the country there is a supply house for biology teachers. It is the variety store. Bits of wire, string, cheese-cloth, plastic, light bulbs, and other odds and ends that are so important in teaching a laboratory course are found there. We describe two pieces of "apparatus" below, but there are many others.

For less than $1 you can purchase a "dissecting kit" from a dime store. The razor blade, scissors, pin, pencil, and forcepts in Figure 8.18 actually cost 97¢, yet they will do the same job that a set costing four times that much will do.

A DNA molecule is illustrated in Figure 8.19. It is inexpensive to make and can be used to teach the structure of nucleic acids as effectively as models costing 10 times as much.

Printed Tools

A growing number of teachers maintain a room library of current periodicals and texts. Magazines such as the *Scientific American, American Biology Teacher, Bioscience,* and *The Science Teacher* are included, along with standard reference works on a wide range of biological subjects. Frequently the books can be checked out of the school library for room use, but the periodicals usually must be purchased outright. A finance plan to accomplish this can be incorporated into the student laboratory fee, which the school assesses. The amount of the assessment depends on how many periodicals you wish to buy, but a fee of 50¢ to $1 per semester usually provides enough revenue to establish such a library for student use.

A room library is useful because it is convenient. The articles and instructions in the magazines that pertain to student projects or serve as resource material for panel discussions and short papers and reports are almost within arm's reach. Students do not have to run to another part of the building to locate the material. Because the room library is used only in the biology classroom, there is no need for a "circulation desk" to check the materials in and out.

Some of the commercial supply houses and certain industries publish monthly magazines that contain articles on science. These can be valuable addi-

Fig. 8.19—This DNA molecule model was made by these two students out of doughnut-shaped breakfast food, straws, string, and toothpicks. How does the model differ from real DNA?

tions to your room library, and they will widen its use as a reading and reference center.

Periodicals, texts, and other publications that could form the nucleus of an effective room library are listed in Appendix E 283 to 287.

THE PROBLEM OF SUPPLY

Where Does the Money Come From?

It is up to the teacher to become involved with the planning of the science budget. In many systems a portion of the cost of science equipment is borne by student laboratory fees, and the balance is paid by a supplementary amount from the schools' general operating fund. While operating costs can vary widely, a simple rule of thumb to follow is to multiply the number of students you have in class for *the full year* by $5.

For example, if you taught five sections of high school biology, with 30 students in each section, a reasonable amount to spend on expendable supplies for the year would be $750. In many schools each student pays a laboratory fee for the year of about $5, so the biology program pays for its own equipment and supplies without supplementary help from the overall budget. Larger, permanent items like microscopes, kymographs, room furniture, and certain specimens are normally purchased by the school district as capital outlay items and are not considered part of the yearly expense for operating the department.

The kind of biology you teach will determine what kinds of tools you use to teach it. This is a question that must be answered by each teacher individually, but to help you formulate a basic

142

list we have included some suggestions in Appendix A 209. The larger the variety of teaching materials you have at your disposal, the more versatile your biology course will be. Because biology is laboratory-centered it relies heavily on the use of materials that must be used by students as they make their study of life.

CREATING CHAPTER CONCEPTS

1. Consult the teacher's manual for the BSCS Green, Yellow, and Blue Versions to determine why the writers of these biology texts felt it was essential that students work with living materials in the laboratory. Summarize your findings in a paragraph or two.
2. Make a list of living tools, available within a 25-mile radius of your present school, that could be kept easily in the classroom. Include a brief statement about each specimen you list, indicating how you would use it in teaching.
3. List 5 specimens, available locally, that you could preserve by freezing for later use in teaching.
4. Devise a student project that would be most effectively carried out by making use of the microscope train that was mentioned in the chapter.
5. Draw plans for making an apparatus for microdissection. Include instructions for drawing out glass rods to make fine needles, and details on making the movable base to hold the glass rods in position. Try to specify the use of easily obtained, inexpensive material.
6. Using a tape recorder, make a recording of a bird song, or the sound of mice or crickets. Record the sounds at $7\frac{1}{2}$ inches per second, and then play it back at $3\frac{3}{4}$ inches per second. Record the playback on a second tape recorder at $7\frac{1}{2}$ inches per second, and then slow down the second playback. Using this piggyback process for slowing sounds, with two tape recorders you can bring the sounds down to a lower pitch and slower speed, and easier analysis and study. Play your finished recording for the class.
7. Make a list of 5 LP records that you would like to use in teaching biological sounds.
8. List the titles and sources of 5 folk song recordings that have biological themes.
9. Write a script that could be used by a class in recording an audio version of the evolutionary time scale. For help with this, see *Laboratory and Field Studies* by Lawson and Paulson.
10. Visit a camera store and get information on two different brands of single-lens reflex cameras. Write a one-page description of each, including the accessories available for the camera. Include a discussion of the uses you hope to make of the camera and the reasons you favor one brand over another.
11. List 5 uses that can be made of a camera in teaching biology, in addition to those listed in the chapter.
12. Draw up plans for building a school darkroom. Include specific equipment that should go into the darkroom, its source of supply, and cost.
13. Go to the Curriculum Library for science in your college or university library, and photograph the stacks that contain books in biology. Stand as far back as you can from the shelves in order that you can cover as large a field as possible.
14. Prepare an annotated list of single-concept films that you would find helpful in teaching biology. Include source of supply and cost.
15. Take a series of 5 or 6 pictures that illustrate one aspect of a general biological theme. Stages in the life cycle of an insect, the differentiation of a stem tip as it becomes a flower, or a group of pictures showing types of ecological niches could be used as subjects. Enlarge the pictures to 8 x 10, put them on a bulletin board, and make appropriate labels for the finished display.
16. Send to several major television suppliers (see Appendix F 294) for information on closed-circuit television. Then compile a brief list of equipment that you could use within your classroom for more effective teaching.
17. Design and cast a plaster model of a

biological subject. Paint the model and label the parts with India ink.

18. In the last part of the chapter we listed tools that should be purchased. Using that list, indicate sources of supply and cost for the tools.

19. Discuss the merits of the biology teacher being given a 10- or 11-month contract for the purpose of maintaining stock, gathering materials, and constructing teaching aids.

EXPANDING THE CHAPTER

BRILMAYER, BERNICE G. 1962. *All About Vines and Hanging Plants.* Doubleday & Co., Inc., Garden City, N.Y. A reference source that contains descriptions of plants that can be grown inside, including specific hints for overcoming problems in culturing them.

BROWN, J. W., and LEWIS, R. B. (eds.). 1957. *Audiovisual Instructional Materials.* The Spartan Book Store, San Jose, Calif. This helpful book lists a large number of additional references on audiovisual methods and equipment.

BROWN, RANDOLPH R. 1965. Microprojector, Microscopy, and Science—A New Look. *The Science Teacher* 32(5):39–42. Some unique techniques for biology teaching are discussed.

DAVIS, H. M. 1964. *Scientific Instruments You Can Make.* Science Service, Inc., Washington, D.C. A book of helpful suggestions on construction of a number of inexpensive teaching materials.

DRUMMOND, AINSLIE H. 1965. Information Bridges—Laboratory to Classroom. *The Science Teacher* 32(2):19–22. An enlightening article on the use of periodicals in teaching biology. A good list of useful periodicals is found in this article.

FREE, MONTAGUE. 1958. *All About House Plants.* Doubleday & Co., Inc., Garden City, N.Y. The care and feeding of house plants, including those difficult to grow inside.

HAAS, K. B., and PACKER, H. Q. 1955. *Preparation and Use of Audiovisual Aids.* 3rd Ed. Prentice-Hall, Inc., Englewood Cliffs, N.J. A reference source that is designed to encourage the use of audiovisual materials.

HABER, L. 1964. Technique To Clear Specimens and Stain Bone. *The Science Teacher* 31(7):41–43. An article that supplies an excellent technique for the preparation of some attractive display specimens.

KLINGE, PAUL. 1965. In My Opinion: Responsibility. *The American Biology Teacher* 27(4):244. An article by the editor of *The American Biology Teacher,* pointing out some of the dangers faced by students who work with viruses, use live material, or take field trips.

RASKIN, ABRAHAM. 1961. Science Toys in the Classroom. *NEA Journal* 50:41. A short article that brings to light the usefulness of simple toys in teaching scientific ideas.

RICKERT, FRANCIS B. 1965. Experiments for the Greenhouse. *The American Biology Teacher* 27(2):106–12. The use of the school greenhouse is enlarged by this article that describes 16 excellent experiments that can be carried out.

SANDS, LESTER B. 1956. *Audiovisual Procedures in Teaching.* The Ronald Press Co., New York. A book of basic audiovisual practices as applied to teaching.

STONE, WARREN B., and MANWELL, REGINALD D. 1965. The Use of Albino Laboratory Mice for the Demonstration of Parasites. *The American Biology Teacher* 27(3):208–13. This article is only one of several in the March issue that dealt with animals in biology laboratories.

STRICKLAND, W. L. 1964. Teaching by What Authority? *The Science Teacher* 31(6):23. A discussion of a philosophy that is helpful when teaching the new biology.

TAYLOR, JOHN K. 1965. Moderation in Instrumentation. *The Science Teacher* 32(3):18–19. A well-written article involving a discussion of equipment and its proper use in the science laboratory.

UNESCO. 1962. *Source Book for Science Teaching.* 2nd Ed. United Nations Educational, Scientific, and Cultural Organization, Paris. A standard reference listing sources and uses of many tools with which to teach.

WITTICH, W. A., and SCHULLER, C. F. 1967. *Audiovisual Materials: Their Nature and Use.* 4th Ed. Harper and Brothers, New York. An updated reference book pertaining to helpful audiovisual methods and materials.

CHAPTER **9**

Planning

Field Trips

JOSEPH WOOD KRUTCH, the naturalist, wrote:

Very recently I had occasion to spend a week on the campus of one of the oldest and most respected of the smaller liberal arts colleges of the eastern seaboard. It was one that prides itself on its exclusive concern with liberal rather than preprofessional education. A benefactor gave a beautiful wooded tract . . . which is lavishly planted with native and exotic flowering trees and shrubs. When no student or teacher with whom I had been brought into contact could tell me the name of an especially striking tree, I sought out the head of the botany department, who was also its only member.

He smiled rather complacently and gave this reply to my question: "Haven't the least idea. I am a cytologist and I don't suppose I could recognize a dozen plants by sight." The secrets of the cell are a vastly complicated and important subject. But should they be the one and only thing connected with plant life which a student seeking a liberal education is given the opportunity to learn?[1]

STUDY NATURE—OR BOOKS?

Louis Agassiz's famous admonition to "study nature, not books" may sound a little out-of-date today. High school biology laboratories have become places to do experiments in physiology, nutrition studies, and cytology. Teachers are

encouraged to foster individual investigation in the laboratory. The question is, How much biology will students learn from this kind of study? Or perhaps, *What kind* of biology will they learn? Mr. Krutch believes there is a great deal to be gained from a study of nature, and that there is limited value in taking apart preserved specimens in a laboratory—at least as far as appreciation and reverence for life is concerned. He says:

To proceed from the dissection of earthworms to the dissection of cats—both supplied to hundreds of schools and colleges by the large biological supply houses —is not necessarily to learn reverence for life or to develop any of the various kinds of "feeling for nature" which many of the old naturalists believed to be the essential thing. To expect such courses to do anything of the sort is as sensible as it would be to expect an apprenticed embalmer to emerge with a greater love and respect for his fellowman. And an increased love or respect for living creatures is one of the last things many courses in biology would propose to themselves.

A biology teacher, like any teacher, ends up teaching more than subject matter. He teaches more than the facts and concepts that make up his course. He hopes that, after the students digest the facts and assimilate them into concepts, some kind of metamorphosis will take place in their minds. This "change in thinking" is supposed to give them an "appreciation" of the subject—a view of the total scope of biological science.

How can the use of outdoor laboratories lead to an understanding of "total biology"? How can the study of nature, in natural settings, enrich and complement work in the classroom? To answer these questions you must first decide what you want to emphasize through the use of outdoor laboratories. You have to

---

[1] *The Great Chain of Life,* The Riverside Press, Cambridge, Mass., 1956.

consider the importance of taxonomy, basic field techniques, and the recording and reporting of data. Properly meshed, these aspects can make field biology an enriching, vital experience. Improperly used, they can make field work a meaningless drudge.

NATURE STUDY WITH A PURPOSE

With the current emphasis on molecular biology, cytology, and genetics, some teachers hesitate to include field work in their biology teaching. There are other reasons why outdoor teaching may not be popular. Effective field work is often done best in small groups. The high school teacher who has twenty-five or thirty students in his class finds it hard to give individual guidance and help. Class periods of forty-five to fifty minutes often make it difficult to get to and from the site. This leaves little time for setting up experiments, gathering data, and collecting specimens. Places at which field work can be done are sometimes not found close to the school, and transportation becomes a problem. All of these factors discourage the teacher teaching outside the classroom.

On the other hand, each of the difficulties cited above can be overcome by the creative biology teacher. Certain field problems relate to molecular biology. Large biology classes can be divided into small teams for effectiveness in the field. The forty-five- to fifty-minute class period is sufficient if the teacher does some planning and blocking out of project work with the students *before* field work begins.

Before we turn to specific field techniques, let us examine some of the motives for including outdoor work in a high school biology course.

Field Trips Provide a Look at the Diversity of Life

Of course some teachers might maintain that diversity of life can be studied in the classroom through the use of microorganisms and preserved specimens. But the use of microbes, while showing diversity, is not entirely valid, for it shows only a range within a specific group. A well-organized field trip, or series of field trips, can be structured to show diversity of all living things—protists, plants, *and* animals. As far as specimens are concerned, what better way is there to study the nature of life than by examining *living* things? You should capitalize on students' curiosity and interest in living plants and animals. Use living materials in favor of preserved materials.

Field Trips Furnish Starting Material for Further Work in the Classroom

Ron Koble, a scientist for the United States Geological Survey, believes that direct observation is of utmost importance. He says:

Much of the uniqueness of investigating . . . lies in the necessity sooner or later of making direct observations out-of-doors. Investigating the earth is not something that can be done wholly, or even largely, by experiment, although experiments may in some instances provide strong supporting data. Likewise, the earth cannot be investigated mainly by deduction, induction, or by intuition, although each of these can be important when based on, or supported by, data revealed by nature.[2]

Paul Klinge, writing in *The American Biology Teacher*,[3] says, "If we em-

[2] "The Teacher, the Student, and Field Trips," *The Iowa Conservationist,* February 1965.
[3] "We Shall Overcome," editorial in *The American Biology Teacher,* November 1964.

148

phasize the molecular approach in biology in our teaching, we lose sight of the organism." He believes the ecologist, field worker, and laboratory investigator use a number of techniques in common for their quantitative studies. Suppose, for example, that your class is on a field trip to a pond, and you are demonstrating the apparatus used to measure pH, oxygen, and carbon dioxide in the water. The basic ecological procedures may catch the interest of a student who is interested in both biology and chemistry. He can expand the project into a study of indoor aquaria in which the conditions can be carefully controlled. Let him set up six aquaria and adjust the pH of the water in each with buffers. After plants and fish are put in, measurements can be taken to determine the O_2 and CO_2 concentrations. The balance between these gases in each of the six setups is an index of the photosynthesis/respiration ratio in an aquatic environment. It is a project that could begin during an ecology field trip, then be expanded into a laboratory investigation. Appendix G 319 contains more information about projects like this.

Field Trips Show Ecological Relationships

One common denominator in biology courses concerns the food web. Regardless of the emphasis, whether it be biochemical or phylogenetic, the idea of plant and animal interdependence usually is a part of most biology courses. Teaching the complex concept of autotroph/heterotroph relationships is not easy, but it is made less difficult if you have powerful examples. Ponds, since they are biomes in themselves, are often used as illustrations. But there are others just as dynamic. Choose a large rock, partly covered with lichens and moss. These two groups of lower organisms,

the first in the soil formation sequence, give off metabolic wastes that break down the rock's surface. Thus, using the rock as a substrate, the bryophytes with the help of lichens produce tiny soil particles. In addition, when their tissues decay, the organic residue becomes a part of the newly formed soil. Soon populations of soil bacteria and insects build up. Seeds of herbaceous plants and of many trees find footing in the soft surface on the rock. The entire process of interdependence occurs on the rock, which beginning biology students can see and study. The concept of a biome, or of a small community of organisms, can be built on the tangible, visual evidence of the rock. Figure 9.1 illustrates some of the related organisms on the type of rock we described.

The teacher who operates against a background of improvisation and creativity will have in mind several addi-

Fig. 9.1—A rock like this or a log covered with plants can be teaching tools. How many of the major plant groups and what types of algae and fungi would you expect to find on the rock?

tional objectives that can be developed through the use of outdoor laboratories. The three we have just mentioned are only a start. Jack Higgins, writing in a state conservation magazine,[4] says that the teacher must define the objectives for the field trip early in the planning stage. "There are many things that can be observed," he says, "however in order to be meaningful, the topic must be narrowed down. In other words, the class should be exclusively interested in just one of the many phases of the course, and not in a broad and meaningless over-view of the entire subject." Perhaps this thought is the most important central theme around which field trips should be designed. (See Appendix A 210 for suggested places to take students on field trips.)

THE STYLE AND DESIGN OF A FIELD PROBLEM

In the introduction of their *Manual of Field Biology and Ecology*,[5] Allen Benton and William Werner summarize what field biologists find important in attacking a problem in ecology:

. . . professional or amateur collectors use similar techniques and require a similar background of knowledge. Each must be able to recognize the organisms with which he works, and must know something of their distribution, their ecological requirements, their way of life. Each must know how to collect and preserve samples of these organisms, where and when to find them, and from whom to get additional information about them. Each must have a sound knowledge of the . . . literature . . . and know how to use varied reference books. Each must know how to observe, and how to take and interpret notes on his observations. Each must know how to use such tools as cameras, binoculars, recording devices, chemical materials . . . as the exigencies of his particular profession or hobby require.

Field work in biology is most successful if the biologists have a *wide knowledge* of the tools and techniques that are necessary for investigation. The more you know about the use of taxonomic keys; laying out transects for population studies; and methods for sampling, collecting, and preserving organisms and gathering data, the more effective you will be as a field biologist and a teacher of field biology. These teaching tools are as much a part of working through an effective field problem as the site of work itself. They must be a part of the design of the field problem, integrated into the learning experience.

Keys to Life

Taxonomic keys are like maps—they help you find unknowns. A road map shows a large number of intersecting roads; to get from one point to another you must make a decision at each intersection. A properly made taxonomic key will be like a road map, with one important exception: a key requires a "yes or no" choice, that is, it presents alternatives at each junction. A *dichotomous key,* presenting only two alternatives at each junction, is the most commonly used type in biology. For example, in Figure 9.2 in which trees are being identified, the first choice is "leaves long and needle-like" or "leaves short, awl-shaped, or flat." This first choice separates spruces, firs, and pines from other gymnosperms.

Unfortunately, because of the highly specific terms and peculiar names that are found in many keys, some teachers feel that too much time is lost in

---

[4] "Observation and Awareness Spell Knowledge," *The Iowa Conservationist*, December 1965.
[5] Burgess Publishing Co., Minneapolis, 1965.

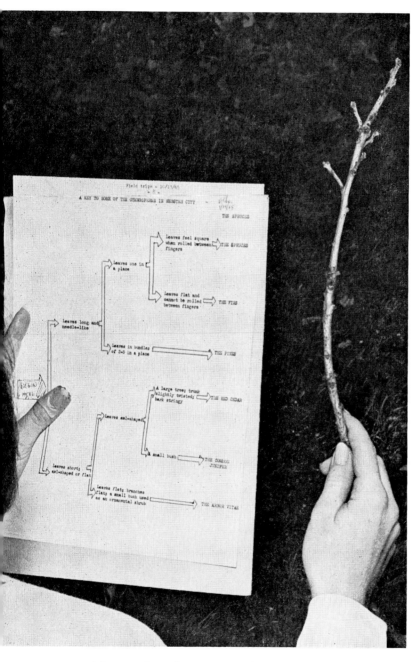

Fig. 9.2—You cannot memorize all the names for plants and animals, but you can identify unknown specimens with a key like this. Biology students should be taught the use of keys early in their study of biology.

teaching students how to use them. The solution to the problem of taxonomic difficulty is simple: choose keys that are nontechnical; that key to class, order, or family only; and that you are familiar with yourself. Do not try to identify everything you find. Concentrate on a few groups, and do not try to key everything in those selected groups to species. The idea of using a key can be taught just as well, and with less frustration to the student, if you identify unknown organisms to class or order only.

When keys and their use are presented to students as *tools,* there is usually a high degree of acceptance. Students see them as learning aids—as necessities in solving problems. But when keys are used as ends in themselves— merely as sources of names of families, genera, and species to be memorized— their purpose is clouded with the memory work that must be done. To make their use successful you must show the students why it is important to use them. Once the reason behind keying is shown to be an integral part of the field problem, identification of specimens through keying becomes as vital as the use of a collecting net or plant press.

The important question is, How can the process of taxonomy be "sold" to the class? Perhaps the following example will illustrate one way to accomplish this. John G. Stoffolano, Jr., a biology teacher at the Oneonta High School in New York, has developed an effective procedure for collecting and studying the arthropods that inhabit leaf litter. In his article in *The American Biology Teacher*[6] he includes instructions for making the apparatus, selecting an area of study, sampling leaf litter, and separating the arthropods. He then determines what organisms are in the collection. This is where a working knowledge of classification and keying comes in. "Using the binocular microscope, various keys, and collected samples," writes Mr. Stoffolano, "one may begin to classify the organisms." He recommends specific keys and furnishes drawings the investigator will find useful. The organisms, once identified, can be compared to each other in terms of population density, number in each genus,

and population fluctuation over extended periods of time. Without a knowledge of the variety of arthropods in the samples, this field problem would be almost meaningless. In the section of the article called "Projects," several investigations are suggested. As you study each of the following suggestions, try to imagine it as a valid biology project if the tool of taxonomy were removed.

1. Compare the kinds and numbers of microarthropods inhabiting different kinds of leaf litter. Example: Compare pine and oak litter.
2. Compare the kinds and numbers of microarthropods inhabiting the leaf litter at various seasons.
3. Compare the kinds and numbers of microarthropods present in relation to the amount of water and/or the amount of leaf litter.
4. Study the microarthropods living in fresh water.

Sampling Life

Effective field workers often devise their own methods of getting samples, as the diatometer in Figure 9.3 illustrates. This simple device, made of an old wooden slide box, microscope slides, and rubber bands, can be used with a high degree of success in lakes, ponds, and rivers. It is especially useful where silt clutters up ordinary collecting nets. While specific methods of collecting and preserving are discussed in Appendix B 220, 231, and D 277, we include some of the important ones here.

Knowing how to make field collections is a little like knowing how to speak a language—first you learn words, then you put them together to make

_____
[6] "Studying Microarthropods Inhabiting Leaf Litter," November 1964.

Fig. 9.3—This diatometer is useful for collecting all kinds of protists. Two slides are inserted in each slot. After about a week the diatometer can be removed from the water and the slides can be scraped. Diatoms and protozoa are the most numerous organisms collected in this way.

a bottle of FAA (10% formalin, 5% glacial acetic acid, 70% ethanol) is a good general preservative. Nets, traps, and tweezers (for small organisms) make the collecting easy.

FIELD WORK IN A CITY

If you live in a city and are unable to reach suitable areas for field biology, there are several ways to study biology outside the classroom. While we list details for suggested field trips in Appendix A 210, the following activities are examples of field work in a city.

1. *Animal Behavior.* Visit a park and observe birds, insects, or small mammals. Watch for behavior patterns that show how the animals have adapted to each other and to man. Examples of such adaptations include pigeons and squirrels that come close to people for small bits of food.

2. *Plant Adaptations.* Make a survey of the vegetation in a park and list the growing conditions, including light, moisture, type of soil, and the effect of other living things. Look for vines on trees and buildings, weeds in cracks in the sidewalk, and lichens on rocks and trees. Compare the vegetation in two different parks, listing possible reasons for differences.

3. *Life in a Fountain.* Study the algae, protozoans, and bacteria in a park fountain or street corner drinking fountain. Often there are small water leaks at the base of a fountain. Take collections from these areas for later study in the classroom.

4. *Soil Microbes.* Take small samples of soil from several locations in a park, forest preserve, or in other

sentences that mean something. Collecting is like that. First, you learn about different methods of collecting, then you choose the techniques to accomplish your aim. The collecting technique you use depends largely on what you are trying to discover through the field problem. For some situations it is necessary to collect specimens and take them alive to the laboratory. Plastic containers, gallon pickle jars with screen tops, and other improvised cages make excellent transporting containers as well as laboratory housing for the animals. If the organisms are to be killed and preserved, areas within a city. In the laboratory analyze the samples for protists and other forms of life. Nematodes and rotifers may also be found in the collections.

On the other hand, much valuable information about plants and animals can be gained without collecting any material. Suppose, for example, your biology class wanted to begin a long-range study of the changes in the local flora. Each student could be assigned an area to survey, with the instruction that he identify all trees and shrubs in that area. After some practice keying these plants he could survey the population of the assigned area. His experience with laboratory specimens would help him identify the plants without taking any leaves or twigs. Even the slower learners can identify common trees and shrubs on sight after suitable experience.

While larger plants, like the trees in our example above, can be examined closely without taking samples back to the classroom, smaller organisms usually require study with a microscope. This makes on-the-spot identification difficult in the field. Furthermore, several reference books are often required. Unless you have facilities for operating a permanent field station, it is more satisfactory for high school students to take specimens back to the classroom for more accurate study.

Suppose your class were to begin a long-range study of the changes in the population of soil insects in a selected plot of ground near the school. With organisms this small it would be almost impossible to make significant identifications in the field. Representative samples from the plot would have to be collected, taken to the laboratory, and processed there.

DEVELOPING SKILL IN COLLECTING

To insure that the field work is done correctly you must emphasize proper technique in collecting and preserving organisms and in recording and interpreting data. Collecting organisms and interpreting data go hand in hand. Data can be skewed in favor of this conclusion or that, depending on how the collecting is done.

For example, if you plan to make a comparative study between the insects in a prairie and those from a wooded area, you must collect from all possible niches in each habitat. In the prairie some insects live on the herbaceous plants. Some hover above the flowers. Others live nearer the surface of the soil, in the grasses. If you use only one type of collecting technique, many of the insects will not be found. For example, some insects in the Order Hymenoptera may be entirely omitted from the collection if you use a beating net exclusively. Forceps and baited traps aid in collecting certain kinds of prairie insects.

On the other hand, collecting in the woods requires different techniques than prairie collections. Insects in woods can be collected with light traps, baiting, or by the process of sugaring trees.

For the best results in using the sweeping or beating net it should be swung in wide arcs about 6 inches above the ground, knocking insects and other arthropods from the vegetation. Figure 9.5 shows how the insects are killed, sorted, and pinned. When using the beating net keep in mind that the net should be emptied frequently into the killing jar, that some attempt should be made to walk into the vegetation at a constant pace, and that this method will

a. The sweeping or beating net.

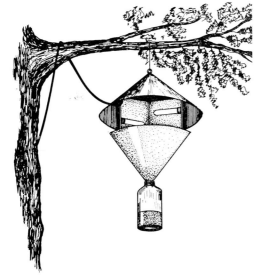

b. An insect light trap.

d. The process of sugaring trees.

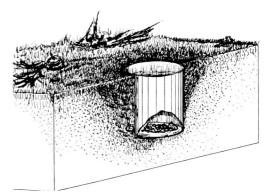

e. A tin-can trap for carrion beetles.

c. A fluorescent light on a sheet, at night.

Fig. 9.4—The sweeping or beating net and other methods for catching insects are shown here.

not catch all of the insects. If you are making a survey of the insect population in a field or woods, you must use a variety of techniques. The sweeping net is only one. (For additional material on collecting insects, see Appendix D 255.)

The aquatic counterpart of the sweeping net is called a plankton net. Made from finely woven silk, with about 150 to 300 openings per square inch, such nets are pulled through the water at trolling speed—about 1 meter per second. Though the mesh is large enough to let many microscopic organisms escape, a considerable number are retained. A small vial at the narrow end of the net catches the organisms. Plankton nets come in many sizes, from the three-foot diameter marine nets that are designed to be towed from ships, to the

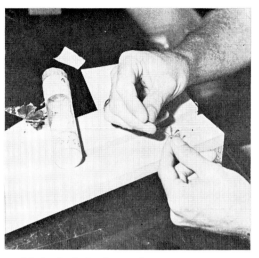

c. Method of pinning an insect.

three-inch size that can be cast from a fly rod.

Aquatic collecting devices are illustrated in Figure 9.6. They include the bottom net (which is excellent for collecting insect larvae), plankton net, and gravy baster. For collecting protists a gravy baster is the most satisfactory. Small fish and crustaceans can be collected with a seine.

Collecting nets and apparatus for surveying aquatic populations are outlined in Appendix D 259.

GUIDELINES FOR FIELD WORK

When you set up a field problem like an insect survey, what planning should you do? Consider the motives mentioned earlier in the chapter: showing diversity, providing material for further investigation, and illustrating ecological relationships.

Use the following four guidelines for planning field work. First, *define clearly and narrow the scope of the field problem.* Keep the problem relatively simple—if there *is* such a thing as a simple problem in ecology. Beginning stu-

a. Killing jar.

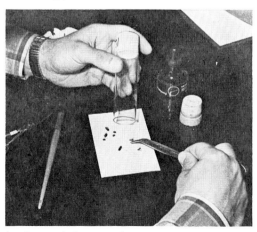

b. Insects being sorted.

Fig. 9.5—A valuable insect collection is not only an asset to the school museum, it reflects important data on the local area. Students can collect insects and make population studies from year to year.

Fig. 9.6—These aquatic collecting devices are useful for collecting small crustaceans, all forms of plankton, and tiny fish.

a. Bottom net.

b. Plankton net.

c. Gravy baster.

dents tend to be confused if the field problem is diffuse and encompasses a wide variety of subjects.

Second, *give the field work a theme.* Is it collecting technique you are trying to teach? Are you teaching the use of a taxonomic key? What about the diversity of life concept? Are you doing field work to show how many genera of arthropods live in fallen trees? These are all valid themes for field work; each defines the scope of the task. However, each could be reduced in scope. In the examples that follow, we have listed some broad, general themes for a few field problems. Each general theme is narrowed to a specific theme.

General Themes	Specific Themes
A study of the insects in McDonald Park	Some factors affecting the distribution of Coleoptera in McDonald Park
Diatoms from a farm pond	Planktonic diatoms from a farm pond, collected during March and April
The effects of minerals on plants	The effect of varying concentrations of ammonium nitrate on the growth in height of *Coleus*.
Vegetation found in Kendall Young Forest Preserve	The ecology of selected angiosperms in Kendall Young Forest Preserve

Third, *identify the concepts* that are to be brought out in the field problem. This stage in the design might seem to be the most nebulous and difficult because it is hard to recognize specific concepts in an area as diffuse as biology. Furthermore, since so many facts are generally required in order to construct the concept, it is easy to let the mass of facts cloud the concept. Certainly collecting, naming, and classifying specimens, and gathering data seem to be rote memory learning. Yet, this kind of study is necessary if we are to make the concepts vivid. F. James Rutherford[7] of Harvard University believes facts and inquiry go hand in hand in the learning process. He says:

When it comes to the teaching of science it is perfectly clear where we, as science teachers . . . stand: we are unalterably opposed to the rote memorization

---
[7] "The Role of Inquiry in Science Teaching," *Journal of Research in Science Teaching*, June 1964.

of the mere facts and minutiae of science. By contrast, we stand foursquare for the teaching of the scientific method, critical thinking, the scientific attitude, the problem-solving approach, the discovery method, and of special interest here, the inquiry method. In brief, we appear to agree on the need to teach science as process or method rather than as content.

Judging, however, by what we can see taking place in many . . . classrooms . . . we might reasonably conclude that there is a large gap between our practices and our convictions. This may well be the result of many factors. . . . *One* of the contributing factors, it seems clear, has been a failure to recognize . . . the close organic connection between process and content in science. The choice is *neither* facts and laws *nor* inquiry and process; it is *both* facts and laws *and* inquiry and process.

Fourth, *condition the students to field work*. Much of this can be done by adequate preparation. Tell them what they will need to look for, what techniques they will need to use, where they will work, and give them a format for working. Use 35mm color slides, or a short film you have taken. Use any device you can to prepare them for the forthcoming work. The students will be more careful workers, and the data they gather will be more valid.

DEALING WITH DATA

You get facts from objective field observations. As the facts are recorded, they should be designated as *qualitative* or *quantitative*. Generally, quantitative data are more useful. For example, suppose you plan to have your class survey two insect populations—one in a prairie, the other in a woods. Qualitative data would list only the orders and genera of insects. Quantitative data would include *quantities* as well as names of the orders and genera.

If you quantify data from a field problem, remember that larger samples make a more accurate survey. Since it often is not practical to take extremely large samples, you may wish to use a mathematical shortcut. Worked out by the Englishman Karl Pearson around the turn of the century, it is called the *chi-square test*. Chi-square helps you determine the probability that chance alone—and not nature—caused the results. The equation below is used to compute the chi-square value.

tails when you present your data. In Figure 9.7 in which data on water chemistry are presented in chart form, how meaningful is it to you? And how meaningful would it be to high school students?

Now look at the same data in Figure 9.8 presented in graph form. Not only does it show results of each test at a glance, it also *shows relationships*.

Presenting data visually is important for clarity. It helps the reader understand the results. As Rudolph Flesch[8]

$$\chi^2 = \text{sum of all } \frac{(\text{expected} - \text{observed})^2}{\text{expected}}$$

chi-square value

	0.0002	0.004	0.016	0.455	1.074	2.706	3.841	6.635
the number of times in 100 that chance alone could have been responsible for the variation	99	95	90	50	30	10	5	1

Note that with a chi-square value of more than six there is less than one chance in one hundred that your results could have been due to chance. Chi-square can be used by many investigators, especially high school students, who have not had courses in statistics or higher mathematics. It is only necessary to solve a simple equation to get the chi-square value. Appendix A 212 gives full details on working out chi-square values you can use to test your data for validity.

Even if your problem is well designed, your results are valid, and you have extensive collections, you must be able to transmit your conclusions to others. And you cannot assume that the others know a great deal about biology. They may know something about it—enough to be able to orient their thinking to your overall problem—but they might not be able to follow the de-

wrote when he discussed how necessary words really are, "People escape from words into pictures, symbols, graphs, charts, diagrams—anything at all as long as it is 'visual.'" But while one picture is worth a thousand words, sometimes you need a thousand words to explain one picture. An important guideline is to make the graph or chart so there is no need for extensive explanations or legends.

How can you make easy-to-understand graphs, charts, and diagrams? Here are some criteria to guide you. These same standards can be used when you evaluate your students' work—as you grade their laboratory and field reports.

1. *Use bar, line, or circle graphs to show relationships between numeri-*

[8] *The Art of Readable Writing*, Harper & Row, Publ., Inc., New York, 1949.

AREA	SAMPLE	DATE	P ALK	T ALK	Ca HARD	TOT HARD	Cl-	SO4=	Fe	PO4≡	NO3-	SiO2
	12	8/9/60	15	185	—	190	—	45	0	0	0.02	18
	13	8/9/60	—	200	—	195	—	60	0	0	0.04	13
	14	8/9/60	—	200	—	200	—	50	0	0	0.04	14
I	25	8/15/60	5	180	7.5	200	7.5	45	0	0.005	—	3.2
	26	8/15/60	1	160	70	200	10	200	0	0.5	0.03	16
	30	9/17/60	20	170	60	200	5	30	0.15	—	—	18
	50	6/11/61	15	210	90	225	10	38	—	—	—	0.5
	15	8/9/60	20	170	—	200	—	43	0	0	0.04	14
	16	8/9/60	—	200	—	195	—	30	0	0	0.04	13
	17	8/9/60	—	190	—	200	—	35	0	0	0.14	14
II	38	4/29/61	20	170	65	190	10	35	0	0	0	1.4
	45a	5/20/61	1	205	75	210	10	40	0	0.005	—	1.5
	46	6/11/61	10	210	70	220	10	35	0	0.005	—	3.2
	48	6/11/61	20	215	90	220	10	20	—	—	—	3.1
	51	6/11/61	12	205	90	220	—	—	—	—	—	—
	10	7/29/60	45	285	80	210	10	55	0.9	—	0.08	1.2
	11	7/29/60	30	100	80	205	10	45	0.5	—	0.08	7.5
	19	8/9/60	—	190	—	200	—	25	0	0	0.05	16
III	20	8/9/60	—	200	—	210	—	25	0	0	0.05	17
	42	5/20/61	0	200	100	240	7.5	30	0	—	0	2.5
	47	6/11/61	10	210	100	225	7.5	38	—	—	—	3.2
	49	6/11/61	20	210	90	235	10	32	0	0.3	—	2.8
	22	8/14/60	30	190	80	220	11	55	0.05	0	0.03	22
	31	9/17/60	30	220	85	200	10	35	—	—	—	20
IV	39	4/30/61	15	160	70	180	10	32	0	0.2	0.05	0.5
	45b	6/16/61	0	215	95	240	15	38	0	0.4	—	3.5
	5	7/29/60	30	220	100	225	12	38	0.15	—	0.06	1.1
V	23	8/14/60	30	220	80	240	10	50	0.07	0	—	22
	32	10/2/60	20	275	120	207	23	42	0	0	0.1	10

Fig. 9.7—The data in this table are hard to interpret at a glance. It is also difficult to see relationships between the different factors to which the data pertain.

cal data. Avoid lists of numbers.

2. *Keep the legend simple.* A well-constructed graph should need only a key phrase or two for clarity.

3. *Pictorialize as much as possible.* Example: On a bar graph of population comparisons of arthropods draw pictures of a beetle, butterfly, and grasshopper instead of writing the words Coleoptera, Lepidoptera, and Orthoptera on the graph.

One unusual way to summarize a study of insect populations is shown in Figure 9.9.

FUSING FIELD WITH CLASSROOM

Dr. Niko Tinbergen, the prominent authority on animal behavior, spends a great deal of time in the field taking movies of birds. He studies their relationships to each other. In a sense he is trying to determine the "sociological relationships" that exist among birds. But Dr. Tinbergen believes that laboratory work is just as important, and he describes its importance in *The Study of Instinct.*[9] In this fascinating documentary on evolutionary processes and natural selection, he discusses the important clues that the study of nature can provide, but he points out that the clues are just that. They are clues. They are not answers. Many of them are only hints that must be investigated further in the laboratory.

As you design your field problems, look at each from the point of view that it is just a start. Try to extract from each field exercise one or two projects that can be continued in the laboratory.

A pond as a site for field work is a good choice. In fact, some teachers prefer aquatic biomes over terrestrial ones.

---

[9] Oxford University Press, 1951.

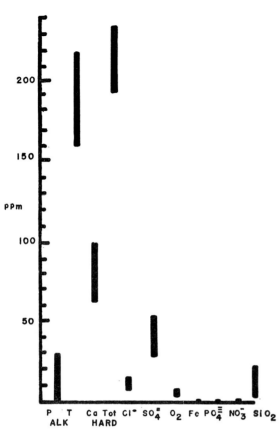

Fig. 9.8—The data in this graph are much easier to interpret at a glance. The graph also shows relationships between the factors on which the data were taken.

William Nutting,[10] a Massachusetts teacher, prefers ponds because:

1. Conditions fluctuate less severely than in a terrestrial environment. (Ponds are superior to brooks because of this.)
2. Samples may be obtained with simple, easily constructed equipment.
3. Specimens from a pond are easily maintained in the laboratory for controlled experiments.
4. The large numbers of different

[10] "Biology of a Pond," *The American Biology Teacher,* May 1966, p. 351.

organisms in a pond impress the student with the interdependence of plants and animals.

Instructor Nutting offers some important guidelines for selecting a suitable site, setting up sample stations, mapping the area, sampling and collecting specimens, and making water tests. In a section called "Laboratory Studies" he cites ways to extend the field work into the laboratory. Among these he suggests that the organisms be collected live, then cultured in the laboratory to obtain precise information on life cycles, food-web relationships and metabolism. Suppose, for example, you discover the larvae of 5 or 6 orders of insects in algae mats among the rocks. With a minimum of care a biology student could collect these larvae, take them back to the laboratory, and raise them to maturity. Although the work in such a project is largely confirmatory, it still makes exciting laboratory work for students. This kind of direct, close observation requires students to form the habits of accurate observation that are so necessary in the work of science.

Balancing Indoor and Outdoor Biology

The use of field work can strengthen biology teaching; however, too much field work may be a weakness. There probably is little justification for teaching ecology an entire semester, just as there would be for teaching a large unit of molecular biology, taxonomy, or any other area. Try small units. Often small problems are as effective as large ones. A walk around the block for a look at trees, or a ride in a school bus to look at a creek or field, might fit your teaching situation best. Try Saturday morning excursions or several weeks of summer field biology. Your facilities, the area of the

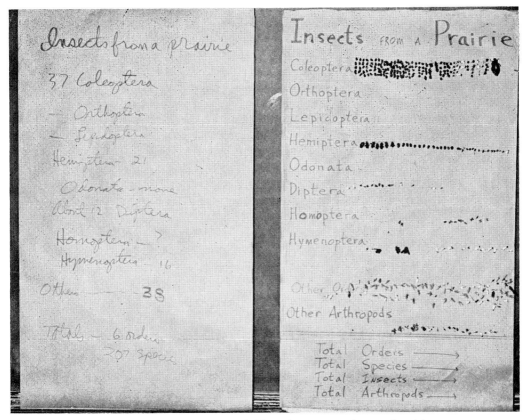

Fig. 9.9—The insects collected in an actual field study were used to make this display on the right. It is easier to study the results of the survey using the insects than it is to interpret the tabulation of insect orders, as in the chart on the left.

country in which you live, and your interests and background all determine the quantity of field work you do. Whatever the choice, well-planned outdoor work is important in creative biology teaching.

CREATING CHAPTER CONCEPTS

1. Do you believe that high school biology should contain a certain amount of field work? What value does outside work in biology have?
2. How can a biology teacher in the center of a city teach field biology?
3. Choose an outdoor habitat and design a field problem that could be worked out in that habitat. Some suggestions include a pond, woods, stream, open

prairie, or river. For teachers who have little access to these habitats, design the investigation for plants and animals that might be found in a city, such as birds that inhabit buildings, or weeds that grow in sidewalk cracks. Structure your problem so that you have:
   a. Pre-laboratory preparation.
   b. Instruction for students on collecting techniques and methods of gathering data.
   c. Methods of writing the report and expressing results. (This might include use of photographs, drawings, or models.)
4. List some objectives you expect to achieve through the use of an outdoor laboratory.

5. Devise a simple taxonomic key that can be used outside. Suggestions are: keys for snakes, trees, twigs, flowers, rocks, reptiles, or leaves.
6. Suggest themes for field work that show students the diversity of life.
7. Construct a graph or chart that is self-explanatory.
8. Design an indoor-outdoor field problem. What aspects of the investigation can be done outside, and which of those must be continued inside?
9. Make a list of units that could be used to teach a summer course in field biology. Base the units on a class of 20, for a 5-week period, meeting only 4 hours a day. Include a list of the equipment you would need to teach the course and the means you would use of evaluating the students.
10. Plan several field problems that students could use for independent study. Use habitats with which you are familiar, and that could be investigated during any season of the year. Block out the project so that it could be worked on by the student over a period of 9 or 10 weeks, on Saturdays or Sundays.

EXPANDING THE CHAPTER

BENTON, A. H., and WERNER, W. E., JR. 1966. *Field Biology and Ecology*. 2nd Ed. McGraw-Hill, Inc., New York. This book is a collection of keys, techniques, and instructions for working through a wide variety of field problems. A "must" for anyone who is designing a field problem in biology.

CHESSMAN, EVELYN. 1952. *Insects: Their Secret World*. Wm. Sloane Associates, New York. An interesting book, giving information about insect adaptations.

CLAUSEN, L. W. 1962. *Insect Fact and Folklore*. Collier Books, New York. A collection that provides little known facts and interesting stories about insect adaptations.

EDDY, S., and HODSON, A. C. 1962. *Taxonomic Keys*. Burgess Publishing Co., Minneapolis. An easy-to-use pictured key that is useful in identifying the major groups of the common animals of the north central states, excluding the parasitic worms, insects, and birds. These keys are especially easy for beginners.

FABRE, J. HENRI. 1949. *Insect World of J. Henri Fabre* (Edwin Teal, ed.). Dodd, Mead & Co., New York. This famous French naturalist has written a number of books that bring the reader face to face with the living world. This book will encourage students to "study nature, not books."

JAQUES, H. E. 1947. *How To Know the Insects*. 2nd Ed. Wm. C. Brown Co., Dubuque, Iowa. This fine little key to the common insects is only one of a series of "How To Know the—" books. While the terms are held to a minimum, it still may be somewhat difficult for the beginning high school student to use. However, the pictures help a great deal to facilitate identification of specimens.

LANYON, W. E. 1964. *Biology of Birds*. The Natural History Press, Doubleday & Co., Inc., Garden City, N.Y. A book that provides an introduction to birds and bird life for the student and bird enthusiast.

LINTON, DAVID. 1964. *Photographing Nature.* The Natural History Press, Doubleday & Co., Inc., Garden City, N.Y. A reference book that will aid students who wish to supplement their field project with photographs.

NEEDHAM, JAMES G., and NEEDHAM, PAUL R. 1962. *A Guide to the Study of Freshwater Biology.* 5th Ed. Holden-Day, Inc., San Francisco. This pictured key allows one to identify the common freshwater organisms found in the United States.

PENNAK, R. W. 1953. *Freshwater Invertebrates of the United States.* The Ronald Press Co., New York. This reference text is a taxonomic catalog of the invertebrates that are found in the United States. Although it is detailed, this text is not too difficult for beginning students. Excellent illustrations make it easier to use.

CHAPTER **10**

Shaping

Biology Units

WE HAVE DISCUSSED the historical background of biology teaching as well as some current ideas. We have talked about the biology setting to teach in, the tools you teach with, and the laboratory you teach from. We have shown the use of outdoor teaching sites and freshly collected material. We have offered some guidelines and examples to follow in teaching structure-function relationships.

Now we are ready to put many of these ideas together. Like a salesman who has been to his company's training school, had a tour through the factory, and spent a good deal of time studying his product, you have studied the ingredients of biology teaching. But, like a salesman, you must put these pieces together to sell your product successfully. This chapter will guide you in assembling important facets of dynamic teaching techniques. It will help you mold biology into teachable units that should motivate and inspire, as well as provide information.

IDENTIFYING THE INGREDIENTS

The Starting Point

What is there in science that is essential and vital to the general education of American children? Answering this question, Laura Zirbes,[1] emeritus pro-

fessor of education at Ohio State University says, ". . . there are different aspects of the question. One is the changing conception of knowledge itself. Another is the changing conception of children who are to be taught."

Key words are *essential* and *vital*. These provide a basis for shaping biology units and can serve as criteria for creativity as you build units.

The particular direction your teaching takes depends on the balance you choose between what you like to teach and feel "comfortable" teaching, and the areas of biology you feel you should teach to provide a well-rounded course.

No matter how you construct a unit, however, there are certain common denominators that should be fundamental to the planning.

Common Denominators

The kind of biology you teach should contain basic threads of relevance to the total education of your students. Even if you decide to give your course an ecological slant, or to emphasize biochemistry, structure-function relationships, or the phylogenetic continuity of organisms, significant overtones should emerge. We feel the following three overtones are significant.

1. *Science should develop the ability to generalize from specific examples.*

It is this *process of generalizing* which should be guided so that particular experiences become the stuff for ever more discriminating comparison, conceptualization, inquiry, and tested inferences. That process is what science teachers should learn to foster and facilitate, but they must also learn that the process cannot be reversed, rushed, or short-circuited without violating the very conditions and values on which the development of scientific attitudes and insights depends. The approach

[1] *Spurs to Creative Teaching.* Used by permission of G. P. Putnam's Sons. Copyright © 1959 by G. P. Putnam's Sons.

which seeks to cover the whole scope of science by such demonstrations is systematically designed to take children over the ground, but it does *not* do what science education should do to *educate* but also to *develop children*.[2]

2. *Science should teach method as well as material.* The art of the experiment, of playing the hunch, and of following the steps from observation and hypothesis to experiment and theory should be a part of our science teaching at all times. Facts are only bare bones if they are separated from the method by which they are determined. In the following example notice how the students discover for themselves the scientific way to solve problems.

A fifth grade was trying to make candles in a study of colonial life, but their "candles" did not "wax," or grow thicker. The teacher said "Isn't it time to sit down and think?" One youngster said, "No, let's go on dipping, dipping, dipping!" And so the teacher let them go on dipping, dipping, dipping a little longer. They went around and around, dipping as they went, but still the candles did not "wax." Finally, one youngster said, "I think we should sit down and think." That was science. When blocked in a problem in life, whatever it is, one does well to sit down and think.

Soon one child said, "I bet I know what it is," and he came up with a probable cause of the difficulty. Almost immediately another child said, "Come on, let's try it and see if that *is* it." As I remember it, this child's idea was that the water on which the wax was floating was not hot enough. They turned up the gas. That was the variable, according to this child's guess, "bet," or hypothesis. After the gas had been turned up, they went around again, dipping, dipping, and dipping. Still the candles did not wax. They could see that the idea was not good.

Somebody said, "Let's think again." Many adults would give up, or ask somebody, instead. . . . Soon someone came up with another suggestion, "Maybe we should

turn the gas *lower* instead of higher, and wait awhile." Another said, "Let's get a thermometer and see exactly how hot it is now." Precision! After a while, one of them looked at the thermometer again and said the heat was going down. "Let's try it now," proposed another, a little impatiently.

Somebody suggested that they go out and play awhile, letting the liquid cool, and then come back. They went, and when they came back, they went around again. Still the candles did not wax! Finally, one little fellow said, "You know, it could be more than one cause. A lot of things are due to more than one cause, and I was just thinking it could be how fast you dip, *and* how hot it is, *and* how long you keep it out before you dip again." They set up a little experiment right there, in which they decided to dip *faster,* to keep their candle strings out *longer* before they dipped again, and to see what happened. And I think that one of the most glorious memories I have is the one of those youngsters when they discovered that they had it. They had solved the problem. But really, they had solved more than that problem. Three or four times, long afterward, I heard them use what they had gotten as a general "hunch" from that experience: "Could be like the candles." "Maybe there's more than one cause." . . . Or they would say, "Well, I guess this is one of those times when we'll have to experiment."[3]

3. *Science should make students aware of the physical and natural world around them.* There is a difficult, intangible quality connected with biology education. Students spend much of their time learning about the matter and mechanics of plants and animals, but we hope they get more than that. We hope students develop an appreciation for nature, that they become aware of the living things around them, and that they cultivate a curiosity for finding out more about trees, insects, and other organisms around them. Excitement with nature and a knowledge of events that take

[2] *Ibid.*

[3] *Ibid.*

place in the world of wild plants and animals vary a great deal, of course. Rutherford Platt wrote an especially vivid description in his book, *The River of Life*.[4] Note the large number of details he is aware of, as he introduces us to the environment of the honeybee.

If you will come with me early on a spring morning to the willow at the foot of my hill, I will show you an exciting animal. The night shadows still lingering in low places are melting away, and the day shadows are shortening in unison. You can see the earth revolve simply by holding out your hand and watching a shadow move across it. . . .

This is the time and place, if you are in the mood, to see events that weave the fanciful fabric of life over land and sea. Every square inch of the landscape is filled with life, which dwells in a soft, gently responsive substance in a multitude of forms. Each is ready to absorb its portion of this day's energy coming in from sky space—at eleven million miles per minute. These receptors of light energy are as different in their sizes, structures, and systems as a green leaf, a flower, an aphid, a caterpiller, a bird's egg . . . a rabbit, or you and I watching a tassel on a willow.

Yesterday's sunlight called out this willow tassel from its bud, by setting in motion water, causing it to flow upward against gravity, and squeeze through the thread of the twig into the bud. . . .

The opening of a bud is similar to the phenomenon of birth which causes an egg to break and a bird to appear at the moment it is ready, a baby to burst out of its mother's womb, fruits to crack when their seeds are ripe. Every living event occurs at just the right time to make ready for the next event and often, as with the bee and the willow tassel, for one individual to meet another precisely at a time and place.

The emergence of the tassel happened at this spot yesterday. This morning we see today's sunlight continue this smooth flow of life where yesterday's left off. The tassel is stippled with red lumps in pinecone spirals, and now these are opening in succession, beginning at the tip and fol-

lowing the spiral course around and around to the base. . . . When the lumps crack open golden pollen grains spill out, turning the tassel yellow before our eyes. This is a signal for the bee. At this spot this moment the Plant Kingdom is summoning the Animal Kingdom to touch it and help itself to these crystals of packaged sunlight.

If you keep these three ideas in mind—generalizing concepts, following the method of science, and increasing the awareness of nature—your teaching will have important holdfasts. Much of what you do in the classroom can be tied to these fundamentals. Of course you must still cope with a large number of terms and techniques, and make a multitude of decisions every day that bear on your effectiveness as a teacher. But we are looking now at the overview of your course, at its design. With the three guidelines that we have discussed acting as a kind of superstructure, let us turn to several specific themes you can use to construct units in biology that will support it.

Themes in Biology

From one standpoint you could argue that there are almost as many themes in biology as there are biologists. But we think some of the themes can be more clearly identified. They include phylogeny, ecology, cytology, genetics, developmental biology, and metabolism. Think back over your own training in high school and college. Which one do you remember best? Which of your science courses contributed the greatest amount to your understanding of, say, metabolism? Did you have an equal amount of training in all phases of biology, or did you choose to concentrate on one or two aspects when you decided upon your electives? In this day of the knowledge explosion it is practically im-

[4] Simon and Schuster, Inc., New York, 1956.

possible to know all areas of biology. How can you overcome academic weakness in one area or another when you are expected to teach "general" biology? Two solutions are possible, and they should be worked together. First, capitalize on your strengths, and second, work to improve your weaknesses.

To capitalize on your strengths you must first analyze what you can do best. If you had to offer a special seminar to other teachers, conduct a short course, or teach a night school program, on which aspect of biology would you base your teaching? Are you better trained in ecology, for example, or physiology? Would you feel more at ease teaching cytology or genetics? Do you prefer doing field work, laboratory investigation, or library research?

Take inventory of your strengths and weaknesses in some of the major areas of biology, using the chart that follows. Analyze your abilities objectively.

	compre-hensive knowl-edge	mod-erate under-standing	little or no back-ground
plant taxonomy			
evolution			
comparative anatomy			
invertebrate zoology			
genetics			
biochemistry			
plant physiology			
field work			
cytology			
phylogeny			
embryology			

Then work to improve your weaknesses by doing outside reading and taking college courses in the summer.

It may sound as if we are advocating a narrow approach by encouraging you to accentuate the positive. Of course we do not want you to neglect weaknesses. What we *are* saying is that you should teach what you know best, for that is how you will be most effective. Then, while doing this, you can be closing the gaps in areas where your background is weak through in-service training, professional reading, and attending conferences.

No matter how strong you are in certain fields of the biological sciences, you will not be able to carry the entire teaching burden alone. This is where a textbook comes in. A text organized like your year's teaching plan makes your work much easier.

Choosing a Text

Most publishers will send sample texts to teachers for evaluation. With several of these texts at hand, you should examine each one, keeping the following points in mind.

MATERIAL COVERED

Does the text cover all the material you feel should be in the course you plan to teach, or are there certain parts and principles omitted? Is the text too detailed, preventing the main issues from making clear impact on the reader? Is the factual material presented in sufficient depth?

If the material is intended for a beginning course but covers much of what should be taught in advanced courses, the student may become discouraged. This often happens if he has to read sec-

168

tions several times before he understands what the author is trying to say. On the other hand, if the text is written in a breezy and shallow manner it will not be sufficiently challenging.

ORGANIZATION OF MATERIAL

This criterion causes disagreement among reviewers since each teacher has his favorite way of organizing subject matter. It may be difficult to find a book that *precisely* meets your needs. If a text does contain most of the material you plan to use, but it is not organized in the way you would like, you may be able to make assignments from the appropriate sections of the book.

THE AUTHOR

The writer's name provides a starting point from which you can evaluate a book. By checking his background and reputation[5] you can get an indication of his stature and experience that qualify him to write. Sometimes the author's reputation may be misleading. For example, it may be based on research or on superior teaching. Neither of these criteria necessarily indicates he can write. Actually, the text should pretty well speak for itself. Some excellent, well-written, easy-to-understand books have been written by authors who have not published a great deal.

PLACEMENT OF EMPHASIS

Frequently an author is more interested in one phase of his writing than he is in other phases. When this is true he is likely to elaborate on certain sec-

[5] See *American Men of Science* or *Who's Who in America.*

tions and give little attention to others. Of course, if you plan to emphasize a *particular theme* in your teaching, you may prefer to use a text that reinforces that theme. Suppose, however, that you narrow your choice to two books. One provides a broad treatment of the subject, is well written, and has only a few minor disadvantages. The other is narrower in scope and stresses the author's "pet" topics. The first text will be more satisfactory because it will give you more latitude in teaching.

STYLE OF WRITING

There is no rule that says texts have to be stuffy, hard to read, and hard to understand. Big words, long sentences, and clumsy constructions make hard reading. If the book is fun to read, or the kind of book that "doesn't seem like a text but more like an interesting magazine article," chances are that its message will get through to students.

Another quality to evaluate relates to the kind of biology you are teaching. If you are teaching by experiment and the laboratory method, evaluate the book's style on that basis. Does it present information in the stepwise, hypothesis-experiment-theory sequence that scientists use? Or, is it didactic, claiming to be factual and ultimate in its presentation of scientific principles? Look for a sense of inquiry in the text. Try to choose a book that discusses the curiosity, tinkering, and hunch-playing that is so often an ingredient in important discoveries.

ILLUSTRATIONS

Some textbook authors feel that the text without the illustrations is almost meaningless because the two are so closely correlated. If one tries to comprehend the drawings and photographs without reading the text portion, there is little meaning to be gained from the book.

Measure the usefulness of the book you are evaluating by this criterion—the dependence of the illustrations on the text. Often an author will write the book and then add pictures and drawings as second thoughts. He sprinkles them throughout to give the book an appearance of being illustrated, but the illustrations have not been a part of the initial formulation. Charts, diagrams, and pictures should clarify the written portion or expand it in the significant places. If the illustrations do not do this they are of little value.

Simplicity of pictures, diagrams, and other types of illustrative material has a big advantage. To a student using the book for the first time the picture must make a strong impact at once. If illustrations are a jumbled mass of extraneous material they will take too much time to decipher.

INDEX

Compare the number of entries in the indexes of several texts for a given topic. The photograph in Figure 10.1 shows what several authors have done in indexing "cells." A complete index makes a book easier to use because a student can find sections of the text that elaborate on terms, processes, or concepts that may be mentioned elsewhere. Often an author will define a term the first time he uses it in the book, and then assume the reader keeps the definition in mind. If the reader happens to skip the first reference to the term, but wants to find the definition later, an index can help locate it. Many books use bold-face type for the page number to show where the term has been defined in the book.

GLOSSARY

Since science has a language usually unfamiliar to the student, a glossary is a very helpful part of the book. Check the glossary for completeness and accuracy. Browse through the text looking for scientific terms that might be unfamiliar to your students, then determine how many of these terms are in the glossary. Also be aware of the wording used in glossary definitions. It should be clear, simple, and written so that a beginning science student can understand it.

BIBLIOGRAPHY

Some bibliographies list only the name of the text, author, publisher, and copyright date. Others, called annotated bibliographies, have a brief description of each entry. The bibliography of *Creative Biology Teaching* is annotated. Short descriptions of each reference make it easier for the student to choose pertinent outside reading. Annotation will catch his interest more effectively than a brief book title.

The Question of Scope

The White Rabbit in *Alice in Wonderland* asked the King for instructions before presenting evidence at the Knave's trial. The King said, "Begin at the beginning, and go on till you come to the end; then stop."

Some textbooks look as if their authors followed the King's advice and kept on writing until they had exhausted the subject. The results in cases like these are not textbooks; they are encyclopedias.

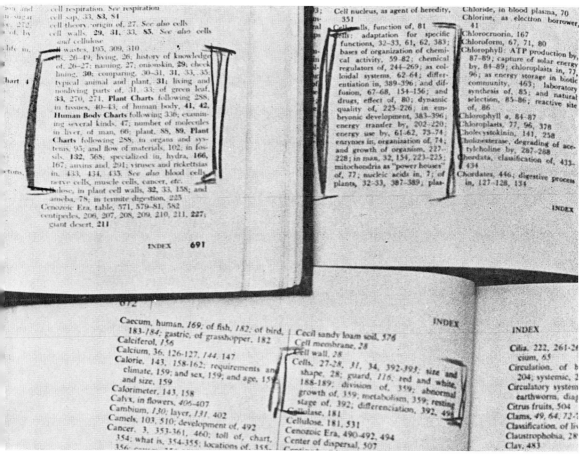

Fig. 10.1—A clue to the extent of coverage a text offers may be found in the index. Note the differences in coverage on "cells" in these textbooks.

An important measure of a book's value is not always what it includes, but what it omits. We have discussed biology themes earlier in the chapter—such as ecology, biochemistry, and phylogeny. Now, using a particular theme as a measuring stick, evaluate a text. How does the book deal with cytology, for example? Is the subject covered in a single chapter, or does the author devote a complete unit to it? Is his treatment confined to a general treatment of the cell and its organelles, or does he spend a great deal of time on the molecules that make up protoplasm? How is cell differentiation brought into the discussion? Does the author favor cells of lower plants, cells of protists, or the variety of differentiated cells found in higher organisms?

One means of determining objectively the scope of subject treatment is to refer to the index of the book. Under the heading "cells" in the index of several biology books, shown in Figure 10.1, note the variation in the coverage of the topic. Not only are there more entries in some texts, but the entries are more widely scattered. The number of entries indicates the extent to which the subject is developed. But the wide distribution of the entries throughout the book indicates the scope of development. If there are pages cited from almost every chapter, the author has apparently used cytological principles as one of the underlying themes of his book. On the other hand, if the page numbers are from a more restricted area of the text, he has probably developed cytology as either a single unit or one chapter.

If you consider the study of cells to be a basic, underlying theme in biology you should choose a text that reflects this view. But if you would rather have your students study cells at one particular time, select a text that does this. A teacher cannot become an expert in every facet of biological science and it is only natural that you develop certain areas as your favorites. But we want to emphasize that the text you choose should match your teaching inclinations as far as possible. If teacher and texts are interlocked in this way biology will seem more unified to the students, and you will have more fun teaching it. In an introductory course such as high school biology, students often become confused if the text and instructor differ a great deal. Rearrangement, reorganization, deletions, and additions to the text should be kept to a minimum. This straightforward approach will be more acceptable to the students.

A student's reaction may be of help in evaluating a text. Richard Trump, a high school biology teacher, has developed a list of questions for students. They look for the answers in three or four different texts, and make a brief write-up. Since the students must study a text carefully to answer the questions, they often point out important strengths and weaknesses.

BIOLOGY TEXT SURVEY: QUESTIONS

1. Concerning osmosis (diffusion), is there a difference between living and nonliving membranes? If so, what is the difference?

2. Narcotics and stimulants have essentially opposite effects on nerve action. Ethyl alcohol is generally considered a narcotic. Is there any biological explanation why it is sometimes called a stimulant?

3. What is the explanation of the bending of a plant toward a source of light?

4. Thiamine is a vitamin whose deficiency is the cause of beri-beri. What does thiamine *do* for the body?

5. Why are virus infections generally harder to control than bacterial infections?

6. One of the more important purposes of the biological classification system is to show the natural relationships between organisms. How do the books you are examining present this idea? (Summarize briefly or quote from the books.)

7. As biologists learn more about the natural relationships of organisms (see #6), they make changes in the classification system. In the past there have been big changes to bring the system into line with what is known about the history of life on the earth. There will undoubtedly be changes in the future too, but probably not as big as the changes in the past. How do the books present this idea? (Summarize briefly or quote from the books.)

8. How is respiration in green plants different from respiration in animals?

9. Why is it important for you to have all of the essential amino acids in the same meal?

10. There is danger in the use of antibiotics for combating minor infections

that the body's natural defenses can handle. How does the study of heredity help explain this danger?

11. Some people believe man should be vegetarian. Others believe he should eat meat as well as vegetables. How does the length of man's digestive tract help to settle this question?

12. There is a possible danger to a baby whose mother has Rh negative blood and whose father has Rh positive blood. What happens that causes this danger relating to the Rh factor?

13. Australia has many kinds of marsupial mammals. The rest of the world has very few. How does Australia's isolation from the continents help to explain these facts?

14. Is there any advantage to the green plant in having the palisade cells near the upper surface of the leaf while the spongy layer is near the lower surface? If so, what is it?

15. Why is the human appendix more easily infected than most other organs of the human digestive system?

16. By injection of gamma globulin it is possible to prevent measles in a person who has been exposed to the virus. Why does the doctor generally inject only enough gamma globulin to make the disease less dangerous?

17. Under what conditions do the plant's guard cells close the stomates?

18. Is there any relationship between the number of young produced by various organisms and the rate of survival of the young? If so, what is the relationship?

Though there are a number of excellent books, perhaps no single text will furnish you the entire "package" of biology. You may wish to synthesize a course from a variety of textbooks, using a multiple-text approach.

In addition to utilizing several textbooks, a teacher can build a useful teaching file. This can be added to and expanded considerably as new materials are found. Appendix A 215 indicates some of the kinds of things that might be included in a teaching file.

One Teacher, Many Books

Evans Clinchy[6] said:

. . . the textbook . . . has enormous disadvantages, even when produced by excellent scholars. . . . In part, because the textbook is a book and is therefore a series of printed pages bound tightly together, the writer tends to take a given body of information and put it in some order, starting with introductory material on page one and ending with advanced material on the last page. In most cases the author becomes involved in a complex process of selecting the important ideas, concepts, or laws that he wishes to discuss, then summarizing the best available evidence that will support or illustrate them, and finally setting down the evidence and the conclusions derived from it in the *most clear, neat,* and *efficient* manner he can devise. The task of the student then, most typically, becomes one of reading the book, remembering what has been said, doing the exercises, at best understanding the material, and then repeating it during an examination which may or may not require him to use the material in some productive way.

One possible alternative to choosing *one* text is to choose several. Then, as you build your teaching units, use the variety of texts as reference material. For example, if your unit is "Respiration," use the resources of several biology texts when you block out the unit. Design student worksheets that utilize three or four text references. You do not have to confine the references to books.

---

[6] *The Revolution in the Schools,* edited by Ronald Gross and Judith Murphy, Harcourt, Brace & World, Inc., New York, 1964.

BIOLOGY TEXT SURVEY: ANSWER SHEET

INSTRUCTIONS: Select a question from list on pages 171–72. Find the answer in each of the two books that you are examining. Fill in the page reference and brief answers on this sheet. Use only the answers that you find in the books, not information from other sources. For each question that you investigate, place a check mark (√) by the name of the text that provided the clearer, more satisfying answer.

Question number____

Text examined_____	Text examined_____
Page reference___	Page reference___
Answer:	Answer:

Question number____

Text examined_____	Text examined_____
Page reference___	Page reference___
Answer:	Answer:

Magazine articles, pamphlets, and instructional aids given free by many industries will help. In Appendix F 295 are some sources from which one may obtain free or inexpensive materials.

ALTERNATE ROUTES TO A COMMON GOAL

After this discussion of biology themes, choosing a text, and the single- and multiple-reference approaches, you may feel confused by the various ingredients in building a unit. We wish there were a single answer to the question, What is the best way to teach? Two popular approaches have already been discussed in previous chapters. They are: (1) the problem-solving, investigatory approach, and (2) the fact-learning, confirmatory approach. Both of these methods could be considered as alternate routes to a common goal. They differ because they emphasize different methods of learning.

Even educational experts disagree about the balance of these two approaches. Dr. A. R. Hibbs[7] believes that a feeling for the discovery and experimentation of science is important to convey to students. He says:

It does not matter whether the student learns any particular set of facts, but it does matter whether he learns how much fun it is to learn—to observe and experiment, to question and analyze the world without any ready-made set of answers and without any premium on the accuracy of his factual results, at least in the field of science.

[7] "Science for Elementary Students," *Teachers College Record*, Vol. 63, pp. 136–42, 1961.

On the other hand, Dr. David P. Ausubel[8] feels that much of the so-called "investigatory activity" practiced by science students is a waste of time. He says:

. . . although laboratory work can easily be justified on the grounds of giving students some appreciation of the spirit and methods of scientific inquiry, and of promoting problem-solving, analytic, and generalizing ability, it is a very time-consuming and inefficient practice for routine purposes of teaching subject-matter content or of illustrating principles, where didactic exposition or a simple demonstration are perfectly adequate. . . . The scientist is engaged in a full-time search for new general or applied principles in his field. The student, on the other hand, is primarily engaged in an effort to learn the same basic subject matter in this field which the scientist learned in his student days, and also to learn something of the method and spirit of scientific inquiry. Thus, while it makes perfectly good sense for the scientist to work full-time formulating and testing new hypotheses, it is quite indefensible . . . for the student . . . to [do] the same thing. . . . If he is ever to discover he must first learn; and he cannot learn adequately by pretending he is a junior scientist.

To acquire facility in problem-solving and scientific method it is also unnecessary for learners to rediscover every principle in the syllabus. Since problem-solving ability is itself transferable, at least within a given subject-matter field, facility gained in independently formulating and applying one generalization is transferable to other problems in the same discipline.

How do these two conflicting views help you make a decision about the "best way" to structure a unit in biology? For one thing, they support the idea that there is no "best way." There are a number of good ways. You must have an adequate general knowledge of the basic teaching techniques, current trends in

[8] "An Evaluation of the Conceptual Schemes Approach to Science Curriculum Development," *Journal of Research in Science Teaching*, December 1965.

biology curricula and texts, and a solid background in biology. Then, using these as tools, you can build the kind of units you feel are best for your particular students.

Teaching With Tests

Tests are not just devices to find out how much students know. They are also instructional tools. As a convincing demonstration of the value of evaluation, try the following experiment in class.

Ask a student to come to the board and draw five lines, each 12 inches long. After each try, measure the line so the student can see how much too long or too short his drawing was. After a few of these "tests" he should be able to draw lines of fairly accurate length.

Now suppose you ask a student to draw five 12-inch lines but do not measure each one until he has put them all on the board. He has no idea of the accuracy of his drawings until the experiment is over. Without the benefit of an evaluation he has no idea of his ability to estimate length.

Tests are important milestones along the learning path. They should enable students to find how much progress they are making. Tests should show up areas of strength and weakness. Once completed, they should be used as study guides to future work because they are yardsticks that measure how well past concepts have been learned. Sometimes students view tests as unpleasant activities to be endured from time to time. Instead of treating them as learning devices, students sometimes see them as interruptions in their classwork, or as times during which they get low grades or show their inadequacies in understanding the material.

One way to overcome a negative attitude about testing is to promote a testing program as an inseparable part of the learning program. Self-tests, study guides, and worksheets help the students condition themselves to taking tests and evaluating themselves. When one is in the habit of testing himself objectively, as a part of his studying, he is less apprehensive about taking a test in class.

Another way to overcome unfavorable attitudes about testing lies in your own attitude about tests and test construction. You should take a straightforward stand and construct the items in an objective, clear, unbiased way. Avoid tricky items or questions on insignificant details. Include a number of questions that are especially easy to answer, putting such items in the first portion of the test. If you do this, the natural fear and nervousness that tend to upset the test-takers will be lessened. Prepare your students for tests and examinations. Tell them the general nature of the questions and explain what you will emphasize in the test. Make an attempt to provide a test breakdown or schedule of the proportion of the test that relates to the work covered in class.

Testing can be accomplished in many ways, including the laboratory test we mentioned in Chapter 7.

INTERLOCKING THE INGREDIENTS

Newspaper editors are somewhat like teachers. They have the daily responsibility of providing information. They must communicate with people of a variety of abilities so that the information is clearly received and understood.

Imagine the chaos if a newspaper editor simply took all of the news, the advertisements, pictures, and announcements and assembled them at random. Without some kind of organizational plan a newspaper would be an unintelligible mixture. By putting top news stories first, human interest and local news next, and classified advertisements in the latter sections, a newspaper editor groups items into an easy-to-read format.

The same kind of organization is important if students are to receive maximum benefit from a biology class. First, they should be told about the plan for the course. The master plan that you follow will then provide a matrix in which to place lecture-demonstrations, audiovisual aids, laboratory work, field trips, and evaluations. Being aware of the plan will give students confidence in you and your teaching and will provide a reference point around which they can organize their thoughts as they study.

The Master Plan

Using the following general set of guidelines you can add as much variation as you wish, fitting the course you teach to your abilities and training and the type of student you have.

UNITS FOR THE YEAR

We talked earlier in the chapter about themes in biology. These included ecology, taxonomy, metabolism, and others. Now, as you work on the master plan for the entire 36-week course, block out time periods for each unit. Perhaps your choice of units will be based largely on the text you use, or it may depend on your choice of reference materials. (See Appendix A 214.)

TEACHING THE UNITS

Students' learning is conditioned by your method of teaching. For example, if you lecture a great deal, note-taking is an important factor. This is quite different from using the technique of class discussion in which students become active participants. If you illustrate important concepts by demonstrations the students have a different role than if they do extensive laboratory work. Techniques of note-taking, report-writing, and keeping laboratory records should be explained to students. Show examples of the kind of written work you require. As students look over good and not-so-good examples of lecture notes, rough drafts of reports, and laboratory notebooks, they can see for themselves what criteria you will use to judge their work.

You will want to mention course requirements such as essays, amount of field work, length and form of individual laboratory reports, and other special aspects of the biology course. In summary, you should inform them early in the school year of their responsibilities, and give them specific instructions on how to carry out these responsibilities. This means, of course, that as you build units you must decide on which of the teaching techniques you are going to rely.

EVALUATIONS

Your students will want to know what kind of tests you plan to have, and how often. If you are going to have a full-period test or examination once every two weeks, once a month, or only a few times a semester, announce it well in advance. Having adequate preparation time is important. You may wish also to have frequent short quizzes which may or may not be announced in advance. Even if you do not give students advance notice of these, they should be informed of your intention to have quizzes.

Your method of compiling grades should also be made clear to students. While the mechanics of grading "on the curve" or by straight percentage are not within the scope of this book, we want to mention the importance of a clear-cut grading system. Some students favor a system of numerical evaluations. That is, they prefer to have their grade expressed in numbers. Suppose you give the class a one-hour test that has a total of 100 points. As you check the papers you subtract points for incorrect answers so that the number of points that remain represent the grade. Numerical grades like this can be accumulated from one test to the next, providing a running score of each student's progress. At any time you can make a list of the accumulated points for all your students, arranging the totals in a column, with the highest at the top and the lowest at the bottom. Then you can split the column into groups of A, B, C, D, and F.

The Object of Your Work

After you have set up the master plan for the year, have explained it to the students, and have worked out means of evaluating their work, you are ready to give attention to the individual units you will teach. In other words, you are ready to make lesson plans, to split a large unit into its smaller parts, and to decide which part should be taught first, how it should be taught, and with what teaching devices.

As you grapple with the mechanics of making a teachable unit, you must have the student foremost in mind. According to some educators, there are concepts that simply do not make any sense to youngsters of certain age groups. It appears that a certain maturity level must be reached before some concepts can be comprehended. Harold Massey and Edwin Vineyard[9] said:

An essential component of readiness is the mental maturity or the mental ability required by and appropriate to the complexity of the learning to be mastered. If one tries long and hard enough, he may teach an average six-year-old pupil to recite parrot-fashion the multiplication tables up to twelve . . . but the likelihood of teaching such a youngster to fully comprehend the meaning of multiplication . . . is very small. A second-grader may learn borrowing in subtraction, but understanding comes much later. . . . The ability to do abstract thinking, to manipulate symbols, to retain the pattern of a sequence or chain of responses, to see the relatedness of parts to parts and parts to wholes, to apply old knowledge selectively to a new situation— all these and more are a part of intelligence or the ability to learn and hence are an important part of learning readiness. One simply cannot be ready to do that which he is unable to do.

Another important thing to remember in connection with maturation is the fact that even children of the same chronological age and with similar potential ability and backgrounds of experience often develop at different rates. Coaching or instruction is futile, and may even be psychologically harmful, in such activities as walking, doing work involving intricate finger manipulations, and learning sight words, until the child has matured or developed sufficiently. This principle applies to other learning tasks as well. "Late bloomers" sometimes eventually excel those who surpassed them in the beginning stages of a learning experience because of earlier maturation. Training becomes important when the child has matured sufficiently to profit from it.

Now we turn to the procedure of forming a unit lesson plan. The form varies a great deal, and we present only one of many plans that could be used.

---

[9] *The Profession of Teaching*, The Odyssey Press, Inc., New York, 1961.

COMPONENTS OF A UNIT LESSON PLAN

1. *Long-range objectives.* You should decide what overall goals are accomplished by the unit you are planning. For example, if the unit deals with mitosis and meiosis, how will it relate to the remainder of the biology course? Will the knowledge that students acquire here be needed later? What depth must you give the subject matter to insure "print-through" at a later time in the course?

2. *Immediate objectives.* Because of laboratory experiences, short-range experiences, short-range experiments, or upcoming classroom discussion, the unit you are currently working on might represent the next hurdle that students should confront. If you are getting ready to do some experiments with *Drosophila,* for example, there are basic techniques of handling the flies that must be discussed. And the monohybrid crosses that fruit-fly experiments may illustrate would be immediate objectives.

3. *Priorities.* Decide the order for teaching the objectives. This step in unit planning is one of the most important. Without a priority assignment the teaching will not flow smoothly from one concept to another. Without the logical, stepwise arrangement of teaching ideas the classwork will become an unconnected, random collection of activities.

4. *The teaching plan.* Using the priority list of objectives, work out effective means of teaching. The use of the teaching tools we mentioned in previous chapters—audiovisual materials, homemade gadgets, and miscellaneous instructional devices—can be a part of your teaching plan. A time schedule should be included, indicating approximately how many minutes you will use to develop each idea. Short quizzes and longer exams should be indicated too, as they are a part of the teaching program. We have illustrated a typical teacher's plan book in Figure 10.2.

In Appendix A 214 we have outlined a lesson plan for a day and in Appendix

178

Fig. 10.2—This teaching plan could be followed by a substitute teacher without much difficulty. It also serves as a guide to the teacher, reminding him of important demonstrations and visual aids. Are important aspects of the unit omitted from the plan above?

A 214 we have given suggestions for making a unit of study. In addition, in Appendix A 217 we have included a number of ways for introducing the same subject or unit to a class. In Appendix A 215 we have made suggestions for building a teaching file.

Practice With Science

The football coach attempts to tell his players how to play the game. The music teacher tries to explain how to bow a violin. But the talk-through technique is effective only up to a point—from there on the student must go it alone. The athlete must compete on his own merit and ability just as the musician must make his reputation on what he is able to do for himself. Years of teaching cannot provide the practice so vital to perfecting a skill. Practice must be coupled with expert coaching from a master teacher.

Practice is the key to your performance as a teacher. If you have had good instruction and a variety of learning experiences in science, you have the foundation for teaching. But practice is as vital to good teaching as it is for any other skill. Practice all aspects of teaching, including choosing outdoor laboratory sites, making tests, choosing a teachable text, and writing discussion guidelines for your students. Above all, practice is necessary if you are to build good units. Through the trial-and-error process you can retain those ideas that work, and discard those that do not. Before long you will have a technique for unit-building that can become the important backbone in the biology course you offer.

A teacher's appearance, manner, and attitude have a great deal to do with the success of putting across subject matter. Even if your planning is excellent, your personal approach to teaching can ruin the plan. The biology teacher before the class sets the "tone" for the course. He can make it exciting, or a drudge.

CREATING CHAPTER CONCEPTS

1. Select one of the following topics and investigate it. Note ways that it could be taught effectively, how audiovisual tools could be used in its teaching, and what kind of laboratory work could be done with it. Then write a brief outline for a unit that could be taught, based on the topic.
   a. Space biology
   b. Adenosine triphosphate
   c. Deoxyribonucleic acid
   d. Viruses
   e. Photosynthesis
   f. Tissue culture
   g. Electron microscopy
   h. Electrophoresis
   i. Biological clocks
2. Visit the Curriculum Library at your college or university and compare three similar units from three current biology texts. Evaluate them using the nine criteria we mentioned in this chapter (see "Choosing a Text," p. 167).
3. Determine several themes you would *delete* from a modern biology course for high school students. Discuss reasons why they should be removed.
4. Examine several biology texts and determine which themes they have in common. Are these themes the most important, in your estimation?
5. Select 25 free and inexpensive ma-

terials from Appendix F 295 or from any other sources that you may choose. These may be magazines or periodicals of your choice or other materials you have read or heard about. Choose the items to go along with the unit you write for (1).
6. Select at random a unit or topic you wish to include in your teaching of high school biology. Then suggest five different ways to introduce the topic to the class. Try to think of unique methods of introduction; do *not* use demonstration, filmstrip, or film.
7. Develop a "recipe file" of 3 x 5 cards for term paper and project topics. You will find several magazine articles helpful, including "The Amateur Scientist" in *Scientific American*.
8. Make a list of ways to use music, art, social science, history, drama, and mathematics in teaching biology.
9. Develop a script for a four-minute 8mm film to illustrate an idea that you wish to teach in biology.
10. List the chapter headings from three biology texts. Then discard some of the chapters and reorganize the others to make the type of biology course you wish to teach.
11. Choose a unit you wish to teach and develop 20 illustrations for it. This might involve photographs, diagrams, charts, 35mm slides, short movies, models, or collections of material from nature.
12. Develop a method of using an overhead projector to aid you and the class in evaluating term papers and special projects.
13. Design a test for your fellow classmates that uses several methods of evaluation. Determine which of the methods is most effective.
14. Work out a two-page "Information for Students" sheet that you plan to use to inform your students about your biology course.
15. Check the *Journal of Research in Science Teaching* and other periodicals to determine if data exist on the optimum frequency for giving tests and other evaluations in high school biology.
16. Some questions evaluate a student's re-

call of facts. Others evaluate understanding of concepts. In your opinion, which of the following is most effective in evaluating understanding of important biological principles?

a. Name the mouthparts of the crayfish.

b. Which of the crayfish mouthparts

are thought to be vestigial?

c. Discuss the evolutionary origin of the mouthparts of the crayfish.

d. Compare the crayfish mouthparts to those of other crustaceans.

e. What adaptations for survival do crayfish mouthparts show?

EXPANDING THE CHAPTER

BELANGER, MAURICE. 1964. The Study of Teaching and the New Science Curricula. *The Science Teacher* 31(7):31–35. The article discusses how the teacher presents his subject matter in light of new curricula developments.

HIGHET, GILBERT. 1950. *The Art of Teaching*. Alfred A. Knopf, Inc., New York. As a book on general teaching methods, this interesting work reads more like a novel than a text about teaching. The author uses many examples from his teaching experiences.

ROUCEK, JOSEPH S. 1959. *The Challenge of Science Education*. Philosophical Library, Inc., New York. This collection of essays on science education provides helpful background for the high school biology teacher.

THURBER, WALTER A., and COLLETTE, ALFRED T. 1959. *Teaching Science in Today's Secondary Schools*. Allyn & Bacon, Inc., Boston. In this book, Chapter 14, "Unit Planning," is especially helpful in building a unit of study.

TRUMP, J. LLOYD; and BAYNAHAM, DORSEY. 1961. *Focus on Change—Guide to Better Schools*. Rand McNally & Co., Chicago. Extensively illustrated with charts and diagrams, this small book will give the science teacher an overview of the total high school program. Background knowledge like this helps in building units in science.

WILES, KIMBALL; and PATTERSON, FRANKLIN. 1959. *The High School We Need*. National Education Association, Washington, D.C. As a summary of several studies carried on by the NEA, this pamphlet will help the biology teacher understand trends in secondary education.

PART 4

APPENDICES

APPENDIX A

Teaching

Aids

LEADERS AND SCIENTISTS WHO CONTRIBUTED TO THE BIOLOGICAL SCIENCES

ALCMAEON (520 BC–?). Greek physician. First person known to have conducted dissections of the human body. He recorded existence of the optic nerve and the tube connecting the ear and mouth, distinguished arteries from veins, and believed that the brain was the center of intellectual activity.

DEMOCRITUS (470 BC–380 BC). The Laughing Philosopher. Best known for his atomic theory. He believed all matter consisted of tiny, almost infinitesimally small particles, so that nothing smaller was conceivable. Hence, they were indivisible (the word atom means indivisible).

HIPPOCRATES (460 BC–370 BC). Greek physician, today known as the father of medicine. His school believed in moderation of diet, cleanliness, and rest for the sick and wounded.

ARISTOTLE (384 BC–322 BC). Considered the founder of zoology. Studied biology and natural history. He classified animal species and arranged them into groups, dealing with more than 500 different animals. He studied the developing embryo of the chick and the complex stomach of the cow.

THEOPHRASTUS (372 BC–287 BC). Greek botanist. Considered the founder of botany. He described more than 500 specimens of plants.

HEROPHILUS (320 BC–250 BC). Greek anatomist. He was especially interested in the description of the brain. He divided nerves into sensory and motor, described the liver and the spleen, and named and described various other parts of the body.

ERASISTRATUS (304 BC–250 BC). Greek physician who continued the work of Herophilus. He noticed the association of nerves with arteries and veins. He believed that all body functions were mechanical in nature.

PLINY THE ELDER (AD 23–79). Roman scholar with a major interest in zoology. He maintained a sense of the wonder and majesty of the natural world.

GALEN (130–200). Greek physician whose best work was in anatomy. He dissected and described in great detail dogs, goats, pigs, and monkeys. He believed that everything in the universe was made by God for a particular purpose. His works were the ultimate medical authority for Europeans until the time of Vesalius in anatomy and Harvey in physiology.

RHAZES (850–923). Persian physician and alchemist who prepared what we call plaster of Paris for forming casts to hold broken bones together. He described differences between smallpox and measles.

ABANO, PIETRO D' (1250–1316). Italian physician who maintained that the brain was the source of the nerves, and the heart the source of the blood vessels.

MONDINO DE' LUZZI (1275–1326). Italian anatomist. He taught anatomy and in 1316 wrote the first book in history to be devoted entirely to anatomy. Therefore he is called the Restorer of Anatomy. He described many of the internal organs of man. This was the only textbook of its kind for the next two centuries (until the time of Vesalius).

PARACELSUS (1493–1541). Swiss physician and alchemist who said the purpose of alchemy was not to manufacture gold but to prepare medicines with which to treat diseases. He wrote on the problems of mental disease, and associated head injury with paralysis, and cretinism with goiter.

FUCHS, LEONHARD (1501–1566). German botanist who was a physician interested in natural history and wrote in great detail about plants. A genus of plants is named for him (*Fuchsia*).

SERVETUS, MICHAEL (1511–1553). Spanish physician who described the circulation of the blood in the lungs.

VESALIUS, ANDREAS (1514–1564). Flemish anatomist who lectured and did his own dissections. He wrote *De Corporis Humani Fabrica* ("On the Structure of the Human Body"), the first accurate book on human anatomy. It was illustrated. His work marked the beginning of modern anatomy. He believed that the brain and nervous system represented the seat of life, mind, and emotion.

GESNER, KONRAD VON (1516–1565). Swiss naturalist who collected plants and animals and described them.

EUSTACHI, BARTOLOMEO (1520–1574). Italian anatomist whose most successful work was done on the sympathetic nervous system, the kidney, and the ear. The name Eustachian tube has been given to the tube connecting the ear and throat. He first described the adrenal glands and studied the structure of the teeth.

FALLOPIUS, GABRIELLO (1523–1562). Italian anatomist who is known today for his descriptions of the inner ear and the reproductive organs of the female. The Fallopian tubes in the female reproductive system carry his name.

FABRICUS (1537–1619). Italian physician who was the teacher of Harvey. He discovered the one-way valves in the veins and described them accurately.

ALPINI, PROSPERO (1553–1617). Physician to the Venetian consul in Cairo, Egypt, who described sexual differences among plants, such as the date palm. Linnaeus, a century and a half later, used Alpini's work as a basis for his classification of the plant kingdom. Alpini was the first European to describe the coffee plant.

HARVEY, WILLIAM (1578–1657). English physician whose book, *On the Motions of the Heart and Blood,* became a great scientific classic, as it described circulation of the blood. He did this without benefit of a microscope.

SYLVIUS, FRANCISCUS (1614–1672). Dutch physician who abandoned the idea that the health of the body depended upon the relative proportions of the four chief fluids or "humors" the body contained (blood, phlegm, black bile, and yellow bile), which theory dated back to the early Greeks. Instead he believed acids and bases had properties of neutralizing each other. He studied digestive juices and correctly believed digestion involved a fermenting process.

MALPIGHI, MARCELLO (1628–1694). Italian physiologist who used a microscope to demonstrate that "hair-like" vessels (now known as capillaries) connect the tiniest visible arteries with the smallest visible veins. He studied other minute aspects of life, such as chick embryos and the internal organs of the silkworm, and the structure of plant stems.

RUDBECK, OLOF (1630–1702). Swedish naturalist whose best-known discovery is the lymphatic vessels, which he demonstrated in 1653.

STENO, NICOLAUS (1631–1687). Dutch anatomist and geologist who recognized that muscles are composed of fibrils, and described the duct of the parotid gland (still called the duct of Steno).

LOWER, RICHARD (1631–1691). English physician who discovered that dark venous blood was converted to bright arterial blood on contact with air. In 1665 he transferred blood from one animal to another and demonstrated that this might be a technique for saving lives. However, transferring animal blood to man or one person's blood to another person often proved fatal. Two and a half centuries later (1927) Landsteiner demonstrated why different types of blood do not always mix satisfactorily.

185

LEEUWENHOEK, ANTON VAN (1632–1723). Dutch biologist and microscopist who ground a total of 419 lenses that could magnify up to 200 times. Some lenses were no larger than the head of a pin. He was the first to discover the one-celled animals. He observed human capillaries and red blood corpuscles more carefully than had Malpighi and Swammerdam, the original discoverers. He described what are believed today to be bacteria. He competes with Malpighi for the title of father of microscopy. He observed nuclei in blood cells but did not identify them.

HOOKE, ROBERT (1635–1703). English physicist. In the field of biology he was an excellent microscopist. In 1625 he published a book, *Micrographia,* with beautiful drawings in microscopy. All his drawings, including those of feathers and fish scales, are beautiful as well as accurate. He is well remembered for the name "cell" which he gave to structures he observed in a fine sliver of cork in 1665.

SWAMMERDAM, JAN (1637–1680). Dutch naturalist who studied minute structure of insect anatomy. He is considered the founder of modern entomology. He worked with muscle tissue and showed that muscles changed shape but not size. He also worked with blood and discovered the red blood corpuscles.

GRAAF, REGNIER DE (1641–1673). He described the fine structure of the male testes in 1668 and the ovary in 1673. (The Graafian follicles are named in his honor.)

GREW, NEHEMIAH (1641–1712). English botanist and physician who studied sexual organs of plants and observed individual grains of pollen produced by the stamens.

LINNAEUS, CAROLUS (KARL VON LINNÉ) (1707–1778). Swedish botanist who developed a clear and concise style of describing species to show how each differed from other species. He was the first to use the symbols ♂ and ♀ for "male" and "female." He based his classification of plants on their sex organs.

BUFFON, GEORGES (1707–1788). French naturalist. Starting in 1752, he and collaborators wrote 44 volumes of *Natural History.* He had an evolutionary viewpoint concerning the earth and man.

HALLER, ALBRECHT VON (1708–1777). Swiss physiologist who showed that a nerve, when stimulated, would produce a sharp contraction of the muscle to which it was attached. He showed, too, that nerves lead to the brain or spinal cord.

NEEDHAM, JOHN (1713–1781). English naturalist who collaborated with Buffon to show that microorganisms arise from nonliving matter (1748). (Spallanzani, twenty years later, showed that Needham simply had not boiled the mutton broth long enough to kill the organisms that were present.)

BONNET, CHARLES (1720–1793). Swiss naturalist who felt that because aphids could reproduce parthenogenetically, every creature existed, preformed in the egg, and that somewhere within that creature was a smaller creature, preformed within it, and so on without end. Generation nested within generation.

SPALLANZANI, LAZZARO (1729–1799). Italian biologist who performed experiments involved in boiling solutions for a considerable length of time to show that microorganisms do not arise by spontaneous generation.

INGENHOUSZ, JAN (1730–1799). Dutch physician and plant physiologist who showed that green plants take up carbon dioxide and give off oxygen, but only in the light. In the dark, they do as animals do, give off carbon dioxide and absorb oxygen. This was the first indication of the role of sunlight in the life activities of green plants.

WOLFF, KASPAR (1733–1794). German physiologist who showed that specialized organs grew out of unspecialized tissue. He proved that the tip of a growing plant consists of undifferentiated cells from which come flowers, leaves, and the like. He showed that the same thing applies to undifferentiated animal tissue, such as he observed in the chick. He is considered the founder of modern embryology.

GALVANI, LUIGI (1737–1798). Italian anatomist who declared there was such a thing as animal electricity, based on the fact that muscles taken from dead frogs twitched. Volta, in 1794, proved Galvani was wrong.

BANKS, SIR JOSEPH (1743–1820). English botanist who was the first to show that almost all the Australian mammals were marsupials and were more primitive than the placental animals of other continents.

LAMARCK, JEAN BAPTISTE (1744–1829). French naturalist who tackled the job of making order out of the classification of invertebrates. He first used the terms "vetebrate" and "invertebrate." He made popular the term "biology." He was the first to devise a scheme to rationalize the evolutionary development of life, maintaining that species were not fixed but were changing and developing. He used the idea of "inheritance of acquired characteristics." He visualized evolution as the product of attempts by the animals to change.

PINEL, PHILIPPE (1745–1826). French physician who was the first to keep well-documented case histories of mental ailments.

CUVIER, GEORGES (1769–1832). French anatomist, the founder of the science of comparative anatomy. He realized how the various parts and systems of the body related to the rest of the body.

BICHAT, MARIE FRANÇOIS XAVIER (1771–1802). French physician who was the first to draw the attention of the anatomist as well as the physiologist to the organs of the body as a complex of simpler structures. He showed that each organ was built up of different kinds of tissues, and that some organs possessed tissues in common. He identified twenty-one types of tissues and may be considered the founder of histology.

YOUNG, THOMAS (1773–1829). English physicist and physician who was the first to discover the way in which the lens of the eye changes shape in focusing the objects at different distances. In 1801 he described the reason for astigmatism.

BROWN, ROBERT (1773–1858). Scottish botanist who gave the name "nucleus" (little nut) to the small body located within the center of the cell. He recognized this as a regular feature of cells.

CANDOLLE, AUGUSTIN (1778–1841). Swiss-French botanist who invented the word "taxonomy" in 1813 to describe the science of classification. His system of plant classification is largely in use today.

LAËNNEC, RENÉ (1781–1826). French physician who invented the stethoscope.

CHEVREUL, MICHEL (1782–1889). French chemist who isolated sugar from the urine of a diabetic in 1815 and showed it to be identical with grape sugar. This was the first step in the direction of recognizing diabetes as a disease of sugar metabolism (a century before Banting and Best were to complete the research).

MAGENDIE, FRANÇOIS (1783–1855). French physiologist who in 1825 showed that the anterior nerve roots of the spinal cord were motor and the posterior roots were sensory. He showed that life could not be sustained in the absence of nitrogen-containing foodstuffs (protein) and that even some proteins such as gelatin were insufficient. He laid the groundwork for the modern science of nutrition and in particular for the work on essential amino acids.

AUDUBON, JOHN (1785–1851). French-American ornithologist who painted some of the most beautiful natural history studies ever done. He was one of the first American conservationists.

BEAUMONT, WILLIAM (1785–1853). American surgeon who treated a young man who had a gunshot wound in the abdomen. The man retained a fistula, but lived to age 82. Through this and other demonstrations, Beaumont demonstrated changes in gastric juice under different conditions.

PURKINJE, JOHANNES (1787–1869). Czech physiologist who was the first to use a mechanical microtome to prepare thin slices of tissue for the microscope instead of a razor used by hand. In 1839 he referred to the living embryonic material in the egg as the protoplasm.

BAER, KARL ERNST VON (1792–1876). German-Russian embryologist who discovered that mammalian (including human) development was not fundamentally different from that of other animals. He pointed out that a developing egg forms several layers of tissue, each being undifferentiated, but from which various specialized organs develop, with a given set of organs from a given layer (germ layers, as he called them). Baer believed that relationships among animals could be deduced better by comparing embryos than by comparing adult structures. Thus he is the founder of comparative embryology.

MÜLLER, JOHANNES (1801–1858). German physiologist who, with Magendie, was founder of the modern science of physiology. In 1826 he showed that sensory nerves could interpret an impulse in just one direction. However, when the optic nerve was stimulated, it recorded a flash of light, whether light was involved or not.

BORDEN, GAIL (1801–1874). American inventor and food technologist who was interested in preparing some form of concentrated food that would be nourishing and easily prepared. He prepared a dried beef product called pemmican that was useful for pioneers crossing the western lands and also for Arctic explorers. In 1853 he produced evaporated milk. Later on he prepared concentrated fruit juices and various beverages.

SCHLEIDEN, MATTHIAS (1804–1881). German botanist who recognized the importance of the cell nucleus. In 1838 he elaborated the cell theory for plants, which Schwann was to extend to animals the next year.

SIEBOLD, KARL (1804–1885). German zoologist. In 1845 he published a book on comparative anatomy in which he dealt with protozoa. He was the first to study cilia and showed that unicellular animals could use them for motion.

OWEN, SIR RICHARD (1804–1892). English zoologist who discovered parathyroid glands in a rhinoceros. He compared tooth structure of various animals.

MOHL, HUGO VON (1805–1872). German botanist who recognized the granular, colloidal material of the plant cells as living, and called it protoplasm, thus adopting Purkinje's word protoplasm.

AGASSIZ, JEAN LOUIS (1807–1873). Swiss-American naturalist who published a huge five-volume detailed work on fossil fishes between 1833 and 1844. Through his research he concluded that glaciers moved, and that thousands of years ago they had grown and moved out over areas where they no longer were present. He proved to the world that there have been Ice Ages in the past.

DARWIN, CHARLES (1809–1882). English naturalist who collected evidence on his five-year trip on H.M.S. *Beagle* for the cause of evolutionary changes in animals and plants. He realized that natural selection was sufficient to explain evolution. In 1859 his world-shaking book, *The Origin of Species,* was published.

HENLE, FRIEDRICH (1809–1885). German pathologist and anatomist who made many microanatomical discoveries such as Henle's loop, a portion of the kidney tubule.

HOLMES, OLIVER WENDELL (1809–1894). American author and physician. In 1842 he discovered the contagiousness of childbed fever. He suggested the term anesthesia (meaning no feeling) as an appropriate term for the process we know as anesthesia.

SCHWANN, THEODOR (1810–1882). German physiologist who demonstrated that pepsin (Greek word meaning "to digest") could digest meat. Pepsin was the first enzyme to be prepared from animal tissue. Schwann, along with Schleiden who established the cell theory, coined the word "metabolism" as representing the overall changes taking place in living tissue.

GRAY, ASA (1810–1888). American botanist who wrote many popular books on botany. He preached Darwinism in America. Beginning in 1842, he was a professor of natural history at Harvard for 31 years.

SIMPSON, SIR JAMES YOUNG (1811–1870). Scottish obstetrician who was the first to use anesthesia for childbirth.

BERNARD, CLAUDE (1813–1878). French physiologist who did important work in studies of digestion by creating fistulas in the digestive system. He showed that digestion occurs in parts other than the stomach. In 1851 he discovered that certain nerves govern dilation of blood vessels, and others govern constriction of vessels. He showed that sugar is stored in the liver as glycogen. He showed that the poisonous effect of carbon dioxide was in its ability to displace oxygen in its combination with haemoglobin.

REMAK, ROBERT (1815–1865). Polish-German physician who was one of the first to make use of an electric current in treating disorders of the nerves, and was a founder of electrotherapy.

LUDWIG, KARL (1816–1895). German physiologist. He devised a kymograph in the form of a rotating drum on which blood pressure could be recorded continuously. He showed that blood circulation could be explained in terms of ordinary mechanical forces. In 1856 he was the first to remove organs from animals and keep the organs alive by pumping blood through them.

SEMMELWEISS, IGNAZ PHILIPP (1818–1865). Hungarian physician who introduced antiseptic procedures in hospitals with the result that childbed fever dropped markedly.

DU BOIS-REYMOND, EMIL (1818–1896). German physiologist who showed that the nerve impulse was accompanied by a change in the electrical condition of the nerve.

MORTON, WILLIAM THOMAS (1819–1868). American dentist who, in 1846, successfully extracted a tooth from a patient under ether. This, plus a successful operation for a facial tumor, done under anesthesia, made operations under anesthesia popular.

HELMHOLTZ, HERMANN (1821–1894). German physiologist and physicist who invented an ophthalmoscope for studying the interior of the eye. He was first to measure the speed of the nerve impulse. He showed that animal heat was produced chiefly by contracting muscles and discovered that an acid (lactic acid) was formed as the muscle functioned.

VIRCHOW, RUDOLF (1821–1902). German pathologist. He was the first to describe leukemia (1845). He showed that cells of diseased tissue had descended from cells of normal tissue.

MENDEL, GREGOR JOHANN (1822–1884). Austrian botanist and monk who worked on garden peas and laid the foundations for our modern beliefs concerning heredity.

PASTEUR, LOUIS (1822–1895). French chemist who heated wines of France to kill undesirable yeasts that might sour the wines. The term pasteurization thus arises. He saved the silk industry of France by destroying infected worms and mulberry leaves and starting over with healthy worms and food. He proved that disease germs were communicated from one individual to another (germ theory of disease). He used a modified rabies preparation to treat patients who had been bitten by rabid dogs. He is known as one of the greatest scientists in history.

GALTON, SIR FRANCIS (1822–1911). English anthropologist who was the first to stress the importance of studying identical twins. He demonstrated the permanence and the individuality of fingerprints.

LEUCKART, KARL (1823–1898). German zoologist who worked out complicated life histories of tapeworms and flukes. He founded the modern study of parasitology.

BROCA, PAUL (1824–1880). French surgeon and anthropologist who demonstrated through postmortem examination that damage to a certain spot on the cerebrum was associated with the loss of the ability to speak. This was the first proof of a connection between a specific ability and a specific cerebral point of control.

HOPPE-SEYLER, ERNST FELIX (1825–1895). German biochemist who, in 1871, discovered the enzyme invertase that hastens

190

the conversion of table sugar (sucrose) into two simple sugars, glucose and fructose. He also discovered lecithin, a fatlike substance containing nitrogen and phosphorus.

HUXLEY, THOMAS HENRY (1825–1895). English biologist who named the phylum Coelenterata to which jellyfish belong. He popularized Darwin's beliefs.

GEGENBAUR, KARL (1826–1903). German anatomist who specialized in comparative anatomy and showed how embryonic structures which form gills in fish form other organs such as eustachian tubes or thymus glands in land vertebrates.

LISTER, JOSEPH (1827–1912). English surgeon who tried to kill germs in surgical wounds by chemical treatment. At first he used carbolic acid. The deaths due to infections decreased.

COHN, FERDINAND (1828–1898). German botanist who was the first to treat bacteriology as a special branch of knowledge. He attempted to classify bacteria.

MARSH, OTHNIEL CHARLES (1831–1899). American paleontologist who discovered a primitive, extinct, birdlike reptile pterodactyl, one of the flying lizards of the Cretaceous era. He worked on dinosaurs. His work probably did more than anything else to create the atmosphere for the acceptance of evolution.

SACHS, JULIUS VON (1832–1897). German botanist who proved that the process of photosynthesis is catalyzed by chlorophyll, within the chloroplast, in the presence of light.

BERT, PAUL (1833–1886). French physiologist who studied the effect of compressed air on the body and realized the cause of the condition known as bends. He published his findings in 1878, and work with compressed air became safe.

WEISSMANN, AUGUST (1834–1914). German biologist who tried to show that environmental changes were not inherited. He cut the tails off 1,592 mice over twenty-two generations and showed that all continued

to bear young with full-sized tails. He suggested that chromosomes contained the hereditary machinery, and their careful grouping during cell division maintains the machinery intact.

HAECKEL, ERNST HEINRICH (1834–1919). German naturalist who believed that life was derived in an evolutionary manner from nonlife and that psychology was just a branch of physiology, so the mind, too, was evolutionary.

WALDEYER, WILHELM VON (1836–1921). German anatomist who is best known for his work on the nervous system. He maintained that it was built up out of separate cells and their delicate extensions. He called each individual cell plus its extensions a neuron. His views were called the neuron theory. In 1888 he gave the name chromosome to the threads of chromatin material that Fleming had discovered.

KOVALEVSKI, ALEXANDER (1840–1901). Russian embryologist who was a strong evolutionist and introduced Darwinism into Russia. He showed that the same three germ layers found in the vertebrate embryos were also found in the invertebrates.

FLEMMING, WALTHER (1843–1905). German anatomist who pioneered in the staining and dyeing of animal cells. He called the absorptive material within the nucleus *chromatin,* from the Greek word for color. He did not realize the genetic significance of the chromatin.

KOCH, ROBERT (1843–1910). German bacteriologist who, along with Pasteur, was the founder of modern medical bacteriology. He worked on the disease anthrax in cattle, transferring it to mice. He learned how to cultivate the bacteria outside the living body, using blood serum at body temperature. He studied bubonic plague and cholera in Asia and sleeping sickness in Africa. He showed that bubonic plague is transmitted by means of a louse that infects rats, and that sleeping sickness is transmitted by the tsetse fly. He discovered (1882) the causative factor of tuberculosis.

MIESCHER, FRIEDRICH (1844–1895). Swiss

biochemist who isolated a substance containing both nitrogen and phosphorus from the remnants of cells. The substance was termed nuclein and later changed to nucleic acid.

STRASBURGER, EDWARD ADOLF (1844–1912). German botanist, one of the pioneers of plant cytology, who studied the behavior of the plant cells during mitosis.

GOLGI, CAMILLO (1844–1926). Italian histologist who introduced the use of silver salts in staining cells. Knowledge of the fine structure of the nervous system was made available by him. He showed the difference between several varieties of the disease malaria.

MECHNIKOV, ILYA ILYICH (1845–1916). Russian-French bacteriologist who showed that in animals the white blood corpuscles are capable of ingesting small particles. He studied the bacteria infesting the large intestine and became interested by their possible connection with longevity or the lack of it. He believed that the natural life-span of man was one hundred and fifty years and that drinking cultured milk would help him attain it.

LAVERAN, CHARLES LOUIS (1845–1922). French physician. In 1880 he discovered the causative agent of malaria was a protozoan and not a bacterium.

BENEDEN, EDOUARD VAN (1846–1910). Belgian cytologist who expanded on Flemming's work. In 1887 he demonstrated two key facts about chromosomes: (1) their number is constant in the various cells of the body and this number is characteristic for a particular species; and (2) in the formation of the sex cells, ova and spermatozoa, the division of chromosomes during one of the cell divisions is not preceded by a doubling. Each egg and sperm has only one-half the usual count of chromosomes.

DE VRIES, HUGO (1848–1935). Dutch botanist who confirmed Mendel's theories.

BURBANK, LUTHER (1849–1926). American naturalist who developed many new varie-

ties of fruits, flowers, and nuts. He also developed the Burbank potato.

PAVLOV, IVAN (1849–1936). Russian physiologist who worked out the nervous mechanism controlling the secretion of the digestive glands, particularly of the stomach.

BUCHNER, HANS (1850–1902). German bacteriologist who was a pioneer in the study of gamma globulins from which the body produces antibodies with which to neutralize and make harmless the microorganisms that invade the body.

GAFFKY, GEORG THEODOR (1850–1918). German bacteriologist who isolated a bacillus and demonstrated it to be the causative factor of typhoid fever.

REED, WALTER (1851–1902). American military surgeon who worked on yellow fever. In 1901 he proved the causative agent which the *Aedes* mosquitoes carried was a filterable virus discovered by Beijerinck a few years earlier. Yellow fever was the first human disease attributed to a virus.

BEIJERINCK, MARTINUS WILLEM (1851–1931). Dutch botanist who worked on the tobacco mosaic disease that dwarfed the tobacco plants and mottled their leaves in a mosaic pattern. In 1898 he published his observations, calling the disease agent causing the mosaic a "filterable virus." Virus is the Latin name for "poison."

LÖFFLER, FRIEDRICH (1852–1915). German bacteriologist who, in 1884, discovered the bacillus causing diphtheria. He showed that natural immunity to diphtheria existed among some animals. This laid the groundwork for Behring's work in preparing an antitoxin. In 1898 he showed that hoof-and-mouth disease was caused by a virus, the first animal disease pinned to such a cause.

HALSTED, WILLIAM STEWART (1852–1922). American surgeon who was the first to use rubber gloves during operations. This marked the transition from antiseptic surgery (killing germs that are present) to aseptic surgery (not letting germs get there in the first place).

RAMON Y CAJAL, SANTIAGO (1852–1934). Spanish histologist who improved upon Golgi's stain and studied the nervous system. He worked out the connections of the cells in the gray matter of the brain and spinal cord and demonstrated the great complexity of the nervous system. He worked out the structure of the retina of the eye.

KOSSEL, ALBRECHT (1853–1927). German biochemist who isolated two different purines, adenine and guanine, and three different pyrimidines, thymine, cytosine, and uracil. He also recognized that a carbohydrate existed among the breakdown products. [In 1909 Phoebus Levene isolated the carbohydrate and showed the five-carbon sugar, ribose, was present in some nucleic acids. In 1929 he showed that an unknown sugar, deoxyribose (ribose minus one oxygen atom), was to be found in them.]

GRAM, HANS CHRISTIAN (1853–1938). Danish bacteriologist who stained bacteria by one of Erlich's methods, then treated the stained bacteria with iodine solution. Bacteria that retain the stain have been called Gram-positive ever since, and those that lose it are Gram-negative.

EHRLICH, PAUL (1854–1915). German bacteriologist who, in 1908, was awarded the Nobel Prize in medicine and physiology for his work in immunity and serum therapy. Chemical 606 that he produced was found to be a killer for *Spirochaeta,* the microorganism that causes syphilis. He discovered a dye, called trypan red, which helped destroy *Trypanosoma* that causes diseases such as sleeping sickness.

BEHRING, EMIL ADOLPH VON (1854–1917). German bacteriologist who discovered in 1890 that it was possible to produce an immunity against tetanus (lockjaw) in an animal by injecting into it graded doses of blood serum from another animal suffering from tetanus. He also developed diphtheria antitoxin.

GORGAS, WILLIAM CRAWFORD (1854–1920). American army surgeon who, in 1904 after Reed's death, was sent to Panama where he placed the mosquito under such effective control that both malaria and yellow fever were wiped out. This helped make possible the building of the Panama Canal.

RUBNER, MAX (1854–1932). German physiologist who tested energy production, by man, in large calorimeters. He measured the nitrogen content of urine and feces and estimated the quantity of the various types of foodstuffs in the diet he fed his subjects. In 1884 he concluded that no one particular type of foodstuff produced energy—the body uses carbohydrates, fats, and proteins with equal readiness. In 1894 he discovered that the energy produced by the body from foodstuffs was the same amount as it would have been if those same foodstuffs had been consumed in a fire.

KITASATO, BARON SHIBASABURO (1856–1931). Japanese bacteriologist who, in 1889, isolated the bacillus causing tetanus and the one causing anthrax. He worked with Behring. In Hong Kong in 1894 he isolated the agent causing bubonic plague and in 1898 the agent causing dysentery.

FREUD, SIGMUND (1856–1939). Austrian psychiatrist who adopted the method of treatment by hypnotism in 1889. In the 1890's he abandoned hypnotism and began to make use of free association, allowing the patient to talk randomly and at will with a minimum of guidance. With the cause and motivation of his behavior known, the patient could more easily avoid that behavior.

BINET, ALFRED (1857–1911). French psychologist who was interested in measuring human intelligence objectively. The term "intelligence quotient" became popular. This represents the ratio of the mental age to the chronological age.

ROSS, SIR RONALD (1857–1932). English physician. In 1887 he located Laveran's malarial parasite in the *Anopheles* mosquito. He continued to study mosquitoes and their breeding habits.

KOLLER, CARL (1857–1944). Austrian-American physician who used cocaine as a local anesthetic during operations. His first local anesthetic was given to a patient during an eye operation. This marked the beginning of the use of local anesthesia.

SHERRINGTON, SIR CHARLES (1857–1952). English neurologist who had developed a theory, by 1906, of reflex behavior of antagonistic muscles which helped explain the way the body behaved as a unit, under the coordinating guidance of the nervous system. He studied reflex action and how it coordinates behavior.

EIJKMAN, CHRISTIAN (1858–1930). Dutch physician who was the first to pinpoint what we call a "dietary deficiency." He traced disease and death among chickens that had been fed polished rice.

DUBOIS, MARIE EUGENE (1858–1940). Dutch paleontologist who tried to find a "missing link" between man and apelike ancestors. In 1894 he published results of his findings in Java of a creature whose skullcap, a femur, and two teeth he found. He called it *Pithecanthropus erectus* ("erect apeman"). In his later years he switched views and maintained the skeletal remains were only those of an advanced fossil ape.

CURIE, PIERRE (1859–1906). French chemist, husband of Marie Curie, who worked with her in the experiments on radium-producing ore.

LOEB, JACQUES (1859–1924). German-American physiologist who tried to show that the tropisms that govern plant behavior (such as simple reactions toward or away from light, water, and gravity) might explain such behavior in other organisms.

FINSEN, NIELS RYBERG (1860–1904). Danish physician who found that short-wave light obtained from the sun or a concentration of strong electric light could kill bacteria in cultures and on the skin. He was able to cure lupus vulgaris, a skin disease brought on by the tubercle bacillus, by irradiation with strong short-wave light.

BAYLISS, SIR WILLIAM (1860–1924). English physiologist who worked on the heart and who married Starling's sister. He and Starling, working together, discovered hormones.

EINTHOVEN, WILLEM (1860–1927). Dutch physiologist who developed the first string galvanometer in 1903. By 1906 he was correlating the recordings with various types of heart disorders.

BATESON, WILLIAM (1861–1926). English biologist who worked on a wormlike creature, *Balanoglossus,* that has a larval stage resembling that of starfish. This was the first indication that chordates were offshoots of a primitive echinoderm stock. In 1905 his experiments indicated that not all characteristics are inherited independently. Some are inherited together. This gene linkage was eventually explained by Thomas Hunt Morgan in his book, *The Theory of the Gene,* in 1926.

HOPKINS, SIR FREDERICK (1861–1947). English biochemist who, in 1900, discovered tryptophan, one of the amino acid building blocks of protein, and showed that it was missing in gelatin. This proved that some of the amino acids could not be manufactured in the body and had to be present in the diet. So he originated the concept of the "essential amino acid."

CARVER, GEORGE WASHINGTON (1864–1943). American agricultural chemist who developed from peanuts alone more than three hundred types of synthetic material, including everything from dyes and soap to milk and cheese substitutes. One hundred and eighteen by products were developed from sweet potatoes. In 1939 he was awarded the Roosevelt medal with a citation which reads: "To a scientist humbly seeking the guidance of God and a liberator to men of the white race as well as the black."

LEISHMAN, SIR WILLIAM (1865–1926). Scottish physician who discovered the protozoan that causes "kala azar," now sometimes called leishmaniasis. He developed a vaccine against typhoid fever.

LAZEAR, JESSE WILLIAM (1866–1900). American physician who worked in Cuba with Reed and studied the malarial parasite. He permitted a mosquito to bite him and was dead within a week of yellow fever.

WASSERMANN, AUGUST VON (1866–1925). German bacteriologist who developed a diagnostic test for syphilis in 1906, based on Bordet's discovery of complement fixation.

STARLING, ERNEST (1866–1927). English physiologist whose most important work was on secretion of the pancreas. It was thought that the pancreas was nerve controlled, but when nerves were cut it still functioned. He and Bayliss discovered that the lining of the small intestine secreted a substance that they called secretin. Under the influence of stomach acid, secretin stimulated the pancreatic flow. Starling suggested a name for all substances sent directly into the bloodstream by a particular organ and causing another organ or organs to function. The term was "hormone" from Greek words that mean to "rouse to activity."

MORGAN, THOMAS HUNT (1866–1945). American geneticist who worked with *Drosophila,* the fruit fly whose cells possess only four pairs of chromosomes. His experiments, in which crossing-over of parts of chromosomes was noted, established chromosomes as carriers of heredity. He located the spot on the chromosome at which a particular gene might exist. In 1911 the first chromosome maps for fruit flies were being drawn up.

CURIE, MARIE (1867–1934). Polish-French chemist who, with her husband Pierre, in 1898 isolated from uranium ore a small pinch of powder containing a new element that was hundreds of times as radioactive as uranium. It was called polonium. They continued working and in 1902 succeeded in preparing one-tenth of a gram of radium. In 1903 Madame Curie and Pierre received the Nobel Prize in physics for their work on radioactivity. In 1911 she and her husband were awarded the Nobel Prize in chemistry for the discovery of two new elements, radium and actinium.

LANDSTEINER, KARL (1868–1943). Austrian-American physician who discovered that human blood differed in the capacity of the serum to agglutinate red blood corpuscles (cause them to clump together). By 1902 Landsteiner and his group had clearly divided human blood into four blood groups which he named A, B, AB, and O. In 1927 Landsteiner added blood groups M, N, and MN, useful in anthropological studies. In 1940 he was involved in the discovery of the Rh blood groups.

SPEMANN, HANS (1869–1941). German zoologist whose outstanding work was done on embryos. In 1920 he demonstrated that even after an embryo had begun to show signs of differentiating into various parts, it could still be divided into halves with each half producing a whole embryo, even though one half was almost all "potential back" and the other almost all "potential belly." He further showed that an area of an embryo develops according to the nature of the neighboring areas. An eyeball develops originally out of a brain material and is joined by a lens that develops out of the nearby skin. If the eyeball is placed near a distant section of the skin that would normally not develop a lens, a lens begins to form. An area containing an organizer that brought about the development of nerve tissue in a frog embryo could even bring about the development of nerve tissue in a newt embryo. It was then seen that embryonic development is under hormonal control.

BORDET, JULES JEAN (1870–1961). Belgian bacteriologist who, in 1901, showed that when an antibody reacts with an antigen, complement is used up. This is a process called complement fixation and has proved important in immunological work. In 1906 he discovered the bacillus of whooping cough and devised a method of immunization against it.

SCHAUDINN, FRITZ RICHARD (1871–1906). German zoologist. He was the first to show that dysentery was caused by an ameba. In 1905 he discovered the organism *(Spirochaeta pallida)* that caused syphilis. A year later Wassermann devised his diagnostic method for the disease. Three years after this Ehrlich discovered specific therapy for syphilis.

RICKETTS, HOWARD TAYLOR (1871–1910). American pathologist who, in 1906, showed that Rocky Mountain spotted fever was spread by cattle ticks. He located the microorganisms that caused it. He showed that typhus was transmitted by the body louse. After Taylor died from typhus fever, the microorganisms that cause typhus and Rocky Mountain spotted fever were named *Rickettsia.*

TSVETT, MIKHAIL (1872–1920). Russian botanist who worked on plant pigments and who, in 1906, was able to separate plant pigment into color bands. He named the process chromatography which means "written in color" in Greek.

BERGER, HANS (1873–1941). German psychiatrist who, in 1929, was the first to devise a system of electrodes which, when applied to the skull and connected to an oscillograph, would give a recording of the rhythmic shifting of electric potentials, commonly called brain waves.

CARREL, ALEXIS (1873–1944). French-American surgeon who, in 1902, developed a technique for delicately suturing blood vessels with as few as three stitches. He attempted to keep organs and portions of organs alive by perfusion or passing blood or blood substitutes through them by way of their own blood vessels. He kept a piece of embryonic chicken heart alive for more than thirty-four years before he deliberately ended the experiment.

LOEWI, OTTO (1873–1961). German-American physiologist who showed that chemical as well as electrical phenomena were involved in a nerve impulse. In 1921 while working with nerves attached to a frog's heart, he showed that chemical substances were set free when the nerve was stimulated. Fluid containing the substance could be used to stimulate another heart directly. He called it vagusstoff (vagus material). Later, Dale showed this to be acetylcholine.

ERLANGER, JOSEPH (1874–1965). American physiologist who, with Herbert Gasser, determined how different nerve fibers conducted their impulses at different rates, with the velocity of the impulse varying directly with the thickness of the fiber.

DALE, SIR HENRY (1875–   ). English biologist. In 1910 he worked on ergot, a

fungus, and isolated acetylcholine from it. This brought about effects on organs similar to those brought about by nerves belonging to the parasympathetic system. He showed that Loewi's vagusstoff was acetylcholine.

TWORT, FREDERICK WILLIAM (1877–1950). English bacteriologist who discovered, in 1915, a type of virus that infested and killed bacteria. They were later called bacteriophages (bacteria-eaters) and were more complicated in structure than the average virus.

BEEBE, CHARLES WILLIAM (1877–1962). American naturalist who, in the 1930's, built a shell of steel with thick quartz windows, and in 1934 descended in waters near Bermuda to a record depth of 3,028 feet. This was the first penetration by man in depths beyond the surface layer of the ocean.

WHIPPLE, GEORGE HOYT (1878–   ). American physician who followed development of red blood corpuscles of dogs in which he had produced anemia by bleeding them. He was interested in bile pigments, and since bile pigments are formed from hemoglobin in the body, he thought he should find out how hemoglobin was handled by the body, beginning with its formation. He kept dogs on various kinds of diets to see the effect on corpuscle formation. He found liver promoted red corpuscle formation the best.

McCOLLUM, ELMER VERNER (1879–   ). American biochemist who discovered a factor in some fats that was essential to life. He spoke of fat-soluble A and water-soluble B. These soon became known as Vitamin A and Vitamin B. McCollum contributed to the discovery of other fat-soluble vitamins such as Vitamin D in 1922 and Vitamin E later. He also did some work on trace minerals.

ROUS, FRANCIS PEYTON (1879–   ). American physician who examined a tumor from a chicken and passed the tumor through a filter. He found this cell-free filtrate was infectious and would produce tumors in other chickens. He did not call it a virus in his report of 1911, but 25 years later when virus research was exploding, there was nothing else to call it. The "Rous chicken sarcoma virus" was the first of the "tumor viruses."

FLEMING, SIR ALEXANDER (1881–1955). Scottish bacteriologist who, in 1928, discovered the mold *Penicillium notatum,* clearly related to the common mold growing on bread. He showed that it inhibited bacterial growth. Penicillin was the first of the antibiotics to be discovered.

WARBURG, OTTO HEINRICH (1883–   ). German biochemist who showed that the heme groups of hemoglobin carried oxygen to the cells. For this Warburg was awarded the 1931 Nobel Prize in medicine and physiology.

MEYERHOF, OTTO FRITZ (1884–1951). German-American biochemist who showed that there was a quantitative relationship between the glycogen that disappeared and the lactic acid that appeared in a working muscle, and that in the process oxygen was not consumed. This was anaerobic glycolysis or glycogen breakdown without air. He showed that when muscle rested after work had been done some of the lactic acid was oxidized. The energy developed made it possible for the major portion of the lactic acid to be reconverted to glycogen.

ANDREWS, ROY CHAPMAN (1884–1960). American zoologist. In 1916 his expeditions in Central Asia resulted in the finding of fossilized dinosaur eggs.

MINOT, GEORGE RICHARDS (1885–1950). American physician who showed that pernicious anemia could be treated by giving the patients quantities of liver. He was right in suspecting that a vitamin deficiency was the cause of pernicious anemia.

HENCH, PHILIP SHOWALTER (1886–   ). American physician who was interested in relieving rheumatoid arthritis, a painful and crippling disease. In the 1940's when corticoids were synthesized and reasonable quantities were available for the first time, he experimented with Compound E, the name given to cortisone. It worked well

and as a result of his efforts, he, Kendall, and Reichstein were given the 1950 Nobel Prize in medicine and physiology.

KENDALL, EDWARD CALVIN (1886–   ). American biochemist who, in 1916, broke down the thyroglobulin molecule and isolated what he called thyroxine. During the 1930's he isolated 28 different cortical hormones or corticoids from the adrenal cortical glands.

SUMNER, JAMES BATCHELLER (1887–1955). American biochemist whose chief interest was research on the nature of enzymes. In 1926 when he was extracting the enzyme content of jack beans he obtained many tiny crystals. He isolated them, dissolved them, and found he had a solution of urease activity. He prepared more, but could not separate the enzyme activity from the crystals. Actually the crystals were the enzymes and they were also proteins. So urease was the first enzyme prepared in crystalline form, and the first enzyme shown to be a protein.

HOUSSAY, BERNARDO ALBERTO (1887–   ). Argentinian physiologist who showed that the pituitary gland had numerous critical functions in the body. In particular, he showed that it affected sugar metabolism. The anterior lobe seemed to produce at least one hormone that had an effect opposite to that of insulin, first isolated by Banting and Best.

REICHSTEIN, TADEUSZ (1887–   ). Polish-Swiss chemist. One of a group in 1933 who succeeded in synthesizing ascorbic acid (Vitamin C). During the 1930's he and his colleagues isolated the various corticoids (working in Switzerland) while Kendall and Hench were doing this in the United States.

ROSE, WILLIAM CUMMING (1887–   ). American biochemist who did research on amino acids in nutrition. Amino acids which had to be present in the diet were called essential ones. By 1937 Rose had shown that of the twenty or so present in nearly every protein molecule only ten were dietarily essential to rats. In the 1940's he showed that only eight amino acids were essential to man. Arginine and histidine, essential in the rat, were not essential in the adult human being.

GASSER, HERBERT SPENCER (1888–1963). American physiologist who worked with Joseph Erlanger on nerve impulses. In their work Gasser and Erlanger shared the 1944 Nobel Prize in medicine and physiology.

ZWORYKIN, VLADIMIR KOSMA (1889–   ). Russian-American physicist. By 1938 he had developed the first practical television camera, which he called the iconoscope. He modified a crude instrument made by German physicists into a practical and useful electron microscope.

MULLER, HERMANN JOSEPH (1890–1967). American biologist who found, in 1919, he could hasten the rate of mutation appearance in fruit flies by raising of temperature. Later he used X rays for this.

BANTING, SIR FREDERICK GRANT (1891–1941). Canadian physiologist who worked with Charles H. Best to isolate an extract (insulin) from the pancreas of dogs. They supplied their extract to dogs that had been made diabetic through the removal of the pancreas. The extract quickly stopped the symptoms of diabetes in the dogs.

NORTHROP, JOHN HOWARD (1891–   ). American biochemist. By 1930 he had crystallized pepsin in gastric secretions. In 1932 he had crystallized trypsin and in 1935 chymotrypsin, both protein-splitting digestive enzymes from the pancreas.

MURPHY, WILLIAM (1892–   ). American physician who worked with Minot in developing the liver treatment for pernicious anemia. He shared with him and Whipple the 1934 Nobel Prize in medicine and physiology.

SZENT-GYÖRGYI, ALBERT VON (1893–   ). Hungarian-American biochemist who studied how ascorbic acid was used in the body. He found a rich source for it in Hungarian paprika. He also studied oxygen uptake of minced muscle tissue. Szent-Györgyi's research made it possible for Krebs to work

out the Krebs cycle. In the 1960's he studied the thymus gland. He isolated some substances which seem to have some controlling effect on growth.

DAM, CARL PETER (1895–   ). Danish biochemist who, in 1929, studied how hens synthesize cholesterol. In studying this he observed small hemorrhages under the skin and within the muscles of chickens fed synthetic diets. Adding citrus fruit juice to the diet (as for scurvy) did not help, so he decided the lack of an unknown vitamin must be the cause of the nonclotting of the blood. This he called Vitamin K for *Koagulation* (German spelling).

CORI, GERTY THERESA RADNITZ (1896–1957). Czech-American biochemist who worked with her husband on the breakdown and resynthesis of glycogen. For their work the Coris were given the Nobel Prize in medicine and physiology in 1947.

CORI, CARL FERDINAND (1896–   ). Czech-American biochemist. During the 1930's Cori and his wife investigated how glycogen, the carbohydrate stored in liver and muscle, is broken down in the body and resynthesized. They filmed the whole series of changes.

KING, CHARLES GLEN (1896–   ). American biochemist who, in 1932, isolated Vitamin C.

ENDERS, JOHN FRANKLIN (1897–   ). American microbiologist who, in 1948, devised a way of growing mumps virus in mashed-up chick embryos bathed in blood. By adding penicillin to the mashed tissue, bacterial growth was stopped while viral growth remained unaffected. Later he worked on polio virus under similar circumstances.

SCHOENHEIMER, RUDOLF (1898–1941). German-American biochemist who, in 1935, introduced the use of isotopic tracers in biochemical research. He used fat molecules containing deuterium atoms in place of some of the hydrogen atoms. Analysis of the fat of the laboratory animals showed that ingested fat was stored in the body, while stored fat was used. Schoenheimer's work was the first to show body chemistry in action. He was the father of isotopic tracer research in biochemistry.

LYSENKO, TROFIM DENISOVICH (1898–   ). Soviet biologist who maintained that acquired characteristics could be inherited. After Stalin's death in 1953 Lysenko's views became worthless.

BEST, CHARLES HERBERT (1899–   ). American-Canadian physiologist who worked with Banting in isolating insulin.

BURNET, SIR FRANK MACFARLANE (1899–   ). Australian physician who suggested in 1949 that the ability of a human being to form antibodies against the proteins of another human being might not be inborn. Antibodies against disease develop only after exposure to the microorganisms causing the disease. Allergies develop only after sensitization to some particular protein. So resistance to the proteins of another person should be developed only during the course of life, possibly very early in life. Burnet's suggestion was acted upon by others.

KARPECHENKO, G. D. (1899–   ). Russian plant geneticist who achieved the first man-made species of plant. He crossed a radish with a cabbage and got a sterile hybrid. He succeeded in doubling the chromosome number and finally got a new plant with tetraploid chromosome number that was fertile when crossed with itself, but sterile when crossed with either a cabbage plant or a radish plant. The "rabbage" plant had a spindly, inedible root like a cabbage, and prickly, useless leaves like a radish. Karpechenko was among the Russian geneticists who were purged as they had been exploring the mechanisms of heredity and evolution instead of working on what the Soviet government regarded as problems that were practical.

THEILER, MAX (1899–   ). South African-American microbiologist who began research on yellow fever in the 1920's. He cultivated the yellow fever virus, infected monkeys, then passed the virus into mice where it developed encephalitis, a brain inflammation. He passed the virus from mouse to mouse and eventually back to monkeys where it produced only a very feeble attack of yellow fever, but induced full immunity to the most virulent strains of the virus. Further research produced an even safer vaccine.

DU VIGNEAUD, VINCENT (1901–   ). American biochemist whose most exciting work was on the hormones produced by the posterior lobe of the pituitary gland. By 1953 he had worked out the exact order in which the amino acids appeared in the chain and was able to synthesize oxytoxin. It was the first protein hormone ever synthesized, and was found to have all the properties of the natural material.

BEADLE, GEORGE WELLS (1903–   ). American geneticist who worked on a mold called *Neurospora crassa,* which, when subjected to X-ray irradiation, will form mutations. Some mutants lose the ability to form a particular compound necessary for growth and will grow only if a compound such as an amino acid is added to the nutrient medium. Beadle found that it was not always necessary to add the missing compound itself to the medium. A different but similar compound might do. This meant that the similar one could be converted into the necessary one. By trying a number of similar compounds and observing which would promote growth and which would not, he could deduce the sequence of chemical reactions that led to the formation of the necessary compound within the mold. He could also tell where a "break" came in the sequence, at which point a reaction existed that the mold could not handle. As an example, he found that two different mutant strains could not form arginine, but when these strains were crossed they could form it since each member of the cross supplied what the other lacked. Beadle concluded that the characteristic function of the gene was to supervise the formation of a particular enzyme. When a gene was altered, it could no longer form the proper enzyme. When this happened some specific reaction would not take place—a sequence of chemical reactions was broken—and a radical change might occur in the physical characteristics of the organism. After the early 1940's it became more and more clear that the gene was a molecule of deoxyribonucleic acid (DNA).

BITTNER, JOHN JOSEPH (1904–1961). An American biologist who worked on cancer in mice. He found some strains that were highly resistant to cancer and others very prone to developing it. His publications in 1936 showed that if young mice of a cancer-resistant strain were fed from a foster mother of a cancer-prone strain, the young developed cancer. If, however, the young mice of a cancer-prone strain were fed by a foster mother of a cancer-resistant strain, they did not usually develop cancer. It seemed that this particular kind of cancer in these specific animals was of an infectious type and the mother's milk carried the infectious agent. In 1949 the Bittner milk factor was isolated, or at least particles were found in the milk of cancer-prone mother mice that were not found in the milk of the cancer-resistant mother mice.

STANLEY, WENDELL MEREDITH (1904–   ). American biochemist who studied tobacco mosaic virus. He grew tobacco and infected it with the virus, then mashed quantities of the leaves and put them through the usual chemical procedure for crystallizing proteins, since he believed the virus was a protein molecule. In 1935 he isolated fine, needle-like crystals. These possessed all the infective properties of the virus. This meant that a virus can reproduce itself within cells, and this is an important criterion of life. Many people found it difficult to accept the fact that a virus can reproduce itself within cells. It seemed to place the virus on the boundary between living and nonliving things. Since 1935 many other viruses have been crystallized. All have been found to be nucleoproteins.

SABIN, ALBERT BRUCE (1906–   ). Polish-American microbiologist. He believed that only living viruses could be counted on to produce the necessary antibodies against polio viruses over a long period of time. In addition, living viruses could be taken by mouth, since they would multiply and invade the body of their own accord, and would not have to be injected by needle as the Salk vaccine. By 1957 Sabin had satisfactory live vaccines of each polio type. Although used successfully in the Soviet Union and other East European countries,

it was 1960 before it came into use in the United States.

WILKINS, ROBERT WALLACE (1906–   ). American physician who introduced into the United States a drug derived from the root of an Indian shrub. It came from India and had been used there in treating high blood pressure. In 1950 Wilkins used it at the Massachusetts Memorial Hospital. In 1952 he reported on its sedative and tranquilizing effects and named it reserpine. This was the first of the tranquilizers.

TATUM, EDWARD LAWRIE (1909–   ). American biochemist who worked on *Neurospora* with Beadle.

FRAENKEL-CONRAT, HEINZ (1910–   ). German-American biochemist who, in 1955, while working with bacteriophages, developed gentle techniques for teasing apart the nucleic acid and protein of a virus without seriously damaging each part, and then putting them together again.

MARTIN, ARCHER JOHN PORTER (1910–   ). English biochemist who developed a method in 1944 to separate amino acids, using paper chromatography.

CALVIN, MELVIN (1911–   ). American biochemist who, together with his workers, discovered and isolated the intermediate products of photosynthesis, determined how they fit together, and developed a scheme to show the process of photosynthesis.

PALADE, GEORGE EMIL (1912–   ). Romanian-American physiologist. He studied intact cells through an electron microscope. The "microsomes," as the small bodies in the cytoplasm were called, showed that they were more than mitochondria fragments. By 1956 he had shown that the microsomes were rich in ribonucleic acid (RNA) and were therefore renamed ribosomes. It was realized that the ribosomes were the site of protein manufacture.

LI, CHOH HAO (1913–   ). Chinese-American biochemist who studied hormones, mainly of the pituitary gland. He and his group isolated adrenocorticotropic hormone (ACTH). In 1956 he showed that the molecule of ACTH is made up of 39 amino acids in a specific order, and that the entire chain of the natural hormones is not necessary for its action. Fragments that consist of little more than half the chain possess major activity. Li proved that the growth hormone from the anterior pituitary of swine or cattle is not effective in human beings, but that ACTH from these same animals is effective.

SALK, JONAS EDWARD (1914–   ). American microbiologist. After Enders and his group showed the way to cultivate polio virus and make large amounts of it available for experimentation, Salk began with attempts to kill the virus so that it would not cause the disease but would produce antibodies that would be active against living polio virus. By 1952 he had prepared a vaccine he tried on children who had recovered from polio and who were resistant to infection. It increased the antibody content of the children's blood. He then tried it on children who had not had polio and it was successful. It became headline news in 1955.

MEDAWAR, PETER BRIAN (1915–   ). English biologist who worked on Burnet's suggestions. He inoculated embryos of certain mice with cells from another strain, hoping the embryos had not yet gained the ability to form antibodies. He felt that by the time the embryo became independent the "foreign" proteins might no longer be treated as foreign. This proved to be true. Once the embryo mice were born they were able to accept skin grafts from the strains of mice with which they had been inoculated in embryo. For this work, Medawar shared the 1960 Nobel Prize in medicine and physiology with Burnet.

WELLER, THOMAS HUCKLE (1915–   ). American microbiologist who worked with Enders and Robbins on the polio virus, and with them was awarded the 1954 Nobel Prize in medicine and physiology.

CRICK, FRANCIS HARRY COMPTON (1916–   ). English biochemist. In 1953 Crick and Watson suggested that the DNA molecule consisted of a double helix, each helix

being made up of the sugar-phosphate backbone known to exist in the nucleic acid molecule. The nitrogenous bases extended in toward the center of the helix from each of the two backbones and approached each other. It was suggested that in the process of replication, the two strands of the double helix unwound and each single helix served as a model for its complement.

ROBBINS, FREDERICK CHAPMAN (1916–   ). American microbiologist who worked with Enders and Weller and shared the 1954 Nobel Prize with them.

WOODWARD, ROBERT BURNS (1917–   ). American chemist who synthesized strychnine in 1954 and also lysergic acid (which influences mental function). In 1956 he synthesized reserpine, the first of the tranquilizing drugs; in 1960 he synthesized chlorophyll; and in 1962 headed a group that worked for three years and was able to synthesize a tetracycline antibiotic.

HOAGLAND, MAHLON BUSH (1921–   ). American biochemist who studied during the late 1950's how nucleic acids bring about the formation of protein molecules. He located a variety of small RNA molecules (transfer-RNA) in the cytoplasm, each of which could combine with one specific amino acid, and no other. Each molecule of transfer-RNA had as part of its structure a characteristic triplet that joined to a complementary triplet on the messenger-RNA. Since each transfer-RNA clicked into a specific place with a specific amino acid attached, a protein molecule was built up, amino acid by amino acid, according to the design that originally existed in the DNA molecules of the chromosome. In this way, chromosomes of a particular cell produce a battery of enzymes (all of which are protein molecules) which guide the chemistry of that cell and eventually produce all the physical characteristics studied by geneticists.

LEDERBERG, JOSHUA (1925–   ). American geneticist who worked with Tatum and showed that different strains of bacteria could be crossed in such a way as to make the genetic material intermingle. Beadle,

Tatum, and Lederberg earned the 1958 Nobel Prize in medicine and physiology.

WATSON, JAMES DEWEY (1928–   ). American biochemist who worked with Crick in working out the Watson-Crick model of DNA structure. In 1962 he, Crick, and Wilkins shared the Nobel Prize in medicine and physiology.

REFERENCES

ASIMOV, ISAAC. 1964. *Asimov's Biographical Encyclopedia of Science and Technology*. Doubleday and Company, Inc., Garden City, New York.

BERNAL, J. D. 1965. *Science in History*. 3rd Ed. Hawthorne Books, Inc., New York.

DAMPIER, SIR WILLIAM CECIL. Reprinted 1950. *A History of Science*. Cambridge University Press, New York.

GARDNER, E. J. 1965. *History of Biology*. 2nd Ed. Burgess Publishing Company, Minneapolis.

COMMON BIOLOGICAL PREFIXES AND SUFFIXES

a—negative, not or without: asexual, asymmetry.

ab—away from: abaxial, abduct.

ad—to or toward: adaxial, admit.

bi—two: bilateral, binocular.

bi(os)—live: biology, biography.

chlor—green: chloroplast, chlorophyll.

chrom—color: chromatin, chromosome.

coleo—sheath: coleoptile, coleorhiza.

derm—skin: epidermis, endodermis, protoderm.

di—two or second: dihybrid, dimorphic.

dicho—in two: dichotomy, dichotomous.

diplo—double, or folded double: diploid.

epi—upon or above: epidermis, epiphyte, epigynous.

hetera—other or different: heterozygous.

homo—like or similar: homology, homozygous.

hyper—over or beyond: hyperacidity, hypertonic.

hypo—under or less than: hypotonic, hypogynous.

iso—equal: isogamy, isotonic.

lamella—leaf or layer: lamella, lamellate.

log(os)—discourse, "science": biology, zoology, entomology.

macro—large: macronucleus, macroscopic.

mega—large: megagamete, megaspore.

meso—middle: mesophyll.

micro—small: microscope, microspore, micron.

mito—thread: mitosis, mitochondria.

morph—form: morphology, morphogenesis.

oid—like, resembling, the "idea" of: zooid, rhizoid.

photo—light: photosynthesis, phototropic.

phyllon—leaf: chlorophyll, phylloclad.

plasma—mold or form of life: protoplasm, cytoplasm.

proto—first: protoplasm, protoderm.

rhiza—root: rhizoid, rhizome.

scler—hard: scleroid, sclerenchyma.

soma—body: chromosome, autosome.

stom—mouth: stomata, hypostome.

syn(n)—together: synapse, synthesis.

tax(is)—arrangement: taxonomy, ataxia.

tel—far or distant: telophase, telescope, telegraph.

trop—turn toward: phototropism, hydrotropism.

troph—food: autotroph, heterotroph.

zyg—united, tied or yoked together: zygote, heterozygous.

Numbers

uni, mono—	1
bi, di	— 2
tri	— 3
tetra	— 4
penta	— 5
hexa	— 6
septa	— 7
octa	— 8
nona	— 9
deca	—10

SPEAKERS FOR BIOLOGY CLASSES AND SCIENCE CLUBS

NO TOWN is so small that it does not have at least some professional people who can contribute considerably to the biology classes and to the science club. Some people may have developed hobbies which make them specialists or near-specialists in certain areas.

The biology teacher needs to seek out such individuals and ask for their services. Students may know of some of these people and can tell you about them.

The list included here is one compiled by E. D. Keiser, Jr., of the Museum of Zoology, Louisiana State University, Baton Rouge, Louisiana. It is taken from his article "The Utilization of Local Speaker Talent in the School Science Club Program" printed in *Turtox News* in August 1965, Vol. 43, No. 8. Your list may not be as impressive or as long as this list, but it may include others.

People from surrounding areas may also be called upon for assistance.

Although Mr. Keiser's list of speakers includes more than fifty different ones, we are using only those that pertain directly to biology. We suggest you read the entire article if you work with your local school science club.

Barber: "The Growth and Structure of Hair"

Beautician: "Beneficial and Harmful Aspects of Cosmetics"

Bee Keeper: "The Natural History of the Honey Bee"

Biologist, Fish: "Sense Organs in Fishes," "The Embryology and Later Development of Fishes"; Game: "The Role of the Hunter in Game Management," "Wildlife Conservation," "Predation—Its Place in the Natural Community"

Chemist: "The Nature of Explosives," "Safety in the Laboratory," "Chemical Concepts of the Nature of Matter," "Isotope Research"

Clergyman: "Evolution and the Bible," "Science and Religion"

Counselor, Marriage: "Biological Problems of Courtship"

Dentist: "Diseases of the Mouth," "Embryology of Teeth"

Engineer, Sanitary: "The Microflora of Sewage Systems"; Water Supply: "Control of Microorganisms in the Community"

Farmer: "Modern Scientific Farming"

Florist: "Flower Culture—The Art of the Green Thumb," "Photoperiodism and Flowering"

Fur and Hide Dealer: "The Trapping of Furbearers"

Game Warden: "Scientific Detection of Game Law Violators"

Geologist: "Volcanoes," "The Water Table," "Fossils and Earth History"

Horticulturist: "Increasing the Yield of the Home Garden"

Law Enforcement Agent, F.B.I.: "Science in Crime Detection"

Librarian: "Use of the Library in Scientific Writing"

Meat Inspector: "Animal Parasites"

Mortician: "The Science of Embalming"

Nurse, Registered: "The Childhood Diseases," "Danger Signals of the Human Body," "Sex Education"

Nurseryman, Tree: "The Science of Grafting," "Tree Growths"

Optometrist: "The Optics of Eyes and Lenses," "Contact Lenses"

Pathologist, Animal: "Animal Diseases"; Human: "Venereal Diseases"; Plant: "Fungus Diseases of Local Plants," "Athlete's Foot"

Pest Control Agent: "Problems of Insect Control," "Termites, Mosquitoes, and Man," "Rat Control"

Pet Show Owner: "The Humane Treatment of Animals in Captivity," "Tropical Fish Culture"

Pharmacist: "The Role of Pharmacy in Modern Medical Science"

Photographer: "Lenses in Photography," "Photography as a Tool of Science"

Physician, Allergist: "The Causes, Diagnoses, and Treatment of Allergies"; General Practitioner: "Sex Education," "Medical Problems of Teenagers"; Obstetrician: "The Miracle of Birth"; Opthalmologist: "Virus Diseases of the Eyes"; Orthopedist: "Posture Problems and Their Correction"; Psychiatrist: "Hypnotism," "Emotional Problems of Growing Up," "Learning and Memory"; Radiologist: "The Development and Medical Use of X rays"

Politician: "The Allocation of State and Federal Funds To Support Scientific Projects"

Psychologist: "The Role of Hypnosis in Psychoanalysis," "Mental Measurement," "Neurosis and Psychosis"

Salesman, Hearing Aid: "The Ear and Artificial Hearing Devices"

Taxidermist: "Preparing Mammal and Bird Skins for Scientific Study"

Teacher (Nonscience), English: "The Role of Language in Scientific Communication"; History: "The History of Science"; Physical Education: "Exercise, Diet, and Bodily Health"

Teacher (Science), Biology: almost unlimited possibilities; Chemistry: almost unlimited possibilities; Mathematics: "Scientific Measurement," "Mathematical Evaluation of Research Data"; Psychology: "The Measurement of Learning," "The Measurement of Psychological Test Data"

Tree Surgeon: "Tree Surgery"

X-ray Technician: "The Physics of X rays," "Biological Effects of Radiation"

SAMPLE LIST OF GOOD TEACHING PRACTICES

1. Do I stress the problems of science?

2. Does my attitude encourage the students to pose problems so that not all the problems are mine?

3. Do our activities solve these problems?

4. Do I help the students realize that science is more than just memorization of facts?

5. Are the activities both challenging and provocative?

6. Do I allow the students to formulate their own hypotheses and to design and perform their own investigations without knowing in advance what the results will be, or do I tell the students everything they are supposed to know?

7. Do I guide the students to consider many methods of attacking problems?

8. Do I use books as lessons in reading, or are the students encouraged to use books as sources of information for testing hypotheses and to reinforce what has already been discovered?

9. Does every student have an opportunity to investigate problems and to discuss his hypotheses knowing that his ideas and contributions will be treated positively, or do I work with just certain students to the

Prepared by Kenneth D. George for the National Science Teacher's Association, Washington, D.C.

neglect of others?

10. Do I allow sufficient time for every student to work on problems with enough depth to satisfy his curiosity and to comprehend the subject?

11. Do I use a variety of methods and techniques, or do I use one to the exclusion of all others?

12. When I ask a student "What do you think about . . . ?" do I really want to know what he thinks or am I asking him "Can you guess what I think about . . . ?"

SUGGESTIONS FOR INSTANT LABORATORIES

NO LAST-MINUTE HUNTING for materials is necessary if they are collected ahead of time and set aside in boxes as instant laboratories.

The following list may be suggestive of instant laboratories you want to prepare for your laboratory:

1. Osmosis Demonstration

In a shoe box place three glass dishes such as custard cups or beakers (large enough to hold an egg), a small bottle of hydrochloric acid, salt, three glass rods, and three overflow cans. Keep the eggs in the refrigerator. When ready to demonstrate osmosis, fill each glass dish about half full of water, then put an egg in each, and add hydrochloric acid (a small amount at a time) to remove the eggshell. Turn the eggs constantly so the shells are removed evenly. When the shell is nearly gone and the inner membrane exposed, remove the water and hydrochloric acid from the glasses and substitute tap water. Rinse the eggs several times.

Measure the circumference of each egg at its greatest girth. In one glass put

an egg plus tap water; in another, an egg plus a strong sugar solution (two to three teaspoons of sugar in water). In a third glass put an egg, plus a strong salt solution (two to three teaspoons of salt in the water). Measure the circumference of each egg at various intervals of time (about once each hour for several hours). Describe the results.

This experiment can also be done using overflow cans to measure the volume of the eggs. Fill the cans with water before placing eggs in them. Then measure the volume of water that is displaced when the egg is put in.

2. Bone Boxes

Bones from each part of the body may be kept in boxes for ready use— head bones in one, arm bones in another, etc. This will take boxes larger than the suggested shoe boxes, but it will be well worth the effort to keep the bones separated and identified.

3. Blood Typing

In a box place a bottle of 70 per cent alcohol for sterilizing fingers; surgical cotton; sterilized individual hemolets (never the spring type by means of which infection can be spread from one individual to another); glass slides; cover slips; and wax marking pencils. If other tests are to be made (such as hemoglobin determination) at the same time, Tallquist hemoglobin scale and paper may also be kept in this instant laboratory. Keep the antiserum for testing both Rh and ABO in the refrigerator until time for using.

4. Starch Digestion

This instant laboratory requires depression or spot plates; iodine-potassium iodide solution; pyrex test tubes; test tube holders; stirring rod; amylase powder (fresh saliva, class donated, is preferable); alcohol lamps (or other means of heating solutions); Benedict's solution.

To make the iodine-potassium iodide solution, dissolve 3 g of potassium iodide in 25 ml of water, then add 0.6 g of iodine and stir until dissolved. Add distilled water to make 200 ml of solution, and store in a dark bottle. When a drop of this solution is mixed with starch, a dark blue color results.

To make Benedict's solution, use the following: sodium citrate or potassium citrate, 173 g; sodium carbonate (Na_2Co_3)—if you use the crystalline form use 200 g but if you use the anhydrous form, use 100 g; copper sulfate ($CuSo_4$), crystalline form 17.3 g. Dissolve the citrate and the carbonate in 700 ml of slightly warm water to speed the action. Filter. Dissolve the copper sulfate in 100 ml of water and pour it slowly into the first solution, stirring constantly. Cool and add distilled water to make a liter of solution.

Benedict's solution is used to test for the presence of certain simple sugars in food, blood, and urine. In the presence of these sugars a yellow or red precipitate of cuprous oxide forms when the reagent is heated with the "unknown." Benedict's solution does not deteriorate upon standing, but it is advisable to keep the solution in a dark brown bottle.

5. Individual Lessons for Using the Flannelgraph

This could include the parts of a flower with individual sepals; petals; stigma, style, and ovary of pistil; and pollen grain, anther, and filaments of stamen. If the parts of the flower are made of flannel or construction paper with sandpaper attached to the back, they will stick to the flannelgraph as you talk about the parts and demonstrate them.

6. Parts of an Insect for a Flannelgraph

Place in a box a head; thorax; abdomen; a series of different kinds of mouth parts as used for sucking, piercing, and biting; types of legs such as those used for running, swimming, carrying pollen, and digging; kinds of wing types as membranous, sheathing second pair, modified wings as halteres or knobs, scale covered, etc. Thus the "typical" insect may become "specific" as the wings, legs, mouthparts, etc. are interchanged.

All these pieces may be made of flannel or of construction paper with sandpaper backing. Cut the pieces so they are large enough to be seen clearly from all parts of the room.

7. A Flannelgraph of Bacteria of Various Shapes, Plus Those of Other Microorganisms (Cocci, Spirilla, Bacilli, Rickettsia, Paramecia, Amoeba)

In addition to the shapes, the names of various organisms can be printed or written on paper that is backed with sandpaper or felt and placed beside the organism types as the demonstration proceeds.

8. Digestion and Emulsification of Lipids

For this instant laboratory you will need dry lipase (steapsin); litmus paper; vials; glass-marking wax pencils; beakers; water bath (substitute pyrex beakers); heating element (candle); stand for beaker over heat; fresh cream (refrigerated) or commercial Pream (Pream may be kept in the instant laboratory); vial of vegetable oil; dry ox-gall or sodium choleate to be made up as used as 1 per cent solution (.1 g in 100 ml water); and a thermometer. If students work in groups there can be as many instant laboratories made up as there are groups of students to use them.

9. Stains

Materials needed for staining various kinds of organisms may be kept in separate instant laboratory boxes, or an entire series of stains may be kept in each box. If students work in groups, as many boxes of stains and other equipment needed for staining may be made up as are needed.

10. Indoor Insect Collection

This can be kept in an area the size of a few instant laboratory boxes. Pin a series of insects to the underside of bottle stoppers and insert the stoppers in bottles. Then instead of an outdoor field trip, a hundred or more insects may be quickly set out for display. With the bottles keep cards stating the information you wish the students to get.

11. Taste Areas on the Tongue

Keep in the box items such as small jars of sugar, salt, citric acid, and a substance for bitter taste (Epsom salts is satisfactory for this); four small beakers or paper cups for every two members of the class for putting materials into water so they are in solution; toothpicks for putting solutions on different parts of the tongue to determine the position of various taste receptors on the tongue; paper cups for students to drink from in rinsing the mouth after making the taste test; taste-test paper for determining tast-

ers and nontasters of phenylthiocarbamide. This taste-test paper may be purchased from American Genetic Association, 1507 M Street N.W., Washington, D.C. 20005. The cost is $ .04 a sheet or $ .03 a sheet for fifty or more sheets. Each piece can be cut into small squares so that one or two sheets can supply an entire class.

12. Cutaneous Senses

Include dividers with blunt points (one for every student); coins such as ten to twelve pennies; a large, shallow dish; some smooth pieces of flat glass; fifty or more sheets of "onionskin" paper; common pins; small beakers for hot water; and marking pencils.

a. Use dividers to determine how close together the two points can stimulate various areas of the skin and still be felt as two distinct points. Check areas on the back of the neck, back of the hand, palm of the hand, ball of the thumb, tip of the nose, tongue, and lips. Record data and compare results.

b. To show how the receptors become adapted to stimuli place a coin on the inside of the forearm and check with a watch to determine how long the touch sensation lasts.

Repeat this experiment at a new spot on the skin, but after the sensation disappears add two more coins on top of the first one to see if the sensation returns.

c. Demonstrate referred pain by placing your elbow in a shallow dish of ice water until you feel pain. You will feel the pain in an area other than that being stimulated. This is referred pain.

d. To demonstrate pressure sense acuteness place a hair on a smooth, flat piece of glass and cover with several sheets of onionskin paper. Use the index finger tip to determine through how many sheets you can feel the hair. Keep adding sheets until you can no longer feel the hair. Test a knuckle, thumb, little finger, and edge of the palm. Record and compare the results.

e. To determine pain, pressure, and temperature receptors, use a skin- or glass-marking pencil to mark off about an inch square on the back of your hand. Divide this area still further into a number of smaller squares. With the point of a straight pin lightly touch selected areas within the test plot on your hand to see which ones are sensitive to pain. Record these in your notebook on a square that is similar to the one on the back of your hand.

Use the blunt end of the pin to determine the areas that respond to touch. Record as with pain.

Heat some water in a small beaker and warm the head of a pin. Test the area on the back of the hand for heat receptors. The pin does not retain the heat for more than two or three tests, so keep reheating the pin head frequently in making this test. Record results.

For determining cold receptors, cool the head of the pin on an ice cube and test as you did for heat. Cool the pin frequently. Record results.

FORTY WAYS TO SAVE TIME AND MONEY IN THE BIOLOGY LABORATORY

1. Needles stuck into corks or pencil erasers make good dissecting needles. Variation in size of needles is good since both larger and smaller ones are needed for dissecting.
2. Small pieces of broken razor blades inserted in the corks make good dissecting scalpels.
3. Hospitals are helpful about supplying schools with plastic hypodermic needles and small vials and bottles. Once a package of hypodermic needles is opened they are no longer "sterile," and although only one may be used from a package, the rest are discarded. Ask and receive.
4. A set of inexpensive, plastic letters for printing insures uniformity in letters for bulletin boards and saves time in printing.
5. Cottage cheese cartons and tin cans make good holders for test tubes, pencils, and colored chalk.
6. Gallon pickle jars make excellent cages for mice, snakes, and rats, and for aquaria and terraria.
7. A supply of plastic bags for carrying plants and other items may be built up by asking students to bring bags from such things as bread.
8. Small vials and bottles, discarded by druggists (and hospitals), make good display bottles for life cycles of insects, seeds, etc.
9. Brown wrapping paper or newsprint from the local newspaper publisher makes a good background for student murals.
10. Shoe boxes, suit boxes, and boxes discarded by book stores may be used for the "instant lab."
11. Aluminum foil, shaped around a petri dish, makes an inexpensive substitute for a petri dish.
12. Slices of cooked potato make excellent media for growing bacteria and molds.
13. Local florists are usually glad to supply free flowers for class use. Though the flowers may be past their prime they can still be used for dissection.
14. Models of practically any plant or animal may be made of modeling clay, covered with latex, and later removed and filled with molding plaster rather than plaster of Paris. A 100-lb bag of molding plaster may be purchased at your local lumber yard for less than $3.00. This makes several dozen plaster models. Figure 8.17 describes the process.
15. Cans that hold several gallons of canned fruit are usually thrown away by bakeries or stores that make fruit pies. These cans may often be obtained

free. They make excellent storage containers or homemade "ovens" and heat boxes.

16. A pressure cooker (used or new) is cheaper than an autoclave and can be used very successfully in sterilizing whatever needs to be sterilized in the biology laboratory.

17. Clip-type clothespins make excellent cheap substitutes for photographic metal clips for hanging negatives in the darkroom. If dipped in paraffin they leave no pressure marks on the negatives. (They are also good for clipping sets of papers together.)

18. Black and white film for 35mm cameras may be purchased from local firms or from camera supply companies in 100-ft rolls to be rolled by the buyer. This may represent as much as a 90% cut in film cost.

19. Before purchasing books for the science room, check to see if they are printed as paperbacks. By purchasing paperbacks the budget is stretched materially.

20. Red ink makes a substitute stain for eosin when eosin is not available.

21. Plastic dishes used instead of glassware in the laboratory save money and time.

22. A powerful cleaning agent that really gets glassware clean may be made from potassium bichromate, 10 g; commercial, concentrated sulfuric acid, 40 ml; and tap water, 50 ml. Dissolve bichromate crystals in the water by heating. Cool, add acid a little at a time and cautiously, as a great heat is created. Use full strength or diluted. Leave glassware in cleaner 12 to 24 hours. Rinse in tap water. Place in a jar of strong, hot soapsuds for 12 to 24 hours. Rinse. Dry at once or place in alcohol until needed. Save cleaner and reuse. Keep cleaner in a wide-mouthed jar such as a battery jar. Label it! Do not immerse hands in it nor pour it down the sink full strength as the acid burns the hands and destroys the enamel of the sink and the lining of the sink pipe. To get rid of it dilute it at least 100 parts water to one part cleaner before pouring down the sink. Flush with a heavy stream of water.

23. For polishing slides and cover slips,

clean in aforementioned cleaner, or in water or alcohol, then place in emulsion of Bon Ami powder, 5 g; and hot tap water, 100 ml. Bon Ami is unsurpassed for polishing slides and cover slips.

24. Satisfactory insect nets and dipping nets may be made using an old embroidery hoop or a wire coat hanger bent in a circle or some other shape, a broom or other implement for a holder, and nylon netting or an old nylon stocking.

25. Gallon battery jars are good homes for cockroach colonies. Grease the inside part of the jar about 4 inches from the top so the roaches are unable to leave the jar.

26. Old cake pans make good dissecting pans. If you wish these to be covered with wax, this may be obtained from old candle stubs or paraffin from the grocery store.

27. If a number of different texts are used as references in the classroom and you have several of each kind, they may be protected by making book covers out of wrapping paper, then painting a stripe of one color for each kind of text on the end of each book cover. Similar books may then be kept together on the shelf.

28. If there is no equipment for heating test tubes in the lab, alcohol lamps serve the purpose very well.

29. Small plastic rings from the dime store, glued onto slides, make good depression slides.

30. Clean plastic detergent bottles, thoroughly washed, can be used for storing reagents.

31. Mineral oil is a good substitute for immersion oil.

32. Cheesecloth cut into small pieces makes inexpensive cloth wipes.

33. Heavy toenail clippers cost less than regular bone clippers or scissors.

34. Cigar boxes lined with corrugated paper serve well for holding insects.

35. Toilet paper makes a good laboratory wiping paper in place of regular Kleenex or Kimwipes.

36. Eyelashes or hairs placed in the microscope ocular are good pointers.

37. Containers such as cheese boxes, flat

fish cans, paper milk cartons, ice cream containers, egg crates, and paper cups are fine for germinating seeds.

38. Make glass stirring rods by cutting glass tubing various lengths and sealing the ends by heating.

39. Make your own spreading board for insects by using a piece of balsa wood and varying the width of the groove to accommodate bodies of both wide and narrow insects.

40. A pair of ice tongs serves well to pick up hot objects, including test tubes and beakers.

BASIC EQUIPMENT FOR THE LABORATORY

There are a few necessities in every biology classroom. Included are:

Book cases
Chalk board with both white and colored chalk, and erasers
Demonstration table with water and heating facilities
Display cabinets and cupboards
Sources of heating, with adequate hot plates or other means for warming materials or cooking them
Tables with chairs, preferably movable tables
Utility cart on wheels
Water outlets with sink

The following materials, listed by categories, are essential for meaningful teaching.

Apparatus and Equipment

Animal cages	Forceps
Aquaria	Hand microtome
Blender	Incubator
Blood typing set	Insect nets: aerial; dip
Bone saw, shears	Kymographs
Bunsen burners	Microscopes: dissecting; binocular; monocular
Corks, assorted	
Culture tube baskets	
Dessicator	
Dissecting pans, tools	Pinning boards, insect pins, boxes
Filter paper	

Plant presses
Pressure cooker (steam pressure sterilizer)
Razor blades
Refrigerator
Ring stands with clamps
Rubber stoppers
Rubber tubing
Student light sources
Terraria
Test tube holders, brushes
Thermometers
Timers
Tripods

Charts

Amoeba	latory system; digestive system; ear; eye; nervous system
Animal and plant cells	
Clam dissection	
Crayfish dissection	*Hydra*
Earthworm dissection	Liver fluke life cycle
Euglena	Mitosis
Frog, male and female urinogenital systems	*Paramecium*
	Planaria
Frog dissection	*Spirogyra*
Grasshopper	Sponge
Human: blood circu-	Starfish
	Tapeworm

Chemicals

Acetic acid	Formaldehyde (40%)
Acetone	Glycerine
Alcohols: butyl; isopropyl; methyl	Hydrochloric acid
Benedict's solution	Lugol's; iodine-potassium iodide
Calcium carbonate	Methyl cellulose
Carbolic acid crystals	Mounting medium: Canada balsam; euparal; Hoyer's
Carbon tetrachloride	
Chloroform	
Chromic acid	Ninhydrin
Ether	Xylene (xylol)
Fehling's solution, 1 and 2	

Glassware and Plastics

Beakers, Pyrex (50 ml, 100 ml, 250 ml, 1000 ml)	Jars: battery; coplin; display (6 oz to 30 oz); specimen (¾ to 16 oz); storage
Bell jars	
Cotton, absorbant and nonabsorbant	Petri dishes
Filter flasks with side tubes	Pipettes: dropping; graduated
Finger bowls	Slides: hanging drop (concave); ordinary microscope slides; cover slips for slides
Funnels	
Glass tubing	
Graduates, assorted (10 ml to 1000 ml)	Stirring rods
	Vials with stoppers

Greenhouse

Chemicals and fertilizers
Growth chamber
Plants
Potting soil: sand; loam
Pots, all sizes for planting
Tables for plants
Tools for using in greenhouse

Microscope Slides

Bacteria types
Bone, human, ground thin
Flat worms (Fasciola hepatica); worms (round) as hookworm
Hydra, cross section
Leaf, cross and longitudinal sections of dicot and monocot
Mitosis
Muscle, human: cardiac; smooth; striated
Nerve cells
Root tips showing mitosis
Roots, cross and longitudinal sections of dicot and monocot
Skin, human and frog, stained
Stem, longitudinal and cross sections of dicot and monocot

Models

Models listed may be purchased from biological supply houses, or they may be made by students. Construction of these, with careful attention to details, would make excellent projects for many students. Directions for making models are given in Chapter 8, Figure 8.17. They are given as a step-by-step process.

Amoeba
Anatomical human model of trunk with head
Cells, animal and plant
Earthworm, internal and external
Euglena
Frog
Human: brain (also frog, fish, dog); ear; eye; heart; kidney; larynx; lungs; reproductive systems; skin
Hydra
Mitosis
Oogenesis and spermatogenesis
Paramecium
Planaria
Tapeworm

Photographic and Projection Equipment

Cameras:
  35mm single lens reflex
  Polaroid
  Super 8 movie
Camera accessories:
  Close-up lens
  Telephoto lens
  Microscope adapter
  Light meter
Electronic flash
Movie lights
Projectors:
  Slide projector, remote controlled
  Filmstrip projector
  Overhead projector
  Opaque projector
  Super 8 film projector
Film loop projector
Screens
Darkroom:
  Enlarger
  Developing trays
  Timer
  Paper safe
  Paper cutter
  Developing tank

Skeletons

Both Turtox and Ward's leaflets describe how to prepare skeletons for display. These are free to biology teachers. Preparation of such skeletons may be an interesting student project. Skeletons of some of the smaller animals may also be stained with alizarin red S, then left for display in glycerine.

Bird
Fish
Frog
Rabbit or cat
Snake

Stains

Alizarin red S
Azure blue
Carmine
Cresyl blue, brilliant
Congo red
Eosin
Fast green
Giemsa stain
Gram's stain
Litmus solution
Methylene blue
Neutral red
Nigrosin
Phenolphthalein solution
Safranin O
Sudan black B
Wright's blood stain

FIELD TRIPS

A TEACHER may feel limited in the number and kinds of field trips on which he can take his students, so he needs to choose trips carefully with the idea of giving students maximum benefits from each trip taken.

Included here is a list of 42 trips with suggestions as to what could be studied. There may be many opportunities for other trips in your community. In planning your biology program for the year, explore all possibilities in your area and select the ones that seem most appropriate for your students.

1. Biological supply house—how biological materials are prepared; amount and variety of specimens; job possibilities for interested biology students.

2. Greenhouse (commercial or college)—practices used for better flower growth and production; variation in plants.

3. Botanical observatory—variation in flowers; special adaptations; great variety of plants.

4. Fish hatchery—stripping of fish; tanks of same age fish; culture methods.

5. Science fair—projects of biology students from other areas.

6. A zoo—unfamiliar animals and their habits; unusual coloring.

7. A farm—animals and farm practices such as contour plowing, conservation practices, and special features of the particular farm.

8. Food processing plant—raw material to finished products.

9. Milk processing plant—raw milk to completed dairy products.

10. Museum—wealth of material from mammals to birds, insects, and other invertebrates and vertebrates; species now extinct.

11. Tree nursery—variety and kinds of trees for display; best kinds of trees and shrubs to plant (may go with idea of landscaping the school yard).

12. Serum company—what goes into the making of serum, and animals used; need for chemists; research areas; products produced.

13. Dairy farm—care and feeding of animals; milking machines; care of milk.

14. Wildlife preserve—habitats of mammals and birds; feeding; nesting of birds; care of stray animals.

15. Poultry hatchery—care of young; practices to produce early fryers and early layers.

16. Clinic (medical, veterinary)—ideas for future type of work some students may be interested in following (take small group).

17. Crime laboratory—weapons used against criminals; methods used to trace and detect criminals.

18. Cannery—sanitation practices needed for protecting health of individuals.

19. Lumber yard—various cuts of wood; wood grain of lumber made from different species of trees.

20. Mink or other animal farm (beaver, fox, frog)—care given animals; type of food; breeding programs; method of "harvesting" the pelt and other usable portions of the animal.

21. Fossil area—stream bed; area cut through hill; river bed (collect fossils for class study).

22. Ice-cream plant—manufacture and packaging of ice-cream products.

23. Mental institutions—care given patients; responses.

24. Camping trip—ecology of region; flowering plants; fungi; small mammals; insects; trees; other things of local interest.

25. Marine aquarium—habits; feeding; habitat; needs of various species.

26. Cheese factory—processes involved in making various cheeses.

27. Film-processing plant—black and white pictures and colored ones.

28. Sewage disposal plant—processes used by a city in getting rid of wastes.

29. Cavern or cave—flora, fauna, and fossils found within it.

30. Insectary (college and university)—type collections of insects; research projects; career possibilities.

31. Food experiment stations—new foods; food values; acceptance and usage by general public.

32. Erosion areas—how much of the land is being taken away each year; what measures could retain the soil; kinds of soil present.

33. Agricultural experiment farm—different kinds of crops (grains, fruits, vegetables, trees) being tested on different soils; use of sprays.

34. City department of health—facilities for caring for the health of the city's population; sanitary measures taken to prevent disease.

35. Seed laboratory—sorting and grading of commercial seeds; removal of weed seeds; testing seeds for viability.

36. Genetics laboratory (college or university)—results of many crosses of both plant and animal materials.

37. Paint factory—manufacture of

paints and paint pigments; kinds of bases used to get certain colors.

38. Bird sanctuary (area protected for birds where they may be fed or just protected from harm by man)—feeding; nesting; protection of young; aggressiveness; variation in color of same species.

39. Apiary—bees at work; artificial insemination of young queen.

40. Water purifying plant—how a city purifies its drinking water.

41. Lectures or talks (college or university)—areas of general interest to the biology class.

42. Field habitats (plowed field, open meadow, peat bog, quaking bog, shore line of a river, lake region, pond area, stream, wooded area, open plain, prairie, railroad track)—identification of certain plant groups or animal groups; comparisons of flora and fauna of several areas; insects of each area; appearance of certain species compared with same species taken from a different habitat.

DETAILS ON WORKING OUT
USE OF THE CHI-SQUARE METHOD

WHEN the experimental results of scientists differ from the results they expect, an important question arises: How much is the observed difference a matter of accident? For example, suppose that in studying the cross between two kinds of tomato plants it was expected that half the offspring would have green leaves and half would have yellow. (This expectation is based on Mendel's laws and the supposition that a single pair of genes is responsible for the difference between green leaves and yellow leaves.) In one actual experiment it turned out that, of 1240 seedlings, there were 671 with green leaves and 569 with yellow leaves. This is clearly different from the 620 of each kind to be anticipated in a total of 1240. Is it a relatively important difference? Or is it a minor difference that is more a matter of chance than anything else? In this case 620 green-leaved plants were expected and 671 were counted, a difference of 51. Similarly the number of plants with yellow leaves, 569, differs by 51 from the expected 620.

The amount of difference can be expressed in several ways. One way is by expressing it as a percentage of the total: $51/1240 = 4.1\%$. A better way of working out a measurement of the amount of difference was invented by Karl Pearson in England in 1900. He called this measure of variation "chi-square." (It is indicated by the Greek letter χ with the square sign, χ^2.) Chi-square is found by squaring each difference between the number expected and the number observed and then dividing by the expected number in each class; and finally adding the quotients together. The difference for the green-leaved plants, 51, is squared, making 2601, and then divided by the expected number of 620. The quotient is 4.2. Again, the difference for the yellow-leaved plants, 51, is squared, to give 2601, and divided by 620, to give the quotient, 4.2. Added together, the two quotients come to 8.4, which is the value of chi-square.

$$\left. \begin{array}{llll} 671 - 620 = 51 & 51^2 = 2601 & 2601/620 = 4.2 \\ 620 - 569 = 51 & 51^2 = 2601 & 2601/620 = 4.2 \end{array} \right\} = 8.4 \text{ chi-square}$$

This does not tell us very much by itself. Fortunately, mathematicians have solved an elaborate distribution equation which gives us the information to judge whether any particular chi-square value represents the sort of differences which occur by chance alone very commonly, or by chance alone most uncommonly, or by chance alone at probabilities somewhere between. The following table which shows these relationships allows us to see about how often a given value of chi-square could have been produced *just* by chance.

	Chi-square Value (χ^2)							
	0.0002	0.004	0.016	0.455	1.074	2.706	3.841	6.635
The number of times in 100 that chance alone could have been responsible for the variation	99	95	90	50	30	10	5	1

$$\chi^2 = \text{sum of all } \frac{(\text{expected} - \text{observed})^2}{\text{expected}}$$

From the table we see that a chi-square figure of 8.4 goes beyond the table. This means that there is less than one chance in 100 by chance alone of getting as big a difference as this experiment actually showed. Experience has shown, and it is generally agreed, that when the probability of an event occurring by chance alone is as little as one or even five chances in 100, then that difference is said to be *significant* (that is, the differences from expectation are not occurring by chance alone, but are attributable to definite causes).

How consistently will repetitions of the same experiment produce differences from each other? If repeated crosses produce about the same differences from what was expected,

it is time to make a search for a suitable cause for the variation. (In the case of the tomato plants, further crosses did
show about the same results. The difference was caused by
a loss of yellow-leaved plants, constitutionally less sturdy than
the others. Fewer of them germinated and lived.)

A UNIT OF STUDY

A UNIT of study may be planned in much
the same way as the study for a day.
However, it does not need to be as specific as the plan for one day. In working out a unit of study on cells the instructor should list all the concepts
about cells he wishes the students to acquire, references to be used, materials
and visual aids to be presented, kinds of
discussion questions, experiments and
field trips, projects, and resource materials he may want to use. From this
large mass of ideas and suggestions the
instructor sifts materials to meet his
class needs and plans for as many days
or weeks as the unit will take. By adding to and subtracting from this material
from year to year he can keep the unit
functional and up-to-date.

LESSON PLANS

A LESSON PLAN is a guide to something
you wish to accomplish. It may be in the
form of a detailed plan with a budget of
time for all activities and parts to be accomplished or it may be informal notes
listed to help you remember all the
things you hope to cover and get across
to students in the allotted period for
class time. If you teach four or five
classes of biology each day, after about
the first three, it is difficult to remember what you have taught in each class.
With some sort of lesson plan you are
better prepared to know what you have
done each class period. Granted, no two
classes are the same, but there are cer-

tain concepts you hope to get across to
each group, and with some sort of plan
to get the job done you are more likely
to attain your goal.

In making out a plan for a day
these are some of the things you want to
think about:

1. What is the general subject you
   want to teach?
2. What goals do you want the class
   to attain?
3. What demonstrations or experiments do you have in mind?
4. How or by what method or methods will you approach the lesson?
5. How will you tie in the material
   with previous work?
6. How will you use educational media with the lesson?
7. How long will each part of the lesson take?
8. What sequence will you follow for
   materials to be used?
9. How will you evaluate the work?
10. What kind of assignment will you
    make?
11. How will the material tie in with
    what is to follow?

These and other questions may be
asked before you actually make the plan
you wish to follow. You may find after
teaching a class from the plan you make
that it might work better if you altered
it. By all means change it! There is
nothing sacred or inviolate about a lesson plan. Change it any way that seems
to make it more effective. The follow-

ing is a one-day lesson plan based on one teacher's plan.

TOPIC: Cells

OBJECTIVES: To find out what cells do.
             To learn about various kinds of cells.
             To appreciate the importance of cells.

MATERIALS   Film (protozoans to show
TO USE: different kinds of single cells), microscopes, slides, cover slips, eye droppers, cultures of protozoans, pond water, a projector, cheek cells, clean toothpicks, a chart showing different cell types from the human body (or pictures you have collected of types of cells, or a flannelgraph showing kinds of cells).

PROCEDURE: Show film of various kinds of protozoans. Discuss various shapes and kinds of protozoans and how each kind seems to get along in its environment.

             Make a slide of pond water or of special culture of protozoans and have each student note appearance of these cells. How does each move about? How many different shapes are there? Examine cells from the cheek through a microscope. Stain with iodine. Use a chart or a flannelgraph to show cells found in the human body.

DISCUSSION QUESTIONS:

1. About how many cells do you think are found in the human body?
2. What kinds of cells make up the human body?
3. What evidence do you have that cells are being replaced constantly in the human body?
4. How do broken bones and cuts heal?
5. What happens to the body if cells are not replaced?
6. What are some of the largest cells you know about? (Yolk of ostrich egg is a single cell.)
7. Where do cells get material for replacing themselves?
8. How do jobs of cells in green plants differ from those of cells in man?
9. Why can cells from man be seen more easily if stained than if not stained?
10. What is meant by specialization of cells? How does this apply to the cells that make up the human body?

ASSIGNMENT: Read the text material on the cell. Note how the organelles function and how the cell functions as a whole. Select one structure within a cell and find out all you can about it. Be prepared to contribute this information to the class. The list of structures is placed on the bulletin board. No more than two persons are to select the same part of the cell for the report.

BUILDING A TEACHING FILE

A FILE may consist of questions, problems, experiments, "attention-getters," data on interesting articles or pictures, or a variety of other teaching materials. If the contents of the file are well organized, they are exceedingly useful. To be of most value files need to be kept up-to-date.

In building and organizing a file of teaching experiments it is well to use cards of uniform size, preferably 5″ x 8″, which are large enough to permit drawings and explanations of experi-

216

ments. If each experiment follows the same general outline or pattern of title, objective(s), materials, method, procedure, observation(s), questions, conclusions, and drawing(s), or some general plan such as this, a most useful file may be developed and built into an integral part of the teacher's resource background. These experiments may be taken from the file or located easily if they are classified under headings which correspond to the teaching organization of the teacher. They may be performed by the teacher himself or handed to a student to set up and perform. A file of such resource material may provide opportunity for a student who is ahead of the others to perform experiments and develop initiative and dexterity in the use of equipment and in learning new processes.

Whenever you find some experiment which seems likely to prove good for a class, it is well to write it up right away, indicating source and location of the experiment. Then if you need to check the experiment later, it is a simple process to check the source and reread the material.

The following example illustrates the organization to follow in making cards for a teaching file.

EFFECTS OF TOBACCO ON CELLS—BIOLOGY

OBJECTIVE: To study the effects of tobacco on cells.

MATERIALS: Smoking tobacco (a cigarette, cigar, or small quantity of smoking tobacco which would fill a cigarette), glass of water, rich culture of protozoa, glass slide, cover slip, microscope.

PROCEDURE: Soak cigarettes or other forms of tobacco in a glass of water until the water is dark brown. Take a drop of the rich protozoan culture and place on the glass slide. Examine. Add a drop of the tobacco water. Examine both before and after the tobacco water has been added, using microscope.

OBSERVATIONS: What movement did you see among the protozoans before you added tobacco water? What movement did you observe just after and in the time following the addition of a drop of tobacco water? Was the movement changed at all? If so, in what way?

QUESTIONS: What changes do cells undergo when they come in contact with tobacco? Why do you think this may be so? What might cause smokers to have a cough? What other experiments could you devise to show the effects of tobacco?

CONCLUSIONS: Does tobacco appear to affect cells? If so, in what way?

PRACTICAL APPLICATIONS: Why do athletes refrain from smoking? What do the red blood corpuscles that move through lung capillaries and collect oxygen do when they come to smoke-filled lungs?

For building a file of useful and helpful magazines or other publications, the following information may be included on cards:

Name of publication
Address of publication
Cost of publication
How often it is published
Kind of materials found in it
Class in which you could use it
Ways in which the publication may be used

If a file of magazines or publications, such as is indicated here, is kept and classified into various topics or subjects studied in a certain class, it can be of great help in planning bulletin boards, sources of information, or even provide excellent material for student reports. It takes time to build a file, but once it is started, it grows rapidly into a useful and stimulating source of teaching material.

Still another type of material which may be used in teaching is a file of interesting sayings, questions, thought provokers, or stimulating ideas. They may be placed on the bulletin board and used as "attention-getters" or as material which you wish the class to note. Examples of this are:

Law is a form of order, and good law must necessarily mean good order.—Aristotle

Self-confidence is the first requisite of great undertakings.—Samuel Johnson

I will have to associate with myself as long as I live so I want to make myself the kind of a person that I can stand being around. —Barton Morgan

Drudgery is generally work that is badly done by an incompetent person.—Henry L. Doherty

By agreement small things grow, by discord the greatest go to pieces.—Sallust

A task is something to be done—not contemplated. The only work that counts is what you put behind you.—Anonymous

The human race is divided into two classes —those who go ahead and do something and those who sit and inquire, "Why wasn't it done the other way?"—Oliver Wendell Holmes

Genius is eternal patience.—Thomas Carlyle

Trifles make perfection, but perfection is no trifle.—Michelangelo

No man has had a liberal education unless he is in harmony with nature.—Huxley

INTRODUCING A NEW SUBJECT IN BIOLOGY

WITH the great variety of ways available for introducing new material, it is a good idea to select one that seems to be effective and that makes an interesting approach. The same approach does not have to be used in every class, nor does it necessarily need to be used year after year. Try out different ways of introducing the same topic and see which is most effective.

Consider a unit on the cell. The following list includes a number of ways that it can be introduced. Undoubtedly you can add many others to this list.

1. Show a film such as the twenty-minute one put out by the American Cancer Society, called *From One Cell*. It may be borrowed as needed from your nearest American Cancer Society office. It discusses some of the things that cells do. Most of the film is concerned with cell physiology. Very little is shown on cancer cells. It is an excellent film for introducing the study of cells.

2. Use a film, a filmstrip, or 35mm slides that show the variety of cells.

3. Show a film that points out the parts of a cell and how they function.

4. Prepare a quiz of fifteen or twenty-five questions and statements pertinent to cells to see how much students know about cells and what their ideas are concerning them.

5. Have students draw their conception of what a cell looks like and how its contents look. Ask them to list the reasons why they believe cells are important to them. Discuss their ideas and build on them.

6. Ask students to name every kind of cell they can. Discuss functions of these cells, how they are alike and how different. Draw some general conclusions about cells from this discussion.

7. Use a flannelgraph to build a cell. Place various parts of cells on the flannelgraph, using different colors for different parts of the cell. Discuss function of each part as you put it on the flannelgraph.

8. Make a list, with students' help, of internal cell structures. This can include compounds as well as structures. Develop the idea of the cell as a power-

ful unit of structure and of function.

9. Read an interesting passage from a source other than the text, pointing out certain features about cells.

10. Have students use microscopes to examine various kinds of cells such as epithelial cells from inside the mouth, *Elodea,* blood cells and corpuscles, and onion root-tip cells. Let them use available stained slides or prepared slides to get some ideas about size, shape, and appearance.

11. Use a model of a cell to convey the concept that cells are three dimensional.

12. Introduce parts of a cell by means of the chalkboard.

13. As a model for a cell, use a big box or a frame covered with cellophane, waxed paper, or tissue paper. With materials such as paper, rubber, or other suitable things, make parts that may be found within the cell and place them in the box. Use an inflated balloon for a nucleus.

14. Use an overhead projector with overlay transparencies to build the concept of the cell and what it contains. Discuss functions as parts are indicated.

15. Compare the classroom to a cell. Have each student describe part of the contents of the cell, relating the sizes of the organelles and the total size of the room.

16. Examine some cork under a microscope, then compare it with some living cells, such as protozoa. Discuss reasons why Hooke might have given the name "cell" to what he observed through the microscope. By comparing the cork and the living cells, have members of the class give some of their ideas as to what constitutes a cell.

17. List the life activities carried on by a human being. Discuss how they are carried out by a single cell such as an *Amoeba* or a *Paramecium.*

18. Have students examine some pond water or water from an aquarium that is rich in protozoans. Ask students to identify as many different forms of single cells (by shape and size) as possible.

19. Ask each student to write a description of a cell and why it is important. Use this as a basis for class discussion.

20. Have students make slides from epithelial cells lining the mouth. Scrape a few from inside the cheek and place on a clean glass slide. Add a drop of potassium iodide (very weak) to stain the cells. Add a cover slip and examine first under low then under higher power of the microscope. Note size, shape, and any definite parts you can see.

21. Examine a culture of *Amoeba* or *Paramecium* to observe structure and motion. Discuss ability of single cells to meet situations and how they manage to survive.

22. Compare small cells of bacteria with some of the largest cells (egg yolk).

23. Use a model of an animal cell and one of a plant cell to show how they are alike and how different.

24. Have each student look up all he can about one certain structure of a cell such as mitochondria, nucleolus, or ribosomes and report that information to the class.

25. List things the class knows or has heard about cells. Sort out the facts from the fiction.

26. Set up, or have students set up, a series of microscope slides of various tissues of the human body. Have everyone observe them, then discuss them in terms of how they function in the body.

27. Discuss what was known twenty-five years ago about cells with what is known now.

28. From your file of pictures, stories, notes, and other information and literature on cells make a bulletin board display of cell information. Give the class an opportunity to read this before discussing it.

29. Discuss parts of a cell such as *Paramecium* or *Amoeba* and decide what changes a free-swimming protozoan might undergo to become a parasitic form.

30. Hand out a mimeographed sheet with a picture of a cell, its parts, and functions of these parts. Use this as a basis for discussion of cells. Indicate whether all parts of a typical cell are necessary and functional in all cells.

Working

With

Protists

COLLECTING, CULTURING, AND USING PROTISTS[1]

LIVING THINGS in the biology room make teaching more interesting and realistic to both the students and the teacher. Processes that are talked about can be demonstrated. Activities of living things are better understood because they can be observed.

Some living things for your classroom can be collected in the field, then cultured and maintained in the laboratory. Others may be purchased from biological supply houses as they are needed. In many cases they may serve as a source of supply for culturing and maintaining certain species in the laboratory.

Protozoans. Several general considerations should be given protozoan cultures. Most of the free-living types used in the laboratory can be maintained at room temperatures between 18° and 24° C (about 65° and 75° F). Temperatures below 14° C (about 57° F) or above 28° C (about 82° F) are generally harmful to protozoans. They do not mul-

tiply at too cold a temperature, and they tend to disintegrate if the temperature is too warm. A pH of 7, or slightly higher, is desirable for best growth.

Distilled water, or boiled spring or pond water, is desirable for use rather than tap water, which may contain materials toxic to protozoans. If tap water is used, it should stand for several days in a glass or plastic container before being used.

Dishes having a large surface area in relation to volume are preferable to those that are narrow at the top.

Moderate light conditions are generally most desirable for maintaining protozoans. Direct sunlight should be avoided except for organisms containing chlorophyll.

Requirements that meet the needs of an adequate food supply, protection from enemies, suitable environment for individuals as well as for maintaining successive generations, and conditions supporting reproduction must be maintained for successful rearing of protozoans.

The needs of each species differ slightly, but general conditions are rather similar. Some of the specific needs are included in the discussion for individual species.

After a culture is started it can be maintained by adding food supplies as needed. Water may also be added from time to time.

In addition to collecting the different species, you may obtain them directly from a number of biological supply houses as pure or nearly pure cultures. These may be used for class studies or for starting cultures in the laboratory.

To subculture various protozoans a few drops of the old culture containing the desired specimens are placed in

---

[1] The term protist is used to include protozoans, algae, and bacteria.

broth or nutrient material in a clean dish and allowed to develop.

In time, food supply in any culture runs low. Add small amounts of food rather than an oversupply. Overfeeding is frequently the cause of a culture degenerating.

In collecting protozoans from a pond, scoop up scum along with floating and semifloating plants and place in quart jars partly filled with water, cover with a lid, and take to the laboratory.

Scoop plants and debris from the bottom of the pond and treat the same way.

In the laboratory, remove lids and place jars on a table or sill where they receive moderate, but not bright, light.

Some of the protozoans can be located immediately but others may take a day or two to become concentrated in a particular area of the jar.

Keep jars for a period of several weeks for observation. A succession of species, changing from day to day, may be observed. One kind that may be plentiful one day may be absent another day, succeeded by a different species.

Some conditions that cause the day-by-day changes are: the type and amount of decaying plant material; temperature; degree of sunlight; chemical constitution of the water; and the kinds of protozoans present.

Collect *Amoeba proteus* from decaying water plants, especially water lilies. Collect in freshwater ponds and streams that are rich in mineral content and low in organic matter. Place a piece of the plant in about 50 cc of pond water in a finger bowl and observe daily for *Amoeba*. If any are present they should appear within two weeks. Use a microscope to examine a few drops of sediment from the bottom of the finger bowl.

To provide a suitable medium for growing *Amoeba,* heat pond water to 70° C (158° F), cool, and add 4 grains of wheat to each 100 cc of the water. Inoculate with a few drops of fluid from a *Chilomonas* culture to provide food for *Amoeba.*

Cultivate in a dish with water less than one inch deep.

A day or two later, inoculate with *Amoeba.*

For best results in maintaining *Amoeba,* temperature of the water should be about 70° to 75° F, light should be diffused since direct sunlight is harmful to them, and food should be scarce. Finger bowls that can be stacked are excellent for growing *Amoeba.* Each bowl should contain about 75 to 100 cc of distilled water with a grain of wheat cut in two and a 2-inch piece of hay cut into $\frac{1}{4}$-inch lengths, or just 3 grains of boiled wheat per culture dish.

If *Amoeba* are to be placed in this medium, use a clean pipette and transfer them into it. After two weeks discard half the medium and add enough distilled water to bring the volume to about 100 cc. If the water does not look murky, add the same amount of food as used originally. If the medium looks too rich, reduce the feeding. *Amoeba* cultures need to be examined every two weeks. If mold develops on the hay or wheat, this does not appear to harm the *Amoeba.* Cultures should be ready for class use after a period of six weeks. They should be subcultured to keep the colonies going.

Collect *Paramecium* from the oozy debris at the bottom of old, permanent ponds. Keep the collection in a bowl with just a small amount of the pond water in it. The organisms become more abundant as the debris becomes more foul smelling.

221

A simple method of preparing a medium for culturing *Paramecium* is to fill several 4½-inch diameter finger bowls about two-thirds full of boiled pond water, and add 4 or 5 kernels of wheat or rice and 12 to 15 pieces of inch-long cooked timothy hay to the water.

Inoculate in five or six days with *Paramecium*. A few days after inoculation, the *Paramecium* may be seen concentrating near the surface of the water and around the food. Maximum concentration will occur within about two weeks. Food should again be added to maintain the culture. Such a culture can be kept pure for a period of about six weeks. Subcultures may be made from it.

It is preferable to obtain a culture of *Paramecium* that is relatively pure rather than to make a "hay infusion" from a mass of hay placed in water for a few days, since in the latter method the bacteria multiply so rapidly they frequently crowd out the *Paramecium*.

Paramecia are less difficult to culture than *Amoeba*. They grow at ordinary room temperature and in any type of glass jar, but finger bowls are especially convenient to use. However, use only clean dishes in culturing.

Euglena is very easy to culture and to collect. It requires plenty of food, but little attention. Tall jars such as battery jars are excellent for culturing *Euglena*. A window sill in direct sunlight, with temperature varying from normal to 90° F, is satisfactory. Boiled wheat, rice, and timothy make good food.

Euglena may be found in barnyard or pasture ponds containing considerable amounts of organic material, among algae of ponds, in green pond scum, and on green slime of decaying plants in ponds and streams. Lily ponds are good sources for collecting.

Fill a clean battery jar (6 x 8 inches) with distilled water. Add a large handful of timothy hay or about 100 grains of wheat or rice. Place culture on a window sill. After about a week add a few drops of *Euglena* culture which can be collected from stagnant pools. Within two weeks the water will appear green and the surface will be covered with a scum containing large numbers of *Euglena*. This culture should last at least four weeks. If food is added every two weeks it should last several months. To start new cultures, subculture and place in fresh media.

Euglena may encyst on the surface of the dish and be kept for months in this condition. Bring them out of encystment by adding hay infusion.

Stentor can be cultured in any type of dish such as finger bowls or battery jars. This protozoan likes rich cultures containing protozoans such as *Euplotes*, *Colpidium*, and *Chilomonas*. One must prepare media with these protozoans present.

A medium such as that prepared for *Paramecium* may be made, and then a culture containing protozoans, such as those indicated, added to it. After this has become well established, add the *Stentor* culture. Keep it in diffused light with a temperature of 65° to 75° F. Subculture from time to time.

Arcella vulgaris is found in the bottom ooze of many older shallow ponds. Specimens can be separated from the ooze and other protozoans if one examines the ooze through a binocular microscope and picks out individual specimens with a fine pipette. Place in a watch glass for observation. It is slow moving and easily studied. *Arcella* is also collected from the bottom of many old cultures of protozoans and from aquatic plants.

In preparing a medium for *Arcella,* heat pond water as for *Amoeba,* cool, and add 2 grains of wheat and ½ g of hay. Inoculate with *Chilomonas,* and in two or three days add *Arcella. Arcella* will grow by feeding on the infusion alone, but if *Chilomonas* are available, *Arcella* seems to grow more rapidly. *Arcella* may be collected from the bottom of many old cultures or in the bottom ooze of any shallow pond. It may be isolated from the ooze with a fine pipette and placed in the culture in a shallow dish.

Chilomonas may be obtained from a pond when decomposing vegetation is present, especially if the pond water is permitted to stand in the laboratory for a few days with a grain or two of wheat added to it.

To isolate *Chilomonas* use a fine pipette to remove specimens and place in pond water that has been previously heated to 70° C (158° F), cooled, and has had 4 or 5 grains of wheat added to it. Within a week the water appears cloudy with *Chilomonas* which serve as a valuable food source for many larger protozoans.

Cysts of *Chilomonas* seem to be ever present in the air, and if culture jars are left uncovered, spontaneous inoculation of *Chilomonas* occurs.

Colpidium, Euplotes, and *Halteria* are common inhabitants in ponds. A simple medium may be prepared for these protozoans by filling 4½-inch diameter finger bowls about two-thirds full of boiled pond water and adding 4 kernels of boiled wheat or rice and 12 to 15 inch-long pieces of cooked timothy to each bowl. Any of the above-mentioned species may be inoculated directly into the medium after it has cooled.

Colpoda in the active stage usually occur early in cultures of moist soil. Later on they encyst in the culture.

Isolate active *Colpoda* from the soil cultures by using a fine pipette. Place them in culture dishes containing an ordinary hay infusion that has been made by steeping a handful of hay, preferably timothy, in hot water for an hour or more, then straining, cooling, and placing it in culture dishes.

After a day or two, allow the culture fluid in the watch glass to evaporate by exposing it to the air. During this slow drying period the animals encyst.

Leave the dry culture exposed for a day or two and then add new hay infusion. The animals have divided within the cysts in the meantime. They revive and are present in greater abundance than before. The drying and reviving process may be repeated until a concentrated culture of the organism is obtained.

Pipette the concentrated culture into a petri dish containing a piece of ordinary filter paper that is cut to exactly cover the bottom of the dish, and moisten the culture with hay infusion. Leave the petri dish uncovered so the moisture will evaporate slowly. When dry, cut the filter paper containing the encysted organisms into small pieces and keep indefinitely to start new cultures when needed.

To start new cultures from the specimens on the filter paper, place pieces of the filter paper in watch glasses or culture dishes and add hay infusion. The protozoans revive in a short time and new cultures may be obtained.

Didinium is a carnivorous protozoan that lives mainly on *Paramecium.* To locate specimens, examine pools and streams containing considerable amounts of decaying organic matter, including small amounts of sewage. Collect sediment, submerged decaying leaves and plant stems, as well as water from the edges of the pool, pond, or stream. Generally *Didinium* is collected in the encysted condition with cysts lying free in the sediment or attached to submerged objects.

To activate the cysts and stimulate rapid multiplication of active specimens, add strong *Paramecium* cultures, preferably those that have been grown in timothy hay infusions.

The food supply is depleted rapidly as four or five generations of *Didinium* are usually produced in each 24-hour period at room temperatures. At least two paramecia are needed for each *Didinium* to reach full growth before dividing. At higher temperatures than 21° C (70° F) growth is exceedingly rapid and as many as nine generations may be produced within a 24-hour period.

If *Didinium* colonies are to be maintained in the classroom, it is necessary to keep well-fed, thriving cultures of *Paramecium* going, as *Didinium* will not grow well on underfed paramecia. They seem to lose the ability to encyst, suffer a reduced division rate, possess structural abnormalities, and die within two to three weeks. When a culture of *Paramecium* is at a peak, introduce *Didinium*. *Didinium* will exhaust the food in a week to ten days.

Difflugia is found in fresh water, especially where clean sand is present. When examining water for *Difflugia* it may often be overlooked on a slide because superficially it resembles a pile of sand grains.

Cultures of *Blepharisma,* a slow-moving peach-colored protozoan, should be introduced into shallow culture dishes. This species needs more boiled wheat than the amount used for *Amoeba.* After a culture is started, inoc-

ulate with *Blepharisma* in about two weeks. They move slowly along the bottom of the dish and may be subcultured every four to six weeks.

Stylonychia may be inoculated into a hay infusion consisting of one part rice stalks to four parts of pond water. Wheat or hay-wheat infusion may also be used. In 24 hours add a pinch of dried yeast and inoculate with *Chilomonas.* After three more days inoculate with *Stylonychia.* Protozoans such as *Colpidium, Euplotes,* and *Halteria* are found in pond collections and may be grown in pure cultures in the laboratory. The culture medium is such as that suggested for culturing *Paramecium.* When introduced into the medium, these protozoans develop rapidly, and within a week will have increased greatly and may be found in the surface scum and the upper half of the culture. They should be cultured about every third week.

Actinosphaerium may be found in permanent freshwater ponds. Place specimens in boiled pond water that contains 4 grains of boiled wheat or rice per 100 cc. Cultures should result within a period of two weeks. Best results are obtained if paramecia or other ciliates are added to the culture.

Actinophrys feeds on small ciliates. To make a hay infusion, cut up some stems of hay into about 3-mm lengths. Put 10 g into 1000 cc of pond water and boil for 10 minutes. Let stand uncovered for 24 to 48 hours. Bacteria are provided a good source of food. Add *Chilomonas* to the medium as food for *Actinophrys.* Subculture about once every three or four weeks.

Centropyxis may be cultured successfully in boiled pond water to which has been added 2 (3 cm) stems of previously boiled timothy hay and 3 wheat grains. *Centropyxis* may be added as soon as the water is cooled to room temperature.

Spirostomum is obtained from habitats where decaying vegetation is present as well as where rich supplies of protozoans such as *Chilomonas* are abundant. Cultures containing 5 grains of wheat and several 2-inch lengths of timothy stems boiled in 100 cc of water with *Chilomonas* added in a day or two to provide food usually give good results. If the bottom of the culture dish contains old, thoroughly washed sphagnum moss the multiplying *Spirostomum* may be seen as white spots on the moss. Place individual specimens on a glass slide for observing *Spirostomum*.

Stentor thrives in protozoan cultures that are rich in forms such as *Euplotes, Colpidium,* and *Chilomonas.* It may be collected from aquatic plants, and sometimes may be found in gelatinous masses on leaves and roots of the water plants.

A hay infusion such as the one described for *Colpoda* provides a suitable medium. When the culture is ready for inoculation add mixed protozoans. In a few days it should be well established and *Stentor* may be added.

Vorticella attach to aquatic plants, decaying vegetation, and sides and bottom of culture dishes. They may be introduced directly into a well-balanced, thriving aquarium where they will multiply and live for months.

To prepare a medium into which *Vorticella* may be introduced, add 4 to 6 grains of boiled wheat to 200 cc of boiled pond water, cool, and introduce *Vorticella* at room temperature.

USING PROTOZOANS IN THE LABORATORY

Euplotes, Colpidium, and *Chilomonas* may be reared in large quantities as food for *Spirostomum* and *Stentor. Paramecium* is food for *Didinium.*

All protozoan species may be observed to determine means of obtaining food.

Nutritional needs may be studied. Vary the diet. Add to the medium different foods such as amino acids, carbohydrates, and proteins, and observe growth, lack of growth, or malformation of species. Use one culture as a check, and for a similar one change the habitat. For a carbohydrate source, sugar or powdered malted milk may be sprinkled over the surface; for protein, use powdered milk.

Observe *Didinium* ingesting *Paramecium.*

Study the kind of motion peculiar to the various protozoans as they move. Slow the motion by adding a drop of cellulose acetate to the slide, or a few strands of cotton.

Watch *Arcella* turn over. Place it on its "back" (top of shell). Since its pseudopodia are not long enough to turn it over, it produces a gas bubble that raises the edge of its shell and allows a pseudopodium to reach the substrate and turn it over.

Observe cysts of *Euglena* and *Colpoda.*

Shade part of a slide and note whether various species are positively or negatively phototropic.

To observe pellicle pattern, flagella, and cilia of various protozoans, place a drop of 10% aqueous nigrosin (10 drops nigrosin per 100 drops distilled water) in a drop of protozoan culture on a slide. Spread, then permit the slide to dry. Observe under a microscope for patterns indicated.

Add a drop of black or blue water-soluble ink to a drop of *Paramecium* on a slide and note how the trichocysts discharge.

Observe streaming of protoplasm and formation of pseudopodia in *Amoeba*. Cut the amount of light you use to a minimum for best observance.

Obtain two strains of *Paramecium aurelia* (most easily secured from a biological supply house) and observe conjugation of the two strains. (Directions for observations come with the purchase of the two strains.)

Note the functioning of contractile vacuole or vacuoles in various species.

Determine size of different species of protozoans. See Appendix G 319 for measuring with the microscope.

An ordinary student microscope has a low power lens magnification of ten times ($10\times$) and ocular magnification of ten times ($10\times$). Total magnification is 10×10 or 100 times ($100\times$). The field diameter of such a microscope is about 1600 microns (1600 μ). One micron (μ) is $\dfrac{1}{1000}$ of a millimeter or $\dfrac{1}{25,000}$ of an inch.

Therefore, an animal that is about one-half the length of the diameter of the low power field is about 800 microns (μ) long. If one-fourth the length, it is about 400 microns (μ) long, etc.

Using a standard high power objective of $43\times$ and an ocular of $10\times$, the diameter of the field is about 372 μ (1600 μ $\dfrac{43}{10} = 372$ μ). Objects measuring half the diameter of this field are about 186 μ; those one-fourth the length are about 93 μ.

The following references are recommended for further suggestions in collecting and culturing protozoans, for identifying them, and for preparing special media:

Jahn, T. L., and Jahn, Frances Floed. 1949. *How To Know the Protozoa*. Wm. C. Brown Co., Dubuque, Iowa.

Lutz, Frank E., Welch, Paul L., Galtsoff, Paul S., and Needham, James G. 1959. *Culture Methods of Invertebrate Animals*. Dover Publications, Inc., New York.

Schwab, Joseph J. 1963. *Biology Teachers Handbook*. John Wiley & Sons, Inc., New York.

Teachers Outline Series. "BICO Invertebrate Cultures." National Biological Supply Co., Inc., 2325 So. Michigan Ave., Chicago, Ill. 60616.

Turtox Service Leaflet Number 4. "The Care of Protozoan Cultures in the Laboratory." General Biological Supply House, Inc., 8200 So. Hoyne Ave., Chicago, Ill. 60620.

Ward, Henry Baldwin, and Whipple, George Chandler. 1945. *Fresh-Water Biology*. John Wiley & Sons, Inc., New York.

TETRAHYMENA PYRIFORMIS IN THE HIGH SCHOOL BIOLOGY COURSE

Tetrahymena is a small ciliate that is easily observed with the aid of a microscope. It embodies the most primitive type of organization found in the Hymenostomatida (this group also includes the *Paramecium*). It was the first animal ever to be maintained axenically, that is, in a sterile medium completely free from all other organisms, and its usefulness to biologists is reflected in literature estimated at over 1,000 contributions. To the physiologist it is important as an experimental animal used in nutritional studies, especially in regard to biochemical pathways; because of its mating types it is used extensively by the geneticists.

Tetrahymena pyriformis is widely

Prepared by Dr. Benton W. Buttrey of the Department of Zoology and Entomology at Iowa State University and presented at the 1967 Science Teachers Short Course at Iowa State University.

distributed in Europe and America, in fresh water, ponds, streams, and stagnant pools, or in thermal springs. It also occurs in brackish water. It is found in the soil and is coprophilous (living in feces). Some strains may at times become parasitic on various animals (especially worms) but the parasitic existence is not compulsory.

THE ORGANISM

The body of this ciliated protozoan is plastic, constantly changing in shape in response to environmental and local pressures, and even a pure culture may give the impression of consisting of several species. The animal is longer than broad, averaging 50 x 30 μ, rounded posteriorly, pointed anteriorly. The shape varies from that of a pear to that of an egg or even a cucumber. The holotrichous cilia are arranged in meridians, closer together anteriorly than elsewhere. The mouth cavity lies anteriorly in the midventral line. This funnel-shaped cavity sinks inward and narrows to a cytostome. Surrounding the buccal area is a typical undulating membrane which is often seen in a living animal.

The endoplasm is homogeneous in axenic cultures, stuffed with food vacuoles in those containing bacteria. There is a single large contractile vacuole posteriorly and slightly to the right of the midline. Feces are discharged at a permanent posteriorly located cytopyge, but this is difficult to distinguish except during defecation.

The macronucleus is round or oval, about 9 x 11 μ. A micronucleus is present; however, a peculiarity of the species is the widespread occurrence of amicronucleate strains, which of course cannot reproduce sexually.

There are no division cysts, and asexual reproduction occurs in the free stage. There are probably no resistant cysts.

CULTURING IN THE CLASSROOM

T. pyriformis is readily grown in a variety of simple media of which the most obvious is hay infusion.

Better results are derived from axenic cultures and these are easily maintained. A suggested medium is 1% to 2% proteose-peptone solution. To make this put 1 or 2 g Difco proteose-peptone in 100 ml distilled water and boil for 5 minutes. Put 10 ml into each of several test tubes. Sterilize by autoclaving or using a pressure cooker. If equipment for sterilizing is not available you can sterilize the media and glassware by boiling for 45 minutes.

Proteose-peptone can be obtained from any biological supply house and costs approximately $2.70 for 4 ounces.

Cultures may be maintained at room temperature (72° F; 22° C), and at this temperature should be subcultured about once a week. At a higher tempera-

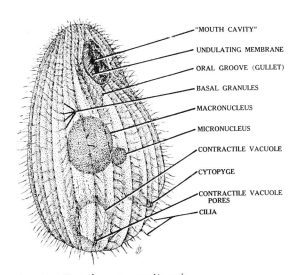

Fig. B.1—*Tetrahymena pyriformis.*

ture (81° F; 28° C) they will reproduce at a faster rate but you may have to subculture them oftener. At a cooler temperature (58° F; 14° C) they do not divide as fast, and thus the cultures will last for a longer period of time without the need for subculturing. In subculturing transfer only a very small drop or use a platinum bacteriological transfer loop.

HOW TO USE IN THE CLASSROOM

This organism is an excellent protozoan to supplement your work on unicellular animals. Study it in the living state as an example of a rather typical ciliate which is quite similar to the pathogenic form *Balantidium* that is normally found in the pig's caecum. By careful observation ciliary action and motion can be observed.

The nucleus can be quickly stained by adding a drop of 1% methyl green (1 g in 100 ml distilled water) to 3 drops of the culture on a slide.

Experimental growth studies could be attempted by growing the cultures at various temperatures, in different media, and at various pH concentrations.

Studies on size and shape could be correlated with all experimental growth studies.

If you are successful in growing this culture it is possible to get mating types in *Tetrahymena*, and conjugation could easily be demonstrated. *Tetrahymena* is one of the very best ciliates for showing conjugation.

USING SIMPLE STAINS FOR *TETRAHYMENA* AND OTHER PROTOZOA

TURTOX CMC-S

Turtox CMC-S is a hard-drying stain-mountant that kills, fixes, clears, stains, and mounts the specimen in just one operation.[1] It stains large protozoa such as *Paramecium* and *Tetrahymena*, and it is very suitable for mounting parasitic worms and ova, free-living nematodes, smaller arthropods or portions of larger ones, small Diptera, lice, mites, hairs, and the like. It is not recommended for tissue sections.

Procedure

Place live specimens in the medium and mix the stain and specimens together on the slide. Immediately, place a cover slip on the mixture. Specimens may also be mounted from alcohol or preservative if they are previously soaked in water 10 to 20 minutes. Specimens containing considerable air are preferably killed in hot water or alcohol and kept in fluid until the air has been eliminated.

CMC-S hardens sufficiently after several hours at room temperature to anchor the edge of the cover slip. The staining action is generally complete after 24 hours. If the microscope slide preparation is to be kept for a long period of time (more than 1 month), it is advisable to ring the cover slip several times on the following day with any quick-drying ringing lacquer (fingernail polish works well).

When thickening occurs in the stock CMC-S from evaporation, thinning is accomplished quite simply by heating the medium in a water bath and stirring in enough distilled water to obtain the desired consistency.

NIGROSIN RELIEF STAIN

The nigrosin relief stain is an ex-

---

[1] Turtox CMC-S can be obtained from the General Biological Supply House, 8200 So. Hoyne Ave., Chicago, Ill. 60620.

cellent stain to show the general outline and surface pattern of *Paramecium* and *Tetrahymena* and other ciliates. The arrangement of the cilia and mouth areas are clearly revealed.

Technique of Preparation

Dissolve 10 g nigrosin in 100 cc distilled water. Nigrosin goes into solution very slowly. Stir and shake over a period of a day or two.

Procedure

A drop of culture containing numerous *Paramecium* or *Tetrahymena* is mixed with a drop of the 10% nigrosin stain. It is then spread out very thinly and allowed to dry. After drying it is best to cover with any mounting medium (such as Kleermount or Permount) and a cover slip.

BRILLIANT CRESYL BLUE SIMPLE STAIN FOR PROTOZOA

Here is a simple, rapid technique using a single solution which is ideal for protozoa in culture such as *Paramecium* and *Tetrahymena*. It is also ideal for human feces to aid in detecting cysts and trophozoites of intestinal protozoa. Another excellent source of protozoa is the large intestine of the frog. This stain works well with this raw material in revealing the various ciliates, flagellates, and amoebae one finds in the frog's large intestine.

FORMULA FOR STAINING SOLUTION:

Brilliant cresyl blue, certified biological (NF)	2	g
Methanol	75	ml
Formaldehyde (USP)	5	ml
Acetic acid, glacial (USP)	2.5	ml
Phenol (USP)	2.5	ml
Distilled water	15	ml

Technique of Preparation

Dissolve 2 g of the dye in 75 ml of methanol. To this solution add the formaldehyde, glacial acetic acid, phenol, and distilled water. After the solution is thoroughly mixed it is ready for immediate use. This solution is stable and may be kept for long periods without deterioration if stored in tightly stoppered brown-glass containers.

Procedure

Mix a small portion of the material to be stained in a drop of the staining solution on a clean slide until a smooth emulsion is obtained. A cover glass is then applied. Immediate microscopic examination may be performed.

The trophozoites and cysts stand out clearly against a dark blue background. The nuclei of intestinal amoebae are distinct. The cytoplasm is pale blue, and the chromatin granules of the nuclear ring are blue-black.

Permanent wet preparations may be made by sealing the cover slip with a ringing lacquer (or fingernail polish). Specimens preserved in this manner show little distortion and no apparent depreciation of the diagnostic features even after a year or two.

HEIDENHAIN'S IRON HEMATOXYLIN

For general purposes, iron hematoxylin is the most important stain for preparation of protozoa which are to be critically studied. It is particularly good for nuclear details.

While the results are superior to many stains, the overall procedure is very time-consuming.

Technique of Preparation of the Reagents

SCHAUDINN FLUID

Saturated mercuric chloride in water	100 ml
95% or 100% alcohol	50 ml

Above stock Schaudinn fluid is kept without acetic acid. At the time of its use, add 1 ml glacial acetic acid to each 19 ml Schaudinn fluid.

Warning: Schaudinn fluid is corrosive to metal instruments and jewelry.

NISSENBAUM FLUID

Saturated solution of mercuric chloride	10 ml
Acetic acid, glacial	2 ml
Formalin	2 ml
Tertiary butyl alcohol C.P.	5 ml

The above reagents are mixed just before using.

HEMATOXYLIN STAIN

Two methods commonly used to prepare hematoxylin stain are:

(1)
Hematoxylin, certified	0.5 g
Distilled water	100.0 ml

Mix and allow to ripen in bright light for 6 weeks. It then remains good for approximately 3 months.

(2)
Hematoxylin, certified	5 g
95% or 100% alcohol	50 ml

Mix and allow to ripen in bright light for 4 to 6 months. This reagent keeps indefinitely. When staining protozoa, mix 1 ml of this reagent with 19 ml of distilled water.

IRON ALUM

Ferric ammonium sulfate	4 g
Distilled water	100 ml

Use fresh purple crystals of ferric ammonium sulfate and dissolve without heat.

Procedure

1. Secure raw material containing protozoa and add enough saline solution so that the mixture spreads evenly.
2. a. Schaudinn fixative method:
   (1) Spread a thin, even film containing the organisms on a cover glass. Avoid too much fluid, and avoid drying.
   (2) Carefully drop the cover glass smear side downward onto the surface of the fixing fluid in a petri dish. Allow it to float for one minute (approximately) and then carefully turn it smear side upward and sink it to the bottom of the fixing fluid. Fix a total of 5 minutes for Schaudinn fluid.
   (3) Place in 50% alcohol 5 minutes.
   (4) Leave in 70% alcohol at least 12 hours. Smears may remain in 70% alcohol indefinitely.
   b. Nissenbaum fixative method:
   (1) Place a large drop of culture on a slide (the droplet must be free of debris).
   (2) Pick up a pipette full of fixative and hold the tip about 2 cm above the slide.
   (3) Slowly drop the fixative directly onto the droplet of culture.
   (4) Once the currents have started, lower the tip of the pipette until it is in contact with the slide. Continue to expel the fixative slowly until the slide is flooded.
   (5) After 15 seconds drain the slide and place in 70% iodine alcohol 3 to 5 minutes.
   (6) Wash in 70% alcohol at least 12 hours. Smears may remain in 70% alcohol indefinitely.

When ready to stain the material, follow this procedure:
3. Place in:
   a. 50% alcohol 1 to 5 minutes.
   b. 30% alcohol 1 to 5 minutes.
   c. Distilled water 1 to 5 minutes.
4. Mordant in 4% iron alum 6 to 18 hours.
5. Rinse in distilled water with 2

rinses of 15 seconds each.

6. Put in hematoxylin (0.5%) for 12 to 24 hours.

7. Remove slides and wash off excess stain with water.

8. Differentiate in 2% iron alum. Check this destaining carefully under a microscope but never permit the slide to become dry.

9. Leave in:
   a. Running tap water 1 hour or longer.
   b. 10% alcohol 1 to 5 minutes.
   c. 30% alcohol 1 to 5 minutes.
   d. 50% alcohol 1 to 5 minutes.
   e. 70% alcohol 1 to 5 minutes.
   f. 85% alcohol 5 minutes.
   g. 95% alcohol 5 to 10 minutes.
   h. 100% alcohol 5 to 10 minutes.
   i. Xylol 5 to 10 minutes.

10. Place in mounting medium with a cover slip over the material.

DIATOMS

THIS MATERIAL represents an introduction to the study of diatoms and a technique for making slides of them.

Confusion of diatoms with desmids is unnecessary since desmids are green while diatoms have a yellowish-brown color. The cell walls of diatoms consist of two overlapping halves which fit together like the parts of a petri dish. These walls are "glasslike" and are variously ornamented with rows of dots and ribs. By way of contrast the two symmetrical "half-cells" of desmids do not overlap and are not glasslike. The siliceous walls of diatoms do not decay as the protoplasm does after death and thus may accumulate on the bottoms of deep

Prepared by Dr. John D. Dodd of the Department of Botany and Plant Pathology at Iowa State University.

lakes and oceans, forming deposits known as diatomaceous earth.

Looking at a diatom from the top means seeing it in *valve view*, while looking at it from the side means seeing it from the *girdle view*. These two views are so different that one is not at first aware they are examples of the same species. When cell division takes place the cell swells somewhat, forcing the valves slightly apart. Cytoplasmic division takes place in a plane parallel to the valves and new walls are formed inside the old ones. One of the new cells is the same size as the parent but the other is usually slightly smaller. Thus, after many divisions, the population of a given species will consist of individuals of a wide range of sizes.

IDENTIFICATION OF DIATOMS

Since the classification of diatoms is based in large part on the shape, symmetry, and markings of the cell wall, it follows that identifications are not feasible unless the wall can be studied carefully. Some of the terms used in descriptions of diatoms are these:

PUNCTAE—Minute pores in the wall, appearing as dots.

STRIAE—Fine lines in the wall which may be resolved as rows of punctae.

COSTAE—Rafter-like thickenings of the wall which appear as coarse lines across the valve surface.

RAPHE—A longitudinal slit in a valve surface which appears as a fine line. It may be continuous or interrupted in the middle.

PSEUDORAPHE—An apparent longitudinal line in the valve surface representing the clear space between parallel rows of striae.

CENTRAL NODULE—A thickening in the

central area of a valve, occurring where the raphe is interrupted.

POLAR NODULES—Thickenings in the valve at the ends of a raphe.

KEEL—A ridge of the valve surface occurring along the edge. Usually the raphe is associated with the keel in such cases. When the keel is not pronounced the raphe may appear to be in a canal, the so-called canal raphe.

TRANSVERSE SYMMETRY—Exists if the surface being viewed (valve or girdle) may be cut in two transversely, forming symmetrical halves.

LONGITUDINAL SYMMETRY—Exists if the surface being viewed (valve or girdle) may be cut in two lengthwise, forming two symmetrical halves.

FRUSTULE—Term applied to the entire cell wall of a diatom.

STAUROS—A laterally expanded central nodule which reaches, or approximates, the edges of a valve.

INTERCALARY BAND—A replica of the girdle band occurring within and parallel to the girdle band.

GIRDLE BAND—A straplike, open-ended band attached to the valve and forming the side of the frustule seen in girdle view.

VALVE MANTLE—The edge of the valve surface where it curves downward.

SEPTUM—A partition extending across the cell in a plane parallel to the valve. It often has perforations of varying size.

The first step in preparing diatoms for identification is to "clean" the material of organic matter so that only the glasslike walls are left. This has been done traditionally by boiling in 50% nitric acid (under a hood) to which a pinch of potassium dichromate has been added. However, in recent years, another technique has become popular (Van der Werff's method). This involves draining off as much of the water as possible, placing the material in a 1000-ml beaker, and adding enough 30% hydrogen peroxide to form a layer about ½ inch deep in the beaker. A small amount (tip of a knife) of potassium dichromate is added. In a few moments an exothermic reaction begins, resulting in a violent boiling of the mixture. (A glass rod should be kept handy during this reaction to "beat it down" and keep it from boiling out of the beaker.)

Upon completion of this aqueous combustion reaction distilled water is added, the material is mixed and allowed to settle for at least three hours. The diatomaceous material will settle to the bottom slowly and form a delicate, flocculent layer which is easily disturbed. Decanting should be done carefully in order not to lose the sediments. The washing process should be done three times and the cleaned diatoms may then be poured into a storage vial. Enough alcohol should be added to make at least a 30% solution to inhibit growth of fungi.

In order to make slides, a drop of water containing suspended diatoms is placed on a clean cover slip and allowed to air dry at room temperature or, perhaps, on a low temperature hot plate. The choice of mounting medium is important as the wall markings do not become clear unless the refractive index is suitable. We use a resinous material dissolved in toluene which is sold as Hyrax (available from Arthur H. Thomas Co., Philadelphia). Another popular medium is Styrax.

When the cover slip is dry it is heated to 300° to 400° C and is then inverted into a drop of Hyrax on a slide. The slide is then heated for a few minutes in the same temperature range until the Hyrax stops bubbling under the cover slip. This allows time for penetration of the diatom frustules and evaporation of the solvent. The slide is allowed to cool but the cover slip should first be pressed down firmly so that it lies flat on the slide. The Hyrax hardens rapidly and the excess along the edges can be scraped off with a knife or razor blade. The mounts are permanent.

Some identifications are possible with high dry objectives, but for critical work oil immersion lenses of high quality are needed and the light source should also be of high quality.

Of the currently available keys to genera, the one by Patrick in the 2nd edition of Ward and Whipple is perhaps the easiest to master.

Attempts to identify specimens to the species level will be generally frustrating except for a few common forms with distinctive characteristics, since several references, most of them in other languages, plus a great deal of experience are required.

To aid somewhat in learning to find your way about in a diatom key, the following breakdown and grouping of some of the more common genera in this area are presented. (The groupings used are not to be interpreted as formal taxons as most of them are merely for convenience here.)

Centric diatoms (recognized by radiate markings on valve surfaces)
  Melosira (cells joined by valve faces to form filaments)
  Cyclotella (markings on valves include ribs, or costae)
  Stephanodiscus (valve markings include puncta, but not costae, and spines are often present along margins)
Pennate diatoms (markings arranged with respect to a longitudinal line)

Group 1—no raphe on either valve
Group 2—true raphe on one valve, pseudoraphe on other
Group 3—only vestigial raphes present
Group 4—true raphes present on both valves
(The above distinctions are critical and must be made only with specimens having both valves intact.)

Group 1—includes *Tabellaria, Diatoma, Asterionella, Fragilaria, Synedra,* and *Meridion.*
Group 2—includes *Cocconeis, Achnanthes,* and *Rhoicosphenia.*
Group 3—the major genus is *Eunotia.*
Group 4—includes a large number of genera and needs to be broken down further, using such features as symmetry and position of raphe on valve:

A. Raphe median—frustules symmetrical longitudinally and transversely in both valve and girdle views.

   Navicula, Pinnularia, Caloneis, Diploneis, Amphipleura, Gyrosigma, Mastogloia, Stauroneis, Anomoeneis, and *Neidium*

Note: in *Pinnularia,* the apparent costae are actually tubules in the wall. Each tubule has a small internal opening near the valve margin and the openings in adjacent tubules are so arranged that the impression of a pair of longitudinal lines is created.

B. Raphe median—valves and girdles symmetrical longitudinally but asymmetrical transversely.

   Gomphonema

C. Raphes median or displaced somewhat but not in keels or canals. Valves transversely symmetric but longitudinally asymmetric.

   Cymbella, and *Amphora*

D. Raphes marginal—sometimes in canals, which may in certain cases be slightly elevated in keels. (This grouping is a convenient but most unnatural one.)

Epithemia, Rhopalodia, Nitzschia, and *Hantzschia*

E. Raphes marginal—inconspicuous, elevated, winglike keels.

Surirella, Cymatopleura, and *Campylodiscus*

STUDY OF LIVING PROTOZOA BY USE OF VITAL STAINS

WHILE permanent preparations of these one-celled forms are essential for an understanding of many facts of cell structure, a more dynamic appreciation can be gained by the study of living animals.

To facilitate such study, Ward's Natural Science Establishment, Inc., has developed a *Vital Stain Set* (Dv 16) that stains structures without killing organisms. This set supplies five of the most commonly used vital stains—Neutral Red, Janus Green B, Methylene Blue, Bismarck Brown, and Brilliant Cresyl Blue. In addition, copper acetate and copper sulfate solutions are provided for immobilizing the protozoa.

For these studies, the best quality microscope slides and No. 1 cover glasses are recommended. The thinness, optical clarity, and colorless quality of the best materials provide the best conditions for observation. In general, a 7/8-inch square cover glass will prove to be most serviceable.

Before use, slides and cover glasses should be clean and free of all traces of grease or any other material that might interfere with observation or proper functioning of the dyes. For thorough cleaning, glassware should be placed in concentrated nitric acid for ten to fifteen minutes, then thoroughly rinsed in distilled water and stored in 95% ethyl alcohol. Before use, the slides and cover glasses are dried with clean cheesecloth. In handling cleaned slides and cover glasses, always use forceps, or if fingers must be used, grasp slides and cover glasses by the edges only.

When studying living protozoa, care must be used to prevent the pressure of the cover glass from distorting the appearance of the animals. Therefore small bits of debris from the culture medium are usually included in the mounting to support the cover glass.

In cases where it is desired to observe a preparation for a considerable period of time, depression slides may be used to advantage. In using these slides, a ring of vaseline is placed around the depression cavity. A drop of the culture is placed on a circular cover glass 1 cm in diameter. The cover glass is inverted over the depression, and the vaseline ring forms an air-tight seal. Unfortunately, this method of mounting does not allow as careful an examination of vitally stained materials as does the use of a regular slide and cover glass.

In examining many protozoa it is desirable to slow down their movements to facilitate observation. While this may be done by using small bits of filter paper or cotton to entrap the animals, it is better to use some method that permits a freer examination. For this purpose, solutions of copper acetate and copper sulfate are included with the Vital Stain Set. Either of these immobilization solutions may be used in either of two ways: (a) a drop is added to the drop of material to be examined or (b) a drop is dried on a slide and the drop of culture medium placed on it. Use whichever method appears more convenient in a given procedure.

The vital stains supplied in this set are made up as 1% solutions in absolute ethyl alcohol. This allows the dyes to remain stable over a relatively long period of time. However, they must be diluted before use. The immobilization solutions—copper acetate and copper sulfate—are prepared for direct use.

The vital dyes may be diluted with absolute alcohol to ten times or more their concentration as supplied. In most cases, the more dilute solutions such as 1:10,000 to 1:100,000 are most suitable. After dilution, the alcohol stain may be placed on the slide as a thin film and allowed to dry. A drop of fluid containing the organisms to be studied is then placed on the slide and covered with a cover glass, which may be sealed with vaseline or melted paraffin. The time required for staining varies from 5 to 30 minutes.

Another way of using vital stains is to make very dilute aqueous solutions. A drop of the dilute solution is added to the drop of culture fluid.

Some specific notes on the dyes included in Ward's Vital Stain Set are:

Neutral Red may either be dried on the slide or added to the medium as an alcoholic or aqueous dye. Neutral Red in dilutions of 1:3,000 to 1:30,000 is an indicator. In alkaline solutions it is yellowish-red, ranging to cherry in weak acids and blue in strong acids. Variations in the pH of food vacuoles may be studied by this means. Neutral Red also will stain the nucleus slightly. The so-called "neutral red granules" received their name from their affinity for this dye.

Janus Green B is used to demonstrate mitochondria (sometimes called chondriosomes). In addition to use as a dye film, Janus Green B may be diluted with either water or physiological salt solution (see *Ward's Culture Leaflet No. 3*) at dilutions of 1:10,000 to 1:500,000. This dye may be used in combination with Neutral Red. This is done by mixing one part saturated dye solution in absolute alcohol and two parts Neutral Red solution (10 parts absolute alcohol and 1 part 1% Neutral Red in absolute alcohol).

Methylene Blue is one of the best vital stains to demonstrate the nucleus, cytoplasmic granules, and cytoplasmic processes. It may be used as described for the other dyes above.

After familiarity has been gained with vital staining technique, *Bismarck Brown* and *Brilliant Cresyl Blue* may be employed in the procedures outlined above to provide an additional variety of staining patterns.

As a general guide for dilutions, the following information from Ball (1927) with respect to *Paramecium* may be helpful. According to this author, structures in *Paramecium* will stain best with Janus Green B at dilutions of 1:800,000; with Neutral Red at dilutions of 1:110,000; and with Bismarck Brown at dilutions of 1:150,000.

For additional information and suggestions on the use of vital dyes with living protozoa, the following references are suggested:

BALL, G. H. 1927. Studies on Paramecium. III. The Effect of Vital Dyes on *Paramecium caudatum, Biological Bulletin* 52:68–78.

KIRBY, HAROLD. 1960. *Materials and Methods in the Study of Protozoa.* Univ. of California Press, Berkeley.

KUDO, R. R. 1954. *Protozoology.* 5th Ed. Charles C Thomas, Springfield, Ill. *Ward's Culture Leaflet No. 4.* Ward's Natural Science Establishment, Inc., P.O. Box 1712, Rochester, N.Y. 14603; or P.O. Box 1749, Monterey, Calif. 93942.

Using

Plant

Materials

TECHNIQUES FOR ISOLATING AND CULTURING FUNGI

I. Culturing a *Physarum polycephalum* plasmodium from sclerotia (crustlike form that is not in the active stage).

1. Place two petri dish halves open side up in a crisper.
2. Place a layer of moist filter paper over the petri dishes.
3. Lay the sclerotium on its substrate on the filter paper. Allow the plasmodium to begin creeping before feeding it.
4. Feed the plasmodium with old-fashioned (*not* quick-cooking) rolled oats. Begin with one flake, then feed two or three as the plasmodium increases in size, until you are feeding several flakes or a small handful to a plasmodium that covers the bottom of the crisper. Feedings do not need to be regular, but always wait until all of the previous feeding is covered by the plasmodium before feeding again. If the oats become excessively contaminated with bacteria or mold, remove them with forceps and feed at another area for a time.
5. Keep a little water in the bottom of the crisper (10 ml is sufficient) and keep the filter paper moist. After the plasmodium is started it may be moistened with a saturated solution of calcium carbonate in water.
6. To subculture the plasmodium, place a piece of filter paper with the plasmodium on it in a crisper set up as before.
7. To maintain the sclerotia, line the edges of the crisper with discs of moist filter paper and allow the plasmodium to creep up on the paper. Place 2 or 3 discs of dry filter paper in a clean, dry petri dish and place the filter paper with plasmodium on top. Leave the lid to the petri dish ajar and allow the plasmodium to dry slowly for 2 to 3 days. When dried, cut off the excess paper and store in petri dishes in the refrigerator. This keeps for months.

II. Isolating myxomycete plasmodia. Myxomycete plasmodia can be obtained for laboratory use directly from field materials or from sclerotia from biological supply sources. Even in the winter, plasmodia can be cultured from bark of living trees, from well-rotted wood or branches in moist places, or from leaves in moist pockets. The field material should be thoroughly moistened with distilled

This section was prepared by Dr. Lois H. Tiffany, Department of Botany and Plant Pathology, Iowa State University, and used in the 1967 Iowa Science Teachers Short Course at Iowa State University.

water, then placed in closed containers lined with moist filter paper. Petri dishes or larger similar containers may be used. Materials held at room temperature, but not in direct sunlight, have been most productive. Several different plasmodia may be obtained from a single source. Plasmodia are usually not obvious in less than 2 weeks, and may develop even after 4 or 5 weeks incubation. Additional moisture can be added as needed, but the materials should not be kept with much excess water in the container.

III. Miscellaneous hints for obtaining fungi for class use without maintaining stock cultures.

1. Place slices of bread in a plastic bag with some moist toweling. Remember that commercial bread has a mold inhibitor added, and fungi will develop more slowly on such a substrate. Local bakery products may not have a mold inhibitor and may develop a fungus flora more readily.

2. Plant parts such as sweet potatoes, oranges, and grapefruit may be placed in closed plastic bags with moist toweling. If citrus fruits with soft spots are selected, *Penicillium* species will usually fruit on the area in a few days. *Rhizopus nigricans* (black bread mold) typically develops on the sweet potatoes. It will fruit much more profusely if the potatoes are broken or cut open before they are placed in the moist situation.

3. Place a layer of moist toweling in a covered container or plastic bag and sprinkle with flakes of oatmeal. *Rhizopus nigricans* (black bread mold), species of *Penicillium* and *Aspergillus*, plus yeasts and bacteria will develop. Typically there are enough air-borne spores that a variety of fungi are obtained, although *Rhizopus* may overgrow everything initially.

4. Dung of small herbivorous animals, such as squirrels or rabbits, is an excellent source of a variety of fungi. Especially in winter this material is easy to collect, and not offensive. Place the dung pellets on moist toweling in a closed container and observe over a period of time. A rather definite ecological sequence of organisms will fruit. Many of these fungi have mechanisms for flinging their spores or spore-containing structures, and often these are sticky and will adhere wherever they land. Also many of these fungi will orient their spore-producing organs toward a light source and spores will be shot toward a definite target area.

STAINING AND CLEARING PLANT MATERIAL

STAINING TECHNIQUES

Permanent Preparations of Free-hand Sections

PERMANENT SLIDES of professional quality can be made by the simple procedure outlined here. The method

This section was provided by Dr. N. R. Lerston and Dr. W. R. Shobe, Department of Botany and Plant Pathology, Iowa State University.

237

works for thin sections of any plant organ and can be adapted for certain whole plant structures that are thin, such as hairs, moss leaves, filamentous algae, etc. Commercial slide preparations are produced by basically the same sequence of events outlined here, except that more steps, time, and equipment are involved.

MATERIALS NEEDED

1. Razor blades, small spatula, teasing needle, forceps
2. 6 Syracuse watch glasses or small petri dishes
3. Slides, coverslips
4. Plant material—need pieces large enough to hold between thumb and forefinger to slice

REAGENTS AND STAINS NEEDED

1. 95% ethyl alcohol
2. 100% (absolute) ethyl alcohol
3. Xylene
4. Safranin O (comes as powder)
5. Fast Green or Chlorazol Black E (both come as powder)
6. Piccolyte mounting medium

SUGGESTED PLANT MATERIAL

1. Corn stems
2. Peperomia stem
3. Castor bean stem
4. Epidendron root (aerial)
5. Fern stipes: *Polypodium, Asplenium,* etc.

Procedure (With Chlorazol Black E)

Cut several sections and place in 95% alcohol. Keep thinnest ones, discard others. Follow this outline, passing the plant material from one step to another in sequence.

1. 95% alcohol, 4 + min.
2. 1% Safranin in 95% alcohol, 5 min.
3. 100% alcohol destain, 10–60 sec.*
4. 0.5% Chlorazol Black E in 100% alcohol, 20–40 sec.*
5. 100% alcohol (to rinse off excess stain), 15–30 sec.*
6. Xylene, 3 + min.
7. Put section in a drop of piccolyte on slide.
8. Add cover slip slowly from one side.
9. Add additional piccolyte at side of cover slip as needed.
10. Store horizontally until piccolyte is dry (week or so).

A Clearing and Staining Technique for Leaves or Other Plant Organs

1. Place fresh leaves in 70–95% ethanol until chlorophyll is removed. For dried or preserved leaves, omit this step and proceed to 2.
2. Place material in 5–10% aqueous NaOH. As cell contents leach out (this may be hastened by oven-heating to 60° C) the NaOH may become discolored. If so, replace with fresh NaOH until leaves are fairly clear. If dark areas remain in leaves after several days, transfer to full-strength household bleach (e.g. Clorox) for 2–5 minutes (keep the time to a minimum as some tissues may tend to disintegrate).
3. Rinse with 3 changes of distilled water (5 minutes each).
4. Place in aqueous chloral hydrate (250 mg/100 ml distilled water) for a minimum of several hours. The tissues should become quite transparent except for lignified areas.

---

* Time varies depending on material and section thickness.

The material may be stored indefinitely in this solution.

5. Repeat Step 3. Do not place fragile material directly into pure water, but pass through a dilution series to avoid possible damage from mixing currents.

6. Dehydrate to 95% ethanol. Pass fragile material through a graded series, but most mature material can go directly from water to 95% alcohol (3 changes of 5 minutes each) without harm.

7. Stain at this point. Either of two procedures (8a–e *or* 9a–e) will give good results.

8. a. Stain for a few seconds with fast green (exact concentration not important) in 95% ethanol.

   b. Remove excess stain by rinsing with 95% ethanol.

   c. Dehydrate with 2 changes of absolute alcohol (5 minutes each).

   d. Stain for 1–5 minutes with Safranin O (½–1% in equal parts of xylene-absolute alcohol). Filter this solution before using.

   e. Proceed to 10.

9. a. Stain for 1 minute with Safranin O (½–1% in 95% ethanol).

   b. Remove excess stain by rinsing with 95% ethanol.

   c. Stain for a few seconds with fast green (exact concentration not important) in 95% ethanol.

   d. Dehydrate in several changes of absolute alcohol (1 minute each).

   e. Proceed to 10.

10. Destain to desired intensity in equal parts of xylene-absolute alcohol.

11. Place material in xylene to stop destaining. Change xylene after a few seconds to avoid safranin precipitation. Examine, and if the stain is

too dark repeat 10 and 11. If not dark enough, restain.

12. Cleared and stained material may be stored unmounted in xylene but more commonly is permanently mounted on slides using any xylene-soluble mounting medium. Piccolyte, which may be obtained ready for use from commercial sources, works well. Thick specimens pose some problems, chiefly from solvent evaporation, and at first more mounting medium must be added at the side of the cover slip.

COAL BALL PEEL TECHNIQUE

AROUND COAL MINES you may find chunks of rock that contain fossilized sections of plants such as the leaves, roots, stems, and even the flowers that were living on earth millions of years ago. If these rocks are ground down, the smooth surface can be etched with HCl. Then a cellulose acetate peel can be made of the etched surface.

To prepare the rock, saw it in two with a diamond saw and grind to a coarse, then a fine grain. Once this is done, you may make thousands of coal ball peel slides. These are the steps to follow, once you have prepared the rock for making slides:

1. Grind the flat surface of the coal ball with carborundum for 30–60 seconds. To do this, sprinkle powdered carborundum over a moistened glass, place the coal ball on the glass, and grind, moving the ball over the glass. When finished, the surface of the rock should be glassy smooth.

2. Rinse with tap water.

3. Put about ½ inch of 5% HCl (5 ml

in 100 ml water) in a pan. Hold the rock face down in the pan for 30–40 seconds to etch it, being careful not to let it touch the bottom of the pan.

4. Rinse with tap water. Do not touch the surface of the rock that is etched.

5. Air dry the surface for 2 to 3 hours, or shorten the process by applying acetone 2 or 3 times and blowing on it to dry it.

6. Cut a piece of cellulose acetate a little larger than the surface of the rock.

7. Wet the rock surface again with acetone and apply the cellulose acetate to the wet surface of one side and roll it on. Do not move the cellulose acetate after it has been placed on the rock.

8. Wait a minimum of 10 minutes before peeling the cellulose acetate from the rock. If it starts to tear as you try to pull it off, wait a little longer. It is better to wait an hour or two before trying to remove it.

9. Examine the peel under a dissecting microscope.

10. Sections of the acetate with the fossil remains make good microscope slides if the acetate is cut into small sections and placed on a slide with an agent such as piccolyte. Put a cover slip on the preparation. Each section of the acetate should make several slides.

SPECIAL GREENHOUSE TECHNIQUES

I. A good potting soil mixture.
  A. (1) 2 parts good garden soil.
     (2) 1 part peat.
     (3) 1 to $1\frac{1}{2}$ parts sand.

(Most plants do well in this mixture.)

(This mixture should be sterilized before using for best results.)

  B. Use preceding mixture plus one of the following:
    (1) Sphagnum moss for most tropical plants, orchids.
    (2) Vermiculite to hold water when quick growth is desired for cabbage, celery, etc. Use for only 3 weeks, however.
    (3) Perlite may be used in place of sand to loosen soil.
    (4) Leaf mold is rich in organic matter and is good for most plants, especially begonias.
    (5) Clay, about 2% of total soil, may be used for most plants.

  C. Use a mixture of *Osmunda* (fern), Sphagnum moss, and Silvaback as a special mixture for orchids, *Bromelia,* ferns, *Anthuriums,* and for Christmas cacti and other plants that do not grow well in soil.

  D. Use poor clay and sandy soil for most cacti.

  E. When potting plants, a small piece of a *broken pot* placed in in the bottom of the pot promotes good drainage (not necessary for pots less than 5 inches wide). When repotting it is necessary to take a 1-inch to $1\frac{1}{2}$-inch larger pot, e.g., a 3-inch pot would be repotted in a 4-inch or $4\frac{1}{2}$-inch pot, and so on.

Prepared by Aalt G. Boon, supervisor of the teaching greenhouses in the Department of Botany and Plant Pathology at Iowa State University. This material was presented at the 1966 Iowa Science Teachers Short Course in response to the requests from biology teachers for practical information on raising plants in the biology laboratories and greenhouses.

II. Propagation.
  A. Seeds, cuttings, root-dividing, runners, offsets, and suckers.
    (1) Seed catalogs give adequate information. (Small flats are better than pots for seeding and most seeds should not be planted deeper than the thickness of the seed. Some seeds are planted on the surface.)
    (2) Cuttings.
      (a) Cuttings should be approximately 4-inch lengths of a young, partially hardened twig, or a young stem.
      (b) Cut just below a node.
      (c) If possible, remove first 4 leaves above the bottom node.
      (d) Dip lightly in Rootone (plant hormone). Plant cuttings straight up 1/4 inch or over in soil.
      (e) Root in flats filled with sand, vermiculite, perlite, or any mixture of the three. Sometimes add a little soil to it.
      (f) Water heavily at planting time and set the flat in a warm, shady place. Keep a little moist in the first 10 days.
      (g) In approximately 2 to 3 weeks the roots are developed enough to plant in proper soil mixture.
    (3) Root-dividing.
      (a) Cut off the top of plants.
      (b) Knock soil and roots out of pots.
      (c) Separate root sections longitudinally.
      (d) Replant the individual root sections in same mixture of a new soil.
  B. (1) After good growth has been attained by seedlings and cuttings, fertilize every 2 to 3 weeks with a weak solution of complete analysis fertilizer.
    (2) Analysis fertilizer recipe: Put 1 teaspoon powdered fertilizer of 20-20-20 (20 = nitrogen, 20 = phosphate, 20 = potash) or 15-30-15 to 1 gallon of water. Fill each flower pot with this solution.

III. Temperature, ventilation, and watering are important in any greenhouse.
  A. Poor ventilation and high temperature are conducive to:
    (1) Damping off (rotting of stem at ground line in seedlings).
    (2) Stem rot (too much water may produce these same effects).
    (3) Mildew (especially). (Above diseases may be *controlled* by such things as *Fermate, Captan,* or *Terraclor,* etc.)
  B. Proper temperature for each plant is learned by experience (many books give information on proper temperature).
  C. The optimum water requirement for each plant is learned by experience (or by looking up information in books). For most plants it is good to water heavily but allow soil to dry between waterings. Watering is *not* held to a rigid schedule like eating breakfast each morning, but water whenever the soil looks dry or feels dry. One can even de-

termine if a clay pot is dry by tapping the pot and noting the light sharp sound produced if it is dry.

IV. Greenhouse pests.
   A. Red spiders (difficult to kill).
   B. Scales (difficult to kill).
   C. Thrips (easy to kill with proper chemicals).
   D. Mealy bugs (easy to kill with proper chemicals).
   E. Aphids (easy to kill with proper chemicals).
   F. White flies
      (Red spiders, mealy bugs, aphids, and white flies can be partially controlled by washing foliage with cold water under pressure to forcibly remove the insects.)
   G. Slugs and snails (should be caught by using a flashlight in the evening, or by using poisons).

REFERENCES

BAILEY, L. H. *Standard Encyclopedia of Horticulture.* The Macmillan Co., New York.

FREE, MONTAGUE. *Plant Propagation in Pictures.* The American Garden Guide, Inc., Garden City, N.Y. 11530.

GRAF, ALFRED BYRD. *Exotica 3.* Roehrs Company, East Rutherford, N.Y. 07073.

POTTER, CHARLES H. *A Grower's Guide to Bedding Plants.* Florists Publishing Co., 343 So. Dearborn St., Chicago, Ill. 60604.

MITICIDES			
TREATMENT	STRENGTH NO.	PEST	TEASPOONS PER GALLON WATER
Aramite	15 W.	Red spiders	4
Chlorobenzilate	25 W.	Red spiders	2
Kelthane	18:5	Red spiders	3–6
Kelthane	EC liquid	Red spiders	1–2
Malathion	SE liquid	Red spiders, aphids, thrips, scales, leafrollers, mealy bugs	2
Malathion	25 W.	Mealy bugs	6–12
Parathion	15 W.	Red spiders, aphids, thrips, leaf-eating pests	2
DDT	50 W.	Thrips, leafrollers, and leaf-eating pests	5
DDT—10% dust		White fly	
Lindane	25 W.	Aphids (especially), thrips	3
FUNGICIDES			
Captan		Leaf diseases, leaf spots	3 (by sprinkler)
Fermate		Leaf spots, rust (does not affect mildew)	5 (by sprinkler)
Sulfur Dust		Mildew, rust	
Sulfur Wettable		Put on hot steam pipes under benches	
Terraclor	75 W.	Stem rot	1–1½

To make soil acid, add one teaspoon sulfur to the soil in a 4- or 5-inch flower pot and mix it well. Add water to make it moist.

To make soil alkaline, add about ½ teaspoon lime to the soil of a 4- or 5-inch flower pot. Mix well and add water to make it moist.

To test for acidity and alkalinity, use a testing kit such as one from Sudbury Laboratory, Sudbury, Mass. 01776. This kit costs about $8.00 and has tests for potash, lime, nitrogen, and phosphorus. A color chart is included with the test kit. To obtain the sulfur for increasing acidity of soil, use sulfur NF, a sublimated powder (one that has been heated, then solidified), from the J. T. Baker Chemical Co., Phillipsburg, N.J. 08365.

PLANTS FOR THE CLASSROOM

YOU PROBABLY will use living algae, bryophytes, fungi, and ferns throughout the biology course you teach. In addition, for the study of germination and growth, and structure of stems, roots, leaves, and flowers you will need living material from angiosperms and gymnosperms.

We have offered suggestions in this appendix for maintaining selected vascular plants. The nonvascular plants are equally as important: algae can be kept in the classroom aquarium; mosses and ferns in a terrarium; fungi in small plastic moist chambers.

TREE SEEDS

Seeds from conifers, elms, maples, oaks, ginkgos, and other common trees may be germinated and grown in the classroom.

Plant seeds from common trees just below the surface of the soil. Use some of the soil from the area where the seeds were collected, or use potting soil mixture indicated earlier in the section of Special Greenhouse Techniques. Soil should be kept moist but not wet. A flat piece of glass over the soil helps to keep the soil from becoming too dry until seeds have sprouted. Before planting various nuts, try freezing some of them. This may hasten germination.

Before planting ginkgo seeds place them in a refrigerator at a temperature of 33° to 42° F for a period of 6 to 8 weeks, then file a small opening in the tough seed coat so that water will enter the seed and furnish necessary moisture for germination.

Suspend avocado seeds by toothpicks in a glass of water with part of the seed in the water. After germination occurs, they should be potted in a light loam soil. If kept moist, but not wet, plants usually make fine growth progress and are interesting for the biology room. Comparison can be made of rate of growth of several plants grown under different conditions.

Citrus fruit trees such as orange, lime, lemon, and grapefruit may be obtained commercially or started from seeds. Plant as directed for other trees. Germination may take up to several months.

HERBACEOUS PLANT SEEDS

Grow nontropical plants such as beans, peas, castor beans, corn, squash, sweet peas, and lettuce at ordinary room temperature. Soaking seeds in warm water for a period of 12 to 24 hours before planting will speed germination. Plant seeds in seed flats filled with light sandy loam soil at a depth of from $\frac{1}{2}$ to 3 inches. Keep the soil moist after the seeds are planted. If the air in the classroom is especially dry, cover the flats with a piece of glass. Remove the glass once in a while to prevent growth of mold. Germination usually occurs in 4 to 6 weeks.

BULBS, CORMS, AND TUBERS

Suspend coconut half-shells from the light fixtures in your teaching room. Cut a whole coconut (after the coconut milk is drained out) in two with a hack saw. Put one-half a sweet potato in each shell and add water. In several weeks the potatoes will sprout and produce long vines. The stems may be sectioned and the vascular systems studied. See Chapter 8 for hand microtomes that can be used for sectioning.

Amaryllis bulbs are rather large and grow best if planted one to a pot. They need rich garden soil. Bulbs should be planted so that just the lower two-thirds is below the surface of the soil with the upper one-third exposed to light and air. After planting, water well and set the pot in a dark place with temperature from 60° to 70° F until growth starts. As soon as the flower bud appears, place the pot in a warm, sunny room and give it plenty of water. Yield is best if bulbs are planted after January 1.

Calla bulbs grow best if planted individually in good-sized pots of rich garden soil and put in a warm room. After leaf has formed extensively, flowers appear. They need plenty of warmth and moisture.

Caladium or *elephant's ear* responds well to the same conditions as indicated for the calla.

Dahlia. The dwarf type needs to be planted 2 or 3 inches deep in pots of rich garden soil. It needs plenty of warmth and sunlight. If these conditions are met it will blossom about 6 weeks after planting and continue blooming for months. Tubers may be planted during the winter or spring months.

Narcissus bulbs, including the yellow, the paper-white, and the Oriental, will not stand freezing but are ideal for indoor growth. They grow either in water or in soil. Plant bulbs 2 or 3 inches apart in shallow bowls and cover with coarse gravel, stones, or pieces of broken flower pots up to the tip of the bulb. Pour in enough water to cover the lower half of the bulb. Flowers appear in 4 to 6 weeks. Growing temperature is from 60° to 70° F. After the first 2 weeks they need plenty of sunlight. Best results occur if bulbs are planted after December 1.

Hyacinths may be grown either in water or in soil. If they are planted in water they should be placed in a dark, cool place 50° to 60° F for 1 month to force the root growth. After a month, bring them to a warm room where there is plenty of light. Flowers usually appear about 3 months after planting.

Tulip and *Crocus* bulbs need a cool, dark place with temperature from 40° to 50° F for about 2 to 3 months for their root structure to develop. They should be planted about 1 inch below the surface in shallow pots of sandy loam soil. They should be well watered before being placed in the dark. At the end of 2 to 3 months pots should be placed indoors in a sunny location with temperature ranges from 60° to 70° F. Within a month after being brought in, flowers appear.

POTTED PLANTS

Potted plants to keep in the classroom include *Aloe, Bryophyllum, Peperomia, Coleus, Asparagus* "fern," *Philodendron, Geranium,* English ivy, and *Sedum.*

Tropical plants need a constant temperature of between 75° and 85° F for growth. Perhaps a series of lights turned on above the plants at night will help maintain this temperature. Humidity must be kept high at around 65 to 90 per cent. Some tropical plants that grow easily in the laboratory are banana sprouts, *Ficus aurem* (the strangler fig), and *Kalanchoe.* These grow well in pots.

Clumps of *mint* and of *parsley* may be placed in pots of good garden soil and grown in the classroom. Ordinary care should keep these plants growing during the school year.

Geranium slips are easily started in water in the fall of the year. As they develop a root system they may be placed in rich garden soil and become part of the room "jungle," mentioned in Chapter 6.

Other plants that slip and root easily include *Oleander, Ficus* (the common rubber plant), *Coleus,* all kinds of ivy, *Philodendron,* and *Sedum.*

FERNS

Ferns may be purchased from greenhouses or taken from the woods. They may also be started from spores and grown in the laboratory. Examples of ferns commonly used are: *Adiantum,* the maidenhair fern; *Asplenium,* a spleenwort; *Cystopteris,* a fragile fern; *Dryopterius,* known as the florist fern and found in practically all greenhouses; *Nephrolepis,* the Boston fern; *Onoclea,* a sensitive fern easily killed by frost; *Osmunda,* the royal, the interrupted, and the cinnamon ferns; *Polypodium,* the common Polypody; *Polystichum,* the Christmas fern; *Pteridium,* the bracken fern; and *Pteretis,* the ostrich fern found commonly in gardens.

USING PLANTS IN TEACHING

Plants of the same species grown in soil with different pH values may grow at different rates. Varying intensity and duration of sunlight will have an effect on growth. Plants that are well watered and those grown under near-desert conditions reach maturity at different times. Good and poor growing conditions for various species may be determined. Vary the depth at which seeds are planted. Soak some seeds for a day or longer and plant others without soaking. Soak some in very hot water and others in cold water. Turn some plants around toward the sun each day and leave others with the same side toward the sun all the time. Roots, stems, and leaves may all be used for cell studies. Hormones may be used to cause changes in the plants. Bending of plants toward the sun may be observed, effects of crowding noted, overwatering and underwatering observed. Bulbs and tubers develop rather rapidly into flowers. Any nursery catalog lists many varieties which may be easily grown as laboratory plants. Try looking at pollen through the microscope as flowers form. Have the students transfer pollen to the stigma so that seeds will form. If cross-pollinated, plant the seeds to produce new forms or colors of flowers. Use a fine camel-hair brush to transfer pollen from the anther of the stamen to the stigma of the pistil.

LIVERWORTS AND MOSSES (BRYOPHYTES)

Liverworts and mosses are distributed over the entire United States and are also prevalent in greenhouses. During summer and early fall liverworts can be collected easily. Most species grow in thin, flat masses close to the ground or attached to bark of trees, rotting wood, and damp rocks. Soil mixture for growing liverworts consists of one part sand, one part cinders, one part loam, and one part peat moss, thoroughly moistened. Place the liverworts on top of this mixture and cover the terrarium with glass. Keep in a cool spot in the classroom, not in direct sunlight. Add small amounts of water from time to time. Plants should continue vegetative growth under these conditions. To induce sexual reproduction, increase the amount of light.

Mosses can be found during all seasons of the year in many different kinds of habitats such as in sandy soil, on rocks, logs, bark, in wet swampy places, and in moist wooded areas, in both hot and cold climates. They grow readily in the laboratory and require little care. Planting mixture for mosses should consist of a mixture of two parts humus, one part peat moss, one part loam, and one part of a mixture of sand and pebbles. Mix these materials thoroughly and add enough water to moisten them. Put the moss on this mixture, sprinkle slightly with water, and put a glass top over the terrarium. Set the terrarium in a cool area of the room, preferably at the north side. Mosses grow best in a cool, airy environment. They die quickly if overheated. Water slightly from time to time.

Mosses may be grown from spores if conditions are kept right. Sterilize a clean porous flower pot with boiling water. Inside the pot to within an inch of the top, put broken pieces of flower pots and a layer of rich loam, all of which have also been sterilized with boiling water, or in an oven. Put a layer of very fine, clean sand over the soil. After all is cooled, blow some spores from the moss onto the sand. Cover the pot with a glass plate and set it in a saucer of water. Do not keep soil too wet. Germination should take place in a month or two.

Lichens are made up of two plants in close association—a fungus and an alga. The alga may live alone, but the fungus dies unless it associates soon with another algal plant, the kind with which it normally associates. Lichens form a crustlike covering over rocks, stones, bark of trees, or on the ground. They may be collected at any time of the year and even though they may appear to be dry and lifeless, if placed in water they soon become flexible and green. When brought from the woods into the laboratory, leave them attached to the branch or bark upon which they were growing. If they were growing on rocks or stones,

bring part of that habitat into the laboratory, place in the terrarium, and put the lichens on them. Cover the terrarium with a glass top, and water slightly and infrequently so as not to encourage growth of mold.

FERNS, CLUB MOSSES, AND *EQUISETUM* (PTERIDOPHYTES)

Ferns should be grown in enclosed glass terraria, with glass tops that can be extended. Use several inches of coarse gravel and broken charcoal for a base, with three to four inches of humus, peat moss, and leaf mold over it. A good soil for ferns consists of one part coarse sand, one part thoroughly pulverized peat moss, and two parts of rich loam. Place rhizomes of ferns in soil so that rhizomes are entirely covered with soil, except for the tips where the new fronds develop. These should be only lightly covered.

Club mosses grow well in the laboratory in soil consisting of two parts of sand and pebbles mixed, one part of peat moss, and one part loam, well mixed and thoroughly moistened. Cover terrarium with a glass top and put in a cool, airy window area. Water slightly, occasionally.

Equisetum grows best if a moderately dry sandy soil is used in the terrarium. It should have good drainage. Dormant roots or young plants grow better than transplanted adult plants.

CULTURE AND CARE OF INSECTIVOROUS PLANTS

Species of insectivorous or carnivorous plants are found in both temperate and tropical regions. Common ones most generally studied in biology classes are pitcher plants, sundews, Venus flytrap, bladderworts, and butterworts. All of these need a moist or aquatic habitat.

Pitcher plants grow in bogs or marshes and generally have some of their roots in the water. Sundews grow on rocks a little above the surface of the water but not in direct contact with it. Venus fly-trap needs a moist soil and good drain-age. Butterworts are small plants re-sembling Venus flytrap, with leaves growing in a sticky, basal rosette to which insects stick. All these plants grow in soils that are low in nitrogen. Ap-parently the insects they capture help supply this lack of nitrogen. Common bladderwort grows well in a balanced aquarium. Its submerged stems have small bladders or traps that often contain little crustaceans and aquatic insects.

In preparing a terrarium for the in-sectivorous plants other than the blad-derworts, cover the bottom with an inch to two inches of coarse gravel for drain-age. Place a layer of acid soil (see Ap-pendix C 242) over this. If acid soil is not available, use regular garden soil mixed with peat moss or dried sphag-num moss. Cover this with a top layer of either living or dead sphagnum moss. Put the plants in the soil so their roots are completely in the soil and covered with both soil and moss. Water thoroughly.

Put the terrarium in a place where it receives some direct but not excessive sunlight each day. Cover the entire ter-rarium with glass to keep the area moist. A satisfactory terrarium may be made from a rectangular aquarium tank or a large battery jar with an all-glass cover. Size of container is unimportant but it should be entirely of glass, and the air on the inside should remain moist at all times.

If care is taken, these plants can be grown from seeds. The plants some-times produce seeds, and they may also be secured from biological supply houses.

To prepare a planting place for seeds of insectivorous plants, use shal-low pots or pans. Put one inch to two inches of soil in the container with a layer of several inches of sandy bog soil. Cover with about 1/4 inch of ground-up or chopped sphagnum or peat moss. Sow seeds very thinly. Keep moist at all times. Place a glass cover over the top or put an inverted bell jar over the con-tainer. Venus flytrap seeds germinate in eight to thirty days; sundew and pitcher plants in two to four weeks.

Carnivorous or Insectivorous Plants

SCIENTIFIC NAME	COMMON NAME
Sarracenia purpurea	Common pitcher plant of northern United States
Sarracenia minor	Pitcher plant of south-ern United States
Darlingtonia californica	Pitcher plant of western United States
Nepenthes spp.	Pitcher plant found in many greenhouses
Droseras rotundifolia	Sundew of northern Wisconsin and Mich-igan
Droseras spp	Sundew of swamps of eastern United States
Dionaea muscipula	Venus flytrap
Utricularia vulgaris	Bladderwort
Pinguiculas spp.	Butterwort

Things To Do With Insectivorous Plants

Grow some plants from seeds. Es-tablish a thriving community of plants.

Observe the traps for capturing prey. Try to determine how each works.

Feed the plants a series of different

foods to see which foods are accepted. Suggestions for feeding are: small insects, tiny pieces of hamburger, pieces of a worm thoroughly minced, a bit of cheese, cooked egg white. Try others. Try changing the pH of the soil for some of the plants.

Get a commercial fertilizer that is high in nitrogen and add a small amount to some of the soil to see how plants are affected.

Lightly touch the trap apparatus with a pencil to see if trap closes.

COLLECTING, CULTURING, AND USING FRESHWATER ALGAE

Nearly any area presenting a wet, a damp, or even a moist situation is likely to yield a number of kinds of algae. They may be found in regions where there are permanent banks of snow as well as in water of the hottest springs. One alga lives on the back of snapping turtles, and others thrive among the hair scales on such animals as the three-toed sloth of Central America.

From green pond scum we can collect several species. Trunks of trees and surfaces of leaves in a moist climate may contain semiparasitic algae. Many are cultivated easily in the laboratory.

Because they may be found near every school, they are ideal for biology classroom study.

They can be collected in vials, jars, and containers of nearly any size. However, upon returning to the laboratory after collecting algae all the vials, bottles, and jars should be opened and their contents put into wide, shallow dishes so they are well aerated. Put in an area away from sunlit windows, preferably in a north window. If kept in a shaded area at a temperature of about 20° C (68° F) cultures may be kept alive for long periods of time. Finger bowls or battery jars serve as good culture containers. Large forms grow well in aquarium tanks.

Do not crowd the material by placing too much in the jars. If it is actively growing, the overabundant supply will be choked off, decay will set in, and general fouling will result, killing the culture. If the algae are growing rapidly, excess material should be removed from time to time.

If a culture is healthy keep it, even if the algae seem to disappear. They may be going through a period of dormancy. Examine the bottom of the jar for spores. Allow the water to evaporate, cover the jar, and set it aside. After a month or two add water and the culture will probably appear again.

Sudden changes in water may damage algae, so it is best to use water from which the algae were collected. When more water is needed, gradually add either distilled or pond water.

An easy way to prepare a medium for culturing some of the freshwater algae may be made in the following way:

Place a gram of commercial fertilizer in a liter of pond or distilled water and heat to about 85° C (185° F). Filter, cool, place in culture dishes, and inoculate with desired algae.

Rich garden soil may be used in place of the commercial fertilizer if heated as indicated, for an hour or two on two successive days. This helps dissolve the humus and kills most of the soil and water organisms present.

Green algae such as *Spirogyra, Closterium* (desmid), *Hydrodictyon, Volvox,* and *Cladophora* are easily cultured on this medium.

Oscillatoria, Vaucheria, Diatoma, Zygnema, Ulothrix, Chlamydomonas, Pandorina, Eudorina, and *Gonium* have also been cultured using these methods.

Most freshwater algae may be introduced into a well-aged aquarium and live for long periods of time in this situation.

Probably the most commonly used blue-green algae in the laboratory are the following four:

1. *Gloeocapsa* inhabits the surface of moist cliffs and rocks, soil in greenhouses, and moist cement and bricks. It may be found in gelatinous masses, sometimes floating in ponds or forming a coating over wet rocks. It is easily maintained in the laboratory by providing a moist situation for it.

2. *Nostoc* builds layers on bottoms of pools, wet meadows, lakes, and various swampy situations. It often forms shelflike growths on the sides of stones in brooks and mountain streams when the water flows over them. The best temperature for maintaining *Nostoc* is about 40° F in the refrigerator.

3. *Oscillatoria* is easily located in stagnant water, on damp ground, and on objects such as flower pots or dripping rocks. It is dark blue-green or nearly black. In culturing it place some of the material in distilled or boiled pond water, and cover to avoid contamination. Cultures may be kept almost indefinitely if a little water is added from time to time to replace any lost through evaporation. Pond water, because of its mineral content, is preferable to distilled water.

4. *Rivularia* is found attached to moist rocks, submerged logs, or leaves and stems of some water plants. It is not found free-floating. Frequently it is found in laboratory aquaria.

In addition to these blue-green algae, several different green algae are frequently used in the laboratory:

1. *Chara* includes many different species and is usually found in slow-flowing streams or in alkaline or hard-water lakes where there is an abundance of calcium present in the form of carbonates or bicarbonates. In the course of its development *Chara* causes lime to be deposited in stems and leaves of plants. This sometimes does damage to the plants so affected.

Remove moist or dried mud from the bottom of ponds in which *Chara* has grown, take it into the laboratory, and add water to it. *Chara* will start growing. It grows best where there is plenty of light.

2. *Chlorella* is free-living but may also be found living as a commensal within tissues or cells of protozoans, hydras, and other small invertebrates. If aquaria are untended or become foul, *Chlorella* is usually found. It may also appear in mixed algae cultures taken from pond water.

3. *Cladophora* grows attached to sticks and stones in either quiet or running waters. It is very easy to grow and may be maintained in one- or two-gallon aquaria. Excess growth should be removed from containers as they become too crowded for successful growth of *Cladophora*.

4. *Diatoms* usually occur in large quantities around springs, in pools, and in ponds. They cling to filamentous algae or form gelatinous masses on submerged plants. Surface mud of a pond, ditch, or other quiet pool or lagoon will yield some diatoms.

If the laboratory is warmer than the area from which they were collected, diatoms frequently show considerable activity. They are often found in aquaria along with other algae.

5. *Nitella* is found in places similar to those from which *Chara* may be taken.

The things said about *Chara* also refer to *Nitella*.

6. *Oedogonium* prefers quiet waters of ponds and ditches. Common habitats are on overhanging grass, leaves, or culms of rushes and old cattail stalks. It may occur as attached filaments, or become free-floating and form cotton-like masses near the surface of the water.

7. *Protococcus,* also known as *Pleurococcus vulgaris,* forms a green film on the moist side of trees, rocks, boards, and the like. It may be collected by taking pieces of bark into the laboratory, storing dry, then placing in a moist chamber for 24 hours before needed in class. By the time the 24-hour period is over, it will have started active growth.

8. *Spirogyra* may be taken from quiet waters of ponds, ditches, small lakes, and lagoons where it may form large, green, cotton-like masses on the surface of the water.

It frequently grows well when placed in a balanced aquarium.

9. *Volvox* occurs in water that is rich in nitrogenous substances such as those found in permanent ponds and pools. Sometimes it forms a "bloom" over the surface of the water during short periods in the summer months. It may be cultured in a medium of 0.2 g of commercial fish meal in a liter of spring water that has been heated to a temperature of 80° to 90° C (176° to 194° F) and filtered through filter paper. To this add 0.5 cc of a 1% solution of $FeCl_3$ in water. Inoculate when the medium has cooled, and place in a window at room temperature.

10. *Chlamydomonas* may be grown in a medium of commercial fertilizer in spring water. Add 1 g of fertilizer to a liter of spring water and heat to 80° to 90° C (176° to 194° F). Filter and place in finger bowls. Inoculate after cooling and place in a location to receive moderate light.

11. *Pandorina, Eudorina,* and *Gonium* may be cultured in the same medium as that described for *Chlamydomonas.* They, too, prefer moderate light.

12. *Vaucheria* is found in greenhouses, on rocks in flowing water, on damp soil, or sometimes as wooly mats floating at the surface of ponds where the plants have broken away from their substrates. They are of a dark green color, but are classified as yellow-green algae.

SUGGESTED WAYS OF USING ALGAE IN THE LABORATORY

Measure oxygen release under varying conditions of sunlight by placing an inverted funnel over the top of the container and an inverted test tube filled with water over the funnel. As gas bubbles move to the top of the container they displace the water in the test tube.

Observe reproduction of algae.

Mount diatoms for a permanent slide collection.

Do some paper chromatography to observe various pigments present in the different algae.

Observe effect of adding small amounts of various materials such as powdered dry milk, dry malted milk, gelatin, amino acid, and sugar to the algal cultures.

Try growing algae in different media. Several of the biological supply houses produce media that may be purchased and used immediately upon mixing.

Raise several similar cultures but vary the amount of light and the temperature.

Dry out some of the algae and later

provide a suitable moist or wet situation for growth. Repeat drying and culturing processes a number of times to observe effects on the culture.

Try growing algae at different pH values. Note effect on the plants of acidic and alkaline situations.

Examine algae from backs of various turtles to see if different species of algae may be identified.

REFERENCES

Many excellent references on algae are available. The following list represents a small portion of these.

BICO. *Teachers Outline Series*. National Biological Supply Co., Inc., 2325 So. Michigan Ave., Chicago, Ill. 60616.

BURKE, J. F. Collecting Recent Diatoms. Preparing Recent Diatoms. Mounting Recent Diatoms. *New York Micro-scopic Society Bulletin* 1(3):9–12; 1(4):13–16; 1(5):17–20, 1937.

CHAPMAN, V. J. 1962. *The Algae*. The Macmillan Co., 866 Third Ave., New York, N.Y. 10022.

FLEMING, W. D. Cleaning and Preparation of Diatoms for Mounting, *Bulletin American Society Amateur Microscopists* 5(3):49–53, 1949.

PRESCOTT, G. W. 1964. *How To Know the Fresh Water Algae*. Wm. C. Brown Co., 135 So. Locust St., Dubuque, Iowa 52003.

SMITH, G. M. 1950. *Freshwater Algae of the United States*. 2nd Ed. McGraw-Hill, Inc., 330 West 42nd St., New York, N.Y. 10036.

Turtox Service Leaflet No. 6. 1961. Growing Fresh-Water Algae in the Laboratory. General Biological Supply House, Inc., 8200 So. Hoyne Ave., Chicago, Ill. 60620.

WARD, HENRY BALDWIN; and WHIPPLE, GEORGE CHANDLER. 1945. *Fresh-Water Biology*. John Wiley & Sons, Inc., 605 Third Ave., New York, N.Y. 10016.

Techniques

Using

Animals

USES OF FROGS IN TEACHING BIOLOGY

Round Worms. As the abdominal wall is cut, look for small round worms that may be present. Place in frog saline, and examine under a microscope. (Frog saline: sodium chloride 8.0 g, potassium chloride 0.2 g, sodium bicarbonate 0.2 g, and calcium chloride 0.2 g in 1,000 ml distilled water. Add calcium chloride last to avoid precipitation of insoluble calcium carbonate.)

Protists. Examine various parts of the intestinal tract for several different protists: anterior part, the flagellate *Giardia agilis*; various parts and rectum, *Endamoeba ranarum;* rectum, three genera of flagellates, *Trichomonas, Hexamita,* and *Euglenamorphe,* and several green flagellates resembling members of the genera *Euglena* and *Phacies.* Tease out parts of the intestine containing protists and place in frog saline. Cover with a cover slip and examine.

The ciliates *Nyctotherus cordiformis* and *Opalina ranarum* are frequently found in the rectum also.

Smooth Muscle Contraction. Re- move a part of the small intestine and place in frog saline. Set it up for use with a kymograph to record contractions of smooth muscle. To make frog saline, dissolve 0.7 g NaCl in 100 ml distilled water.

Epithelial Cells. Keep a live frog overnight in a jar with about one inch of water. Next day mount small pieces of the sloughed-off skin in methylene blue (described under Cell Studies, Appendix G 331.) Place cover slip on this and examine epithelial cells.

Nervous Tissue. Remove tiny pieces of the spinal cord of a freshly killed frog and place on a clean slide. Carefully press a second clean slide on this, then, without moving the two slides further against each other, pull them apart so you have two slides, each with spinal cord material on it. Use methylene blue to stain. Examine, using a microscope.

Carefully expose the brain by scraping and gently cutting the top of the skull from the frog. Continue cutting and scraping until you have exposed the entire brain and the spinal cord. Compare the appearance and parts of the brain with those of other vertebrate animals, including man.

Stimulate exposed nerves with a weak electric current to see what part of the frog is affected. Stimulate points on the skin, also, and note the results.

Appearance of Muscle Tissue. Tease apart tiny pieces of the three kinds of muscle tissues of a frog; for example, leg muscle for skeletal, heart for cardiac, and stomach or intestinal for smooth. Place on three different slides. Stain with methylene blue or Lugol's solution (10 g of potassium iodide dissolved in 100 ml distilled water, and 5 g iodine added). Note differences in the three kinds of muscle cells.

Sperm Cells. Remove a testis from a freshly killed male frog and place in a drop or two of frog saline on a slide. Thoroughly crush the organ and put a cover slip over it. Examine for living sperm cells. They will live in this solution for several minutes and may be seen moving about the slide. Use a minimum of light so you can observe the tails of the sperm.

Effects of Chemicals on Heart Beat. Expose the heart. Try stimulating it with different chemicals and with water of different temperatures. Various strengths of acetylcholine, adrenalin, alcohol, and strychnine (cautious use) could be used. Use a second hand of a watch for timing reactions. Note the length of time for contraction and relaxation to take place. Rinse thoroughly with fresh frog saline after each trial.

The heart may be attached by the tip of the ventricle to a needle and set up with a kymograph to get recordings of differences in response.

Electrical Stimuli (source of stimulus, a dry cell battery). Use to stimulate different sections of the anatomy of a freshly killed frog to observe muscle responses. Determine which part reacts as stimuli occur. Cut a leg and some internal organs from the frog and see if reactions still occur when these isolated parts are stimulated. A stimulator attached to one or two dry cells should give sufficient stimulation.

Eyes. Carefully remove an entire eye from the dead frog. Then cut a small opening in the eye to show fluid within. Remove the lens and place it on a piece of paper that has some writing on it. Describe the results.

Cut a small piece from the cornea and put it in frog saline. Note transparency of this structure.

Remove both eyes from a frog. This can most easily be done by cutting off the top of the head by inserting scissors in the mouth and above the head behind the eyes. Then split the part you removed in two, longitudinally, so that each piece contains one of the eyes. Trim off the eyelids, including the lower, nictitating membrane so you can observe the size of the pupils. (See Figures D.1, D.2, and D.3.)

Put each in a small watch glass and flood one eye with a 5% solution of adrenalin (five parts in ninety-five parts water) and the other with 5% solution of acetylcholine (five parts in ninety-five parts water). Watch for changes in size of each pupil.

When each pupil has reached a maximum (or minimum) size, flood with frog Ringer's solution (to 100 ml distilled water add 0.65 g NaCl, 0.014 g KCl, 0.012 g CaCl$_2$, 0.02 g NaHCO$_3$, and 0.001 g NaH$_2$PO$_4$).

Try to reverse the process by putting adrenalin on the eye that formerly had acetylcholine on it, and acetylcholine on the one that formerly had adrenalin. You may wish to try out other chemicals.

Action of Leg Muscles. Cut through the skin entirely around one thigh of the hind leg of a frog. Slip off the skin as you might remove a glove from your hand. If you hold the cut piece with dry paper toweling it slides off easily. (See Figure D.4.) Observe the large muscle in the calf of the leg (gastrocnemius muscle). Pull on this muscle to see how

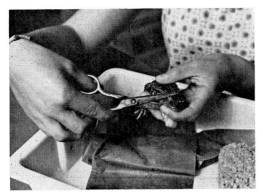

Fig. D.1—Insert one blade of scissors between jaws of frog and cut off head just behind the eyes.

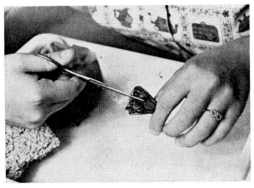

Fig. D.2—Cut head in two, lengthwise.

it moves the foot backward. Pull on the antagonistic muscle in front of the leg (tibialis anterior) and note how it brings the foot forward. Observe the strong tendon (tendon of Achilles) attached to the gastrocnemius muscle and see how it passes over the heel and spreads out to attach to the toes of the frog.

Frog Lung. Examine a small section of the frog lung to see the structure of the alveoli. Tear apart the lung tissue and examine for flat worms or round worms frequently found in this area of the body.

Inflate a lung by placing a medicine dropper in the mouth of the frog, into the glottis, and down into a lung. Alternately squeeze air into the lung and out again to see the lung inflate and deflate. Try blowing through a pipette into the lung to inflate it to maximum size.

Response of Skeletal Muscle to Acid Stimulation. Suspend a freshly killed frog by its lower jaw from a bone clamp attached to a ring stand. Apply a drop of 30% acetic acid (add 30 ml glacial acetic acid to 70 ml water) to one hind toe. Note result and immediately wash the leg with frog saline. Now hold one leg so it cannot move and apply a drop of acetic acid to a toe of the leg you are

Fig. D.3—Two halves of head with eyes intact.

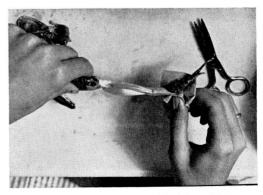

Fig. D.4—Remove skin from hind leg of frog.

holding. Note what happens to the other leg if the one being stimulated and held is unable to respond.

Action of Cilia. Place a frog in a closed jar with a few drops of ether on a piece of cotton. When the frog shows no signs of activity put it on a board or a frog pan, tie it down, and open its mouth. Place some crumbs of cork (obtained by running a nail file over a cork) at the back of the mouth of the frog. Note how the crumbs are swept down the digestive tract by the cilia at the back of the mouth.

Try ways of demonstrating that nicotine from smoking inhibits action of cilia. Put smoke directly on the back of the mouth, or soak a cigarette in water and place some of this water along with cork crumbs at the back of the mouth. Try juice from a cigar soaked in water. Devise other ways to demonstrate inhibitory action of nicotine.

Place a frog in a jar and tilt it from side to side. Note efforts of frog to right itself.

Put a jar containing a frog on a stool and rotate stool. What efforts does the frog make to compensate for this rotation?

How far can a frog jump? Compare the length of the hind legs of a frog to the length of its jump. What part do the front legs play when a frog jumps?

Cover the head of a frog with a small cloth and note the direction the frog jumps, if it jumps. Cover one eye and note the direction of the jump. Change over to the other eye and note the results.

COLLECTING INSECTS

INSECTS are found in nearly all parts of the world, including the Antarctic region. Many are inconspicuous and are not generally seen unless one knows where to look for them. There are more kinds of insects than any other kind of animal life so there is little chance of running out of new material to study.

HABITAT

For studying insects it is helpful to know something of the habitat in which to look for them.

1. Flowers are attractive to butterflies, moths, various flies, bees, wasps, beetles, and bugs. An aerial or "butterfly" net and a killing vial are tools needed for collecting insects from flowers.
2. Shores of ponds or lakes offer sites for collecting dragonflies, damselflies, whirligig beetles, water striders, water beetles, water bugs, mayflies, caddisflies, and water boatmen. In addition to the tools indicated for collecting insects from flowers, an aquatic net may also be of use in this type of habitat.
3. Stems and leaves of plants often harbor galls made by insects. The galls contain insect larvae. Small vials in which to put the galls are needed. For removing insects not in galls from stems and leaves of plants, a beating net or an aerial net may be used. Place insects in the killing tube.
4. Searching the ground may yield many insects, as some spend their lives crawling over and into the ground. A quick hand and a killing tube yield specimens here.
5. Outdoor lights at night attract many insects, some of which are seen only during the night hours. A street light, porch light, or any

light within reach is a good source for collecting insects. A screened window or door, with a light inside the house, attracts many insects. A white sheet spread out with a light focused directly on the sheet attracts many insects, especially moths. These may be picked directly from the sheet and placed in the killing tube.

6. A tin can with a piece of fish, meat, or small dead mammal in the bottom attracts carrion beetles. Bury it so just the top is exposed. Put dirt and leaves around it so there is nothing showing except the open top. Carrion beetles are attracted to the fish or meat and fall into the can but cannot get out again. Remove the insects from the tin-can trap, put them in a killing tube, and then pin them in the usual way.

7. Logs offer hiding places for insects. Some insects are found under stones and rocks, others under the bark of trees.

8. Tear up shelf fungi and mushrooms and look for insects there.

9. Examine rotting wood and see what insects you can find.

10. Slowly heat fungi or wood to drive out the insects.

11. Sugaring trees to attract moths can be done by mixing spoiled fruit juice with equal amounts of sugar or molasses and placing this on the trunks of trees. Place it here and there at convenient heights, using a paint brush. By the time you have sugared the last tree, visitors will already be sampling the first ones. Many will fall directly into the killing tube if it is held just below them. Others that fly nearby may be caught with a net. Sugaring is an excellent way to catch moths,

beetles, and ants. A sultry night when a storm is approaching is an ideal time for sugaring.

12. An aspirator is a convenient way to pick up very small insects. It consists of a wide-mouthed bottle or celluloid vial, some glass or metal tubing, and small rubber hose. A piece of fine gauze is tied over the tube to prevent foreign matter from entering it as the collector draws air through the glass mouthpiece.

13. Collect pupae from a great variety of places. Be sure to keep a record of where each came from. Place them in large containers and watch the adults emerge.

14. Look under dead chickens or other dead animals for carrion beetles.

15. Examine decaying fruit and vegetables for fruit flies or other insects.

16. To find insects that feed on grain, look through flour bins, granaries, and any place that cereals are stored.

17. Examine closets and boxes of stored clothing and old papers for moths, silverfish, and beetles.

18. Mosses and lichens harbor many insects. Examine with a hand lens.

19. Split the stems of dry or green weeds and other plants to find stem borers as well as other insects that may be hiding in them.

20. Dig under trees for the pupae of moths and other insects.

21. After you have driven your car for several miles on a nice day, examine the radiator for many species of insects.

22. Look for tiger beetles on a warm sandy bank along a stream or river.

23. Hold a net outside a car and drive about 25 miles an hour. Empty the net every three or four miles.

24. Examine the debris caught and

tossed up by rising streams during a flood or soon after.

25. Put a piece of bread and jam on a stump on a warm summer day. Come back a few minutes later and collect the insects.

Those are some of the places to look for insects. Not to be overlooked are the insects found on mammals and birds.

SUPPLIES AND EQUIPMENT

In compiling a list of supplies and equipment useful in collecting insects we include the following:

1. Some kind of collecting net that is lightweight but durable.
   a. Aerial nets are used for collecting on land or above water lines and are usually lightweight with the net portion made of nylon. Care must be taken as they tear if caught on barbed wire or thorny bushes.
   b. Sweeping nets are used for brushing over surfaces of plants to collect insects from leaves and small branches. After one has finished sweeping, insects and debris from bushes and other plants are concentrated in the small cone end of the net. Care must be taken not to lose the insects. Work with the killing tube inside the net. The sweeping net is more strongly constructed than the aerial or butterfly net.
   c. Beating nets are strongly constructed so the collector can sweep down into shrubs and rough vegetation. Nets also may be placed under some of the branches and then the collector may shake the branches. Some

insects when thus disturbed fall into the net.
   d. Aquatic nets are useful for collecting aquatic insects. A dip net is made of a frame covered with a shallow cloth and is used for removing insects and other organisms from the water. One type of aquatic net is shaped like a shovel with a coarse screen across the top to keep out aquatic vegetation. It permits insects to go through the screen but keeps out plants. A scraper net may be made of wire mesh, shaped like a deep tray, and mounted at an angle to a handle. It is used in digging and scraping the bottom of ponds and streams.

2. Killing tubes are made with plaster of Paris at the bottom, ethyl acetate over the plaster of Paris, and a small packing of tissue paper pushed down over it. Label *Poison* on the jar.
   a. Take certain precautions when making an insect collection so that insects are damaged as little as possible. Put small delicate specimens in a killing tube by themselves. Larger, heavier specimens may damage the smaller, more delicate ones. Place moths or butterflies in tubes not used by other insects, as the scales from the moths and butterflies tend to collect on other insects, giving them a dusty appearance. Keep the inside of the killing vial dry. Moisture may collect from other specimens inside the jar and this moisture mats the hairs or scales of insects. Use soft paper to dry the inside of the jar. As soon as insects are dead remove them from the killing

tube. Place them in small vials or cellophane envelopes until they can be taken into the laboratory. Do not allow a mass of insects and vegetation to collect at the bottom of the killing tube.

3. Spreading board for spreading wings of butterflies, dragonflies, moths, and the like.

4. Insect pins used in spreading and in mounting insects.

   a. Insect pins vary in size from 000 (very thin and short) to No. 7 (thick and long). Size No. 2 can be used successfully for practically any high school insect collection. Because insects differ in structure, they are pinned through different parts of the body. Pin beetles near the front margin of the right wing cover, close to the midline. Pin grasshoppers through the back part of the prothorax, just to the right of the midline. Pin butterflies, moths, dragonflies, and damselflies through the middle of the thorax at the widest point between or just behind the bases of the front wings. Pin other insects through the thorax just a little to the right of the midline. For insects that are so small it is difficult to put a pin through their body, carefully glue them to a paper point. Punches that can make these tiny paper triangles may be purchased from any biological supply house.

   b. If insects become too dry before they are mounted, they need to be placed in a relaxing jar. For this you need a wide-mouthed jar that has a practically air-tight lid. Fill the bottom of the jar with about an inch of sand. Add enough water to keep the sand moist. Add a few drops of carbolic acid to keep molds from growing. Cover the sand with a piece of heavy cardboard or wood. Place a sheet of paper containing several insects on the cardboard or wood and put the cover on tightly. In 12 to 24 hours the moist air of the relaxing chamber should have softened the specimens so that they can be handled without breaking. If they are large insects and you find they are still too brittle to mount, place them in the relaxer for another 12 hours. Specimens should be watched carefully so they do not get too wet or too soft. When removing insects from the relaxing jar handle them carefully so appendages do not break off.

5. Small vials of "FAAG" are used for preserving soft-bodied insects and larval forms.

   a. FAAG preserving solution:

Commercial formalin	15 parts
95% alcohol	24 parts
Acetic acid (glacial)	5 parts
Glycerine	10 parts
Tap water	46 parts[1]

6. A number of folded paper triangles, glazed paper, or cellophane envelopes for placing insects awaiting preparation.

7. Forceps.

8. A camel's hair artist's brush for handling very small specimens.

9. Field notebook for collecting data.

10. Storage or mounting boxes or cases for housing prepared specimens.

References for working with insects and for ordering supplies for insect studies:

[1] Recommended in *How To Make an Insect Collection*, Ward's Natural Science Establishment, Inc., Rochester, N.Y.

Biological Materials Catalogue. Carolina Biological Supply Co., Burlington, N.C. 21215; or Powell Laboratories, Gladstone, Oreg. 97027.

How To Make an Insect Collection. Ward's Natural Science Establishment, Inc., P.O. Box 1712, Rochester, N.Y. 14603; or Ward's of California, P.O. Box 1749, Monterey, Calif. 93942.

JAQUES, H. E. 1947. *How To Know the Insects.* Wm. C. Brown Co., Dubuque, Iowa 52001.

Turtox Catalogue. General Biological Supply House, Inc., 8200 So. Hoyne Ave., Chicago, Ill. 60620.

Ward's Catalogue for Biology. Ward's Natural Science Establishment, Inc., P.O. Box 1712, Rochester, N.Y. 14603; or Ward's of California, P.O. Box 1749, Monterey, Calif. 93942.

EQUIPMENT FOR SURVEYING AQUATIC POPULATIONS

SEVERAL kinds of aquatic nets were described under Collecting Insects (see p. D 257). These can be used for collecting small invertebrates as well as insects and other aquatic life. However, there are other nets and pieces of equipment that are useful. Some of these can be made and others purchased through biological supply houses. By getting a few pieces of equipment each year one can build a "library" of useful equipment for biology. The pieces described here have limited use in some sections of the country.

1. Fine-meshed collecting seines, made of cotton netting weighted with lead floats, are designed for collecting small fishes, tadpoles, crayfish, large insect larvae, and other aquatic forms. They generally vary in length from 4 or 5 feet to 20 or more feet and go into the water to a depth of 4 feet.

2. Bottom nets are long-handled scraping nets designed to pick up animals and plants from the bottom of an aquatic area. The bag is rectangular and heavily reinforced. The handle is more than 5 feet long.

3. Hand screens may be made of copper wire cloth at least 18 inches deep and 12 inches wide, attached to two handles at least 2 feet long. These are very useful in collecting small animals from rapidly moving streams.

4. Plankton towing nets are cone-shaped with a heavy ring at the wide end of the cone (diameter about 10 inches) and a glass vial at the small end. The net itself is a conical bag about 3 feet long, often made of silk bolting material. This is dragged through the water from a boat and samples taken from the water are placed in other vials, marked, and taken back to the laboratory for study.

5. Dredges are used for scraping along the bottom of rough areas for collecting in marine waters or rough bottom inland waters. The frame is of flat metal and has an open front with protected sides and reinforced bottom.

6. Dredge nets are used for collecting samples from the bottom of a lake, pond, river, or other water area. They are pulled along behind a boat so they scrape the bottom, picking up samples. The net is fastened above two rudders that act like slide runners, moving along the bottom of the water area and carrying the net just above them.

7. Water-glass fishscopes are used for observing below the surface in either fresh or salt water. They are made of aluminum alloy with an

260

inside dull black finish for better viewing. Fishscopes are about 24 inches long and 6 inches in diameter.

8. The visibility disc is used for comparing color and turbidity of different waters. It is made of heavy steel 8 inches in diameter, with each quadrant of the upper surface painted alternately black and white. The disc is fastened to a brass chain or a rope at least 20 feet long.

9. A piling scraper consists of a 16-inch curved blade attached to a triangular bag and is used for collecting specimens from pilings in marine situations.

10. Brass testing sieves are useful in analyzing specimens from sand or mud as well as from water. The sieves are 8 inches in diameter and 2 inches high, and are completely rustproof.

11. The inverting bottom sampler is for use in shallow water and where it is necessary to be accurate in sampling. A 2-quart can with $\frac{1}{4}$-inch mesh hardware cloth is firmly attached to a 6-foot handle for sampling water either from the shore or from a boat. After filling, the can is lifted up above the water for examination.

12. A plant grappling bar consists of a 24-inch bar equipped with 12 strong 6-inch metal teeth that are slightly curved for maximum collecting capacity. The bar is attached to a 10-foot rope for collecting.

13. A plant grappling hook may be a large hook placed at the end of a rope for collecting marine specimens as one tows the hook from a boat, or it may be thrown from shore. Smaller hooks are equally effective in pulling specimens from the bottom of aquatic situations that are less deep.

DEVELOPING A SERIES OF CHICK WHOLE MOUNTS

THE STUDENTS, David Staszak and James Good, who used this technique to develop their slides, each obtained a series of seven good slides of chick embryos varying in age from 12 hours to 108 hours. Each actually produced many more than seven slides, and found it extremely difficult to get a good whole mount of an embryo that was less than 34 hours old. They found it an advantage to obtain at least two embryos at each stage. Each started with two dozen fertile eggs. A set of seven slides costs about $5. If purchased, they would cost around $21. The students developed skill in manipulating the tools needed for obtaining the slides and learned a great deal about chick embryology.

Materials Needed

Sharp dissecting scissors
Tweezers
2 or 3 pipettes
Syracuse dishes (14)
Filters or paper towels
Finger bowl
Concave-center slides
Cover slips
Incubator or a steady heat source
Marking pen
Drying oven

(Other available small dishes may be used in place of Syracuse dishes. Small custard cups substitute nicely for finger bowls.)

Chemicals Needed

Bouin's fixative
Alcohols—50%, 70%, 85%, 95%, and 100%
  (dilute from ethyl alcohol)
Xylene
Acid alcohol (¾% HCl)
Potassium acetate in 70% alcohol
Meyer's paracarmine (stain, diluted 1:10 in
  70% alcohol)
Kleermount or other mounting medium
Chick Ringer's solution

Chemicals Used

Bouin's fixative—may be bought commercially prepared from any of the biological supply houses or may be prepared in the laboratory.
  Use:
      40% formalin 20 ml (or parts of)
      Glacial acetic
        acid   5 ml (or parts of)
      Picric acid (saturated
        solution)   75 ml (or parts of)

Acid alcohol—
      ¾% concentrated
        HCl   0.75 ml
      95% alcohol   100.00 ml

Meyer's paracarmine—
      Carminic acid   1.0 g
      Aluminum
        chloride   0.5 g
      Calcium
        chloride   4.0 g
      70% alcohol   100.0 ml
  Dissolve materials in the alcohol, then filter. Dilute 1:10 in 70% alcohol to use.

Mounting medium—any synthetic resin medium such as Permount or Kleermount.

Chick Ringer's—
      NaCl   9.00 g
      KCl   0.42 g
      CaCl$_2$ 0.25 g
      Distilled water 1000.00 ml
  Dissolve the materials in the distilled water.

Potassium acetate in 70% alcohol—put a pinch of potassium acetate in 70% alcohol.

Procedure

1. Incubate fertile eggs at 100° F in a high humidity area. Rotate the eggs each day. Start timing them 6 hours after starting to incubate to allow for a warming period.

2. When ready to use eggs, place an X on the dorsal side with a marking pen. Be sure to keep the X up until the egg is broken so as to have the embryo oriented.

3. Fill a finger bowl 2/3 full of chick Ringer's solution and break the egg into it, still keeping the X mark upwards.

4. Using bold strokes, make a circular cut around the embryo, keeping about ¼ inch away from the embryo. You should then have a disc of tissue about ¾ inch in diameter, with the embryo attached to it in the center.

5. Gently lift the embryo on the tissue disc to a Syracuse watch glass and clean it by gently swishing in Ringer's. Carefully flatten the embryo on the bottom of the watch glass upside down. Pipette out the yolk and swish back and forth to remove yolk and the vitelline membrane.

6. Wash the embryo in the Syracuse watch glass (still lying flat on the bottom of watch glass upside down). Use circular motion with pipette or you will "suck up" the embryo. Do not try to flatten embryo with forceps.

7. Cut a piece of filter paper or toweling to fit into the watch glass. Next cut a hole in the center of the paper slightly larger than the embryo. The age of the embryo and the date of this laboratory work may be written on the filter paper "donut."

8. Soak the filter paper in the chick Ringer's solution and place it over the embryo, which should still be upside down. The embryo should show through the center hole with a ring of filter paper around it. Press down gently around the outer ring of filter paper with a pipette.

9. Gently pour Bouin's fixative into the watch glass and leave for 24 hours. With a pipette check to make sure that

262

the embryo is sticking to the paper and not to the glass.

10. Wash the embryo in 50% alcohol for about 1½ hours, then transfer to 70% alcohol. The embryo may remain in 70% alcohol for a week or more if necessary.
11. Pour the 70% alcohol out and add paracarmine stain. Leave in the stain for an average of 5 hours.
12. Rinse in 70% alcohol.
13. Destain in acid alcohol for about three days. The embryo will appear much darker after clearing. Check the destaining procedure periodically.
14. Wash with potassium acetate in 70% alcohol to neutralize the acid.
15. Dehydrate by placing the embryo in progressively higher alcohols. Allow 1 hour each in 85%, 95%, and 100% alcohol.
16. Clear in xylene for 1 hour.
17. Use scissors to cut the embryo away from the paper. Be careful not to crack the embryo. You may wish to leave a small piece of paper attached to the embryo to use as a handle in transferring the embryo later.
18. Put a drop of mounting medium on a clean slide and place the embryo in it with its head pointed toward your right.
19. A large embryo (72 hours or older) may be so thick that the cover slip balances on it when placed on the slide. If so, chips of glass broken from a slide, or a glass mounting ring may be used to elevate the cover slip so it does not contact the embryo. Be sure the cover slip is level.
20. Place the cover slip over the embryo and put more mounting medium under it until the entire area is filled. Make sure the cover slip is level.
21. Place in a drying oven in a level spot for one to two weeks.
22. Check periodically to see if the mounting medium has receded. If so, add mounting medium by letting it run under the cover slip.
23. Clean the slide with a razor blade and/or acid alcohol, and label. Label may include date, age of chick, your name, and other data.

SIMPLIFIED METHOD OF BONE STAINING WITH ALIZARIN RED SULFATE

THIS METHOD is best used for staining bones of small animals such as mice, rats, chickens, and small birds. The bones stain a bright red and each may be seen individually as the stain penetrates the bones.

These are the steps to follow:

1. Skin and eviscerate fresh specimens. Try to remove all fat deposits.
2. Macerate by placing the specimen in a solution of 1% potassium hydroxide (KOH) in distilled water. Use about 10 times the volume of the specimen. Change to fresh solution every day or two until the specimen is fairly softened. Baby mice require about two days. Older and larger specimens require a longer period, perhaps up to two weeks. (See Figure D.5.)
3. Add alizarin red S stain, only enough to color the KOH solution light purplish. (Use a few drops of concentrated dye solution in the KOH solution, or

Developed as a simplified form by Joan Sturtevant at the Iowa State University Genetics Laboratory.

Fig. D.5—Skeleton of mouse, age 2½ weeks, stained with alizarin red S.

put a trace of dye power in it. Alizarin red S may be purchased from a biological supply company.

4. Add more stain in a day or so if the bones are not coloring enough.
5. When the bones show red, transfer the specimen to glycerine.
6. After a day or so, transfer to fresh glycerine. (Caution: Do not leave stained specimens in sunlight or they will turn brown.)

To make up a concentrated solution of alizarin red S place a pinch of powder in a small dropping bottle. Add distilled water. If the color is not a deep red, add more powder until the color is quite concentrated. This lasts almost indefinitely. If you prefer, add a small pinch of powder directly to the jar containing the specimen in KOH.

REFERENCES

RUGH, R. 1962. *Experimental Embryology.* Burgess Publishing Co., Minneapolis, p. 417 (Basic Staining).

ST. AMAND, G. S., and ST. AMAND, W. 1951. Shortening Maceration Time for Alizarin Red S Preparation. *Stain Technology* 26:271.

COLLECTING AND BANDING BIRDS

YOU MAY HAVE some students who are interested in collecting birds for making study skins, or who are interested in banding birds. Before picking up any dead birds from any area whether they be in a field or killed along the highway, one must have a collector's license. Check, or have students check, with the local or the state conservation commission or your fish and game department for such a permit.

Each state has a list of birds that are protected by law for that particular state. The list may be the same or it may be different from the birds protected by federal law. The 1967 federal list contains about 125 game birds and 400 nongame birds that are protected by federal law. You can obtain the list by writing to the Superintendent of Documents, U.S. Government Printing Office, Washington, D.C. 20402. The current bulletin number is *Wildlife Leaflet No. 469,* at five cents per copy.

For banding birds, both state and federal permission is necessary. Check with your state conservation commission in your particular state and also with the Chief, Bird Banding Laboratory, Migratory Bird Population Station, Laurel, Maryland 20810.

DIRECTIONS FOR MAKING BIRD STUDY SKINS

SKINNING

1. Make a ventral incision from lower keel to (and usually around) cloaca, as shown in Figure D.6 (arrows), being careful not to penetrate the abdominal muscles.
2. With fingers or blunt instrument, free the skin from the muscle laterally from the incision until the leg is reached.
3. Sever the joint between the thigh and leg ("knee") by bending the leg and pushing this joint into the open, as indicated by arrows in Figure D.7. Then, pull the skin from the lower leg until the base of the leg is reached.
4. Remove the muscle from the leg, clean the bone, and cover it with borax. Wrap with a thin piece of cotton.
5. Repeat this procedure for the other leg. Keep the skin damp at all times with wet cotton or a towel.
6. Cut *around* the cloaca and free the skin in that area and on the sides and back. This is the most difficult part

This material was prepared by Dr. Milton W. Weller of the Department of Entomology and Wildlife at Iowa State University, Ames, Iowa.

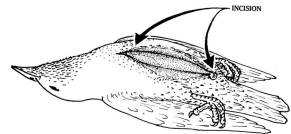

Fig. D.6.

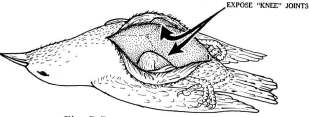

Fig. D.7.

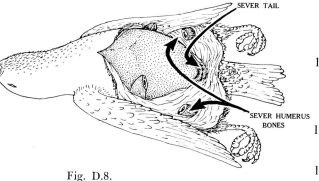

Fig. D.8.

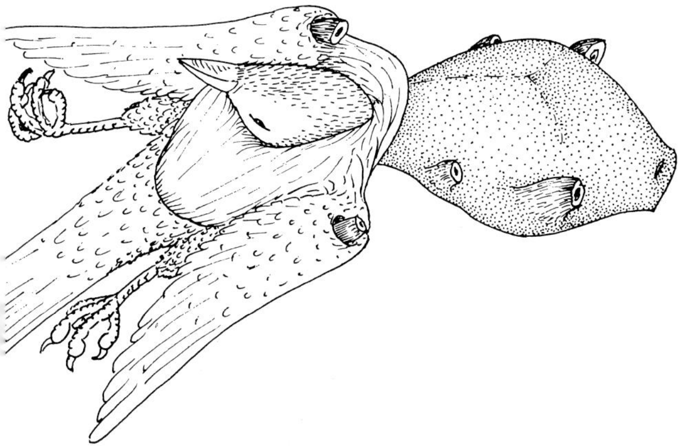

Fig. D.9.

of the skinning process so *proceed slowly.*

7. Skin back until the caudal vertebrae are reached, upon which the tail feathers are mounted. Cut the vertebrae from the body at the area indicated by the broken line in Figure D.8 and free the skin from the dorsal tail region.

8. Carefully remove *all* of the oil gland located dorsally on the caudal vertebrae.

9. Using your thumbnail or a blunt instrument, skin up the back until the wings are reached. Be careful not to stretch the skin. Remove all the fat you can as it will *never* dry as muscle does, and eventually will ruin the skin. Wash in cleaning fluid if necessary.

10. Sever the humerus as close to the body as possible as shown by the arrows in Figure D.8; in small birds this bone will eventually be discarded.

11. Skin out the wing as you did the leg, and carefully skin past the joint of the radio-ulna and humerus and strip the secondaries from the ulna. (Some collectors clean these bones without stripping off the secondaries.)

12. Stop at the distal end of the radio-ulna and remove the small radius. (Again, this is a step not followed by many preparators.) Treat the remaining ulna as you did the leg bones.

13. Leave the skin pushed to the distal end of the ulna as you will eventually place the ulna inside the body.

14. After completing the second wing, skin up the back and neck to the head, as shown in Figure D.9. In small-headed birds you can skin out the head directly. In ducks and other large-headed birds, you must cut the neck at the base of the skull and then make a slit on the throat or crown for cleaning the skull. Keep these head feathers free from blood or moisture.

15. In skinning out the head, keep the cutting instrument as near the skull as possible for you must cut the skin at the very base of the ear and eye openings.

16. Free the skin over all the skull and then remove the eyes and excess muscle.

17. Clean out the brain and also the bony

parts of the lower, inner, and posterior portions of the skull itself by making vertical cuts just inside each jaw and up the back of the skull to get into these bony parts. Remove this ventral and posterior portion of the skull, the tongue, and the brain, and clean out all excess tissue. Dust with borax.

18. Make two larger-than-normal artificial eyes of cotton and place them in the sockets so that the smooth side of the cotton is on the surface.

19. Remove all excess fat from the skin by careful cutting and scraping. Use sawdust or cornmeal to absorb the fat. Rub borax into the skin as a preservative. Borax also can be used for absorbing, but it seems to dull feathers.

20. Make a stitch between the two scapula feather tracts inside the body and between the wings. Draw the two tracts together to one-half their normal distance.

MAKING THE BODY

1. You are now ready to prepare an artificial body of cotton or excelsior. The following method is designed to make a durable as well as attractive study skin, but the procedure is more time-consuming than that of many methods. Find a square stick $1\frac{1}{2}$ times as long as the bird (about the diameter of the bird's neck), sharpen, and force it into the nasal region of the *upper* bill. In large-headed birds, the head may be turned at a right angle to the body but the neck is usually on the same longitudinal axis.

2. Place the bird in a normal position around the stick and mark the approximate length of neck (shorter than normal for appearance) and body.

3. The body and neck should be slightly smaller than the original. Gradually build up the body by using several layers of cotton and *tightly* wrapping each with heavy thread. The body should not *twist* or *slide forward* on the stick. Make the back *flat* and the breast broader and flatter than in life.

4. Slip the sharpened end of the stick into the nasal region of the skull, filling skull spaces with thin cotton wrapping.

Pull the skin over the skull and the artificial neck and body, being sure to place the ulnas inside the body cavity and parallel to the sides of the body. When sewn shut, the ulnas and attached feathers are locked in place.

5. Sew the body shut (and skull incision, if present) and work the feathers over to hide the incisions.

6. Place the wings close to the body and cover their leading edges with the side feathers. Arrange all feathers carefully and pull out minor feathers which "refuse to behave." A good skin is "made" by careful work at this stage.

7. Wrap the body with a very thin sheet of long-fiber cotton. Be sure all feathers are in place. Let dry for a day, open, recheck feathers, and rewrap.

8. Make a label containing at least the following information: species, where collected, date of collection, sex, age, and habitat. Since the color of the soft parts fades or is removed, record color of eyes, bills, and feet. Data on reproductive organs, food contents, weight, and bursa depth (an indication of age) should be noted.

PREPARATION OF MUSEUM-SKINS OF SMALL MAMMALS

SKINNING

1. Record weight in grams, measurements in millimeters, and sex and other desirable data on a specimen label.

2. Carefully make a ventral incision on the posterior one-half (or less) of the abdomen and loosen the skin from the muscle, as shown in Figure D.10.

3. Skin toward the side and expose the thigh of the hind leg, as indicated in Figure D.11. Keep sawdust or cornmeal on moist or greasy areas and keep hands dry at all times.

4. Skin out the leg and sever the bone at the ankle, as indicated by the arrow in Figure D.11, leaving the paw attached to the skin. For convenience, the knee joint may be severed first.

5. Repeat for opposite hind leg.

266

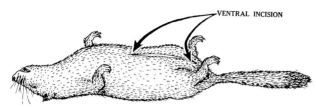

Fig. D.10.

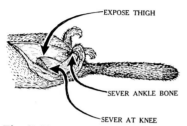

Fig. D.11.

Fig. D.12.

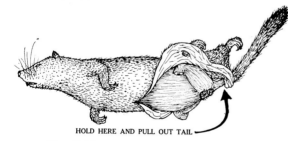

Fig. D.13.

Fig. D.14.

6. Cut around the urogenital openings as shown in Figure D.12 and free the skin from the base of the tail *and the back*.

7. Hold the skin of the tail *firmly* against the tail vertebrae with the thumb and index finger in the area indicated by the arrow in Figure D.13 and pull the tail out *without* everting the skin.

8. Skin forward to the forelegs, removing fat from the skin as you proceed.

9. Skin out the forelegs and cut off at the wrists, as shown in Figure D.14.

10. Skin over the neck to the base of the ears. With the blade at the very posterior part of the ear near the head, cut across the ear opening. Repeat for the eyes, being careful not to damage the eyelids.

11. Cut the lips free from the gums and the nose pad free from the nasal cartilage. (Be especially careful not to damage the skull since this must be preserved.)

12. Make triangular stitching, with two stitches through the inside of the upper lip and one stitch through the lower lip.

13. Clean all fat and muscle from the skin and rub in borax or some other preservative.

14. Keep the skin pliable by wetting with damp cotton.

MAKING THE BODY

1. Tear off a piece of cotton as wide as the head and body of the mammal are long, and sufficiently long to make a tight roll of the same diameter as the mammal. Large specimens are usually made flatter than normal, and very large or delicate skins (like those of lagomorphs) may be supported with a flat board or long leg wires which support both fore and hind feet simultaneously.

2. Using a long forceps, push one end of the cotton into the nose of the skin and fold the other end into the skin.

3. Make four leg wires to insert through the *palms* of the feet. These wires should be long enough to extend from the outstretched foot to the incision so that the end may be grasped easily. Wires should be sharpened with a file

to avoid tearing off the feet when push-ing in the wires.

4. Leaving room on the wires to insert through the palms, wrap a thin piece of cotton around to simulate the leg muscle. Cover the exposed end of the wire so that it will not later rupture the skin.
5. Slip each artificial leg into place.
6. Cut a wire 1½ times longer than the tail so that when inserted through the tail it will protrude into the body cav-ity. Take a long, thin piece of cotton and wrap the cotton around the wire to simulate a tail. Wet the cotton and slip the tail into the skin. The artifi-cial tail cannot be larger than the original and *must* be properly tapered.
7. Sew up the ventral incision with a thread of appropriate color.
8. Attach a completed label to the right hind leg by tying above the foot.
9. Pin the specimen on a board by pin-ning through the feet and around the tail. All four palms or soles should point downward and the feet should be *close* to the body or tail.
10. Open the eyelids, smooth the fur, and allow specimen to dry in a place free of insects.

CARE OF THE SKULL

1. Carefully clean off excess muscle and remove the eyes and brain.
2. Label with a good quality tag bearing the collector's name and collection num-ber, species, and sex of the animal.
3. Tie the string with the tag on it around the lower jawbone.
4. Hang up to dry in an insect-free area.
5. Place in an active dermestid beetle colony until cleaned of "meat." A good colony should take care of a small skull in 12 to 48 hours.
6. Brush off "debris" with a small, soft brush.
7. Immerse in hot (212° F) water for a few minutes to kill clinging larvae and eggs and to wash the skull.
8. Immerse in hydrogen peroxide (3 to 6%) solution for a few minutes to degrease and bleach. For large skulls, boiling in mild ammonia or a 4% borax solution may be substituted.

CARE OF COLD-BLOODED VERTEBRATES

Acquiring animals is one thing, but keeping them comfortable and healthy under laboratory conditions poses a dif-ferent problem.

In handling some animals certain precautions should be taken. Frogs are slippery so must be grasped and held firmly. Large and medium-sized ones should be circled just in front of the hind legs, but small ones are best grasped by the hind legs.

Salamanders should be held firmly, with the head protruding from between the fingers of the hand holding them.

Lizards are quieted by holding their feet. The body should also be grasped to prevent sudden lunges of the animals. Never grab or hold a lizard by its tail because it may break off in your hands.

It is difficult to hold soft-shelled tur-tles. Pressing against the neck with your fingers helps to keep the head from pro-truding far enough to turn around and bite you. The sharp claws can cause serious wounds so should be avoided.

If you capture snakes in the field, hold them just behind the head so they cannot turn around and bite you. Do not try catching poisonous snakes this way, however, unless you have had train-ing and instruction in capturing them. They are best caught by using a snake stick and pinning their heads down, but not hard enough to injure them.

Animals may be transported to the laboratory in bags, bottles, or traps. Deli-cate specimens are best carried in bot-tles. If moisture is needed for the speci-mens, bottles retain it better than other carriers. Plastic bottles are very good.

Snakes are easily carried in bags

tied with a simple overhand knot. String should not be used for tying bags as some snakes may push the string off the bag and escape. If transporting a poisonous snake, hold the bag far enough from your body so you will not be bitten.

Captured specimens may be kept briefly in the laboratory for observation, then released in the same area where they were taken, or they may be maintained in the laboratory for quite some time.

The most versatile cage is a regular aquarium. It should be at least 15 inches long and 10 inches wide and deep. Make a heavy or tight-fitting lid of screen wire tacked to a wooden frame. It is practical to build two hinged doors into the lid so you can get your hand inside the cage to service half the cage at a time. Thus, the specimens need not be caught and removed each time the cage is to be cleaned. This is less disturbing to animals such as lizards and tree frogs. Animals thrive best when left undisturbed.

To prepare the cage, fill the bottom of the aquarium with small clean pebbles to a depth of one inch. Cover with $\frac{1}{4}$ inch to $\frac{1}{3}$ inch of charcoal chips. This helps keep the soil from becoming "sour." Add one inch or more of soil that is sandy or loamy. Small plants can be rooted in the soil, but it is more practical to put plants in flower pots with the base resting in the layer of pebbles or charcoal. When cleaning the cage, lift out the pots and clean rather than uprooting the plants. Some frogs and lizards like to cling to the sides of the pot.

Select any plants for the cage that can be watered daily. If they are sprinkled, the lizards will lap up the drops. Some lizards will not drink water from a dish, so they need this means of obtaining water. However, some terrarium cage residents such as frogs and toads need a shallow water container of glass, plastic, or nonrusting metal for drinking. Many animals like to climb, so you should place a few branches in the cage.

Lizards from dry habitats need the floor covered with dry, clean sand with no charcoal or pebbles. Place cactus plants in their habitat. Since cacti cannot stand daily sprinkling, provide water for the animals by filling a shallow container to the brim and letting some of the water spill over to make a circle of damp sand around the dish.

Snakes, toads, and turtles are usually not good terrarium animals as they burrow out of sight and dig up plants, pebbles, and charcoal. If such specimens are kept in the laboratory they do best on bare pebbles with potted plants and shelter areas. Shelter may include a piece of bark with its concave side down; small flat stones piled up to make a miniature cave; and small, flat cardboard boxes.

Certain animals such as newts, some frogs, and turtles that like water do well if they have a choice of wet and dry environments. To provide for this, insert a partition across the center of the aquarium and pile pebbles on one side with some potted plants and a shelter area. Put water in the other half of the aquarium. Provide a ramp of small, rough stones or bark for small turtles. Young caimans and alligators need the same kind of accommodations but much larger ones. They need an area where they can stretch out full length either in or out of the water.

Only a very few kinds of reptiles and amphibians can live in straight-sided aquariums with no place to crawl out. Included in this group are mud, musk, and soft-shell turtles; rainbow, mud, and swamp snakes; aquatic newts; sirens; Amphiumas; mudpuppies; waterdogs; and hellbenders. These should all have a stone cave at the bottom of the aquarium to serve as a retreat. Aquatic vegetation should be used for resting and hiding areas for these animals.

The simplest cage for a snake is the aquarium. Sheets of newspaper folded and cut to fit the bottom of the cage absorb excess moisture. When soiled they can be lifted from the cage. For convenience, place about 8 or 10 pieces of paper at one time on the floor of the cage. Air holes are not needed if the top is completely screened.

Provide a dish of water large enough for the snake to completely submerge itself yet heavy enough so it will not tip over. Provide a hiding place of bark or stones, or a cardboard box with a hole in its side. Snakes generally thrive best if their living quarters are completely dry except for the water in the water dish.

Both reptiles and amphibians closely approximate the temperature of the room or the cage in which you keep them. When they are cold their body functions slow down; appetites become poor or even nonexistent. If they are too warm they become very uncomfortable. Too much heat will kill them. Temperatures between 75° and 85° F are best for most reptiles. Amphibians will thrive under somewhat cooler conditions.

During summer months reptiles and amphibians may be kept in somewhat cooler places. Lizards and turtles need some sunshine. Snakes can get along without sunshine if they eat well and have good food. All these animals need an area in their cage where they can bask and an area where they can re-treat to the shade as they choose.

During the wintertime, normal room temperature, if the room does not chill at night, is suitable for amphibians. If reptiles are to remain active they need to be kept warmer—around 75° to 80° F. If there is a considerable drop in the temperature at night, supplementary heat such as an electric light with a reflector directed into the cage may be sufficient.

Most reptiles and amphibians are carnivorous or insectivorous. A few eat plant material such as lettuce and fruit.

An excellent ration for alligators, caimans, aquatic salamanders, and turtles is canned dog food that is fortified with vitamins and minerals. You can make your own ration by thoroughly mixing a few drops of liquid vitamin concentrate and a pinch of bone meal or oyster shell flour with raw lean hamburger. Mold this mixture or the canned dog food into small rounded pellets and place them at the edge of the water. Animals will take a piece at a time into the water. Most turtles will eat some vegetable matter such as pieces of lettuce. Some will eat lawn cuttings. Feed them some green material at least once a week.

Tortoises and box turtles should have soft fruits and berries in addition to greens. Occasionally they should be given meat.

Reptiles and amphibians need live insects or other invertebrates for best living and growing. In winter when it is difficult or impossible to collect live insects, meal worms or tubifex worms from tropical fish stores may suffice. Roaches, crickets, and grasshoppers are other foods that have proved to be satisfactory for reptiles and amphibians.

Cold-blooded animals are interesting to have in the laboratory but they take considerable care and attention. Perhaps after they have been observed and studied for a few weeks or days, you may want to return them to their natural habitat. If you do return them, release them in a familiar habitat, preferably the one from which they came.

If baby turtles are purchased from a pet shop and have enamel painted on their shells, remove it by flaking it off with a knife or razor blade. Turtle shells are living, growing parts of the animal and such things as paints or enamels cause malformations of the shell by preventing proper growth.

When preparing an aquarium for fish in the laboratory, use larger rather than smaller aquaria. Some inexpensive all-glass bowl-type aquaria are available up to two gallons. These may be used when just a few small fish will be placed in each bowl.

Metal-frame glass aquaria are available to 50-gallon or even greater capacity. Problems such as leakage are reduced if water is kept in the tanks even when fish are not there.

For tropical fish, water temperature should be maintained at about 80° F. If the laboratory becomes chilly at night use a light for heating the water.

A rule of thumb for calculating the number of fish to place in an aquarium is an inch of fish per gallon of water. If fishes are crowded in an aquarium, both reproduction and growth are inhibited.

Before placing fish in an aquarium, fully prepare the area for the inhabitants. Use spring or well water or rain water collected in a nonmetal container, or water that has stood long enough to develop a population of microscopic and semimicroscopic plants or animals. You may also use conditioned water or water that has had fish in it before. Some fish grow better in water that has had fish in

it before. It has been suggested that water previously occupied by fishes contains a bacteriophage that controls the growth of undesirable bacteria.

Many small fishes will reproduce in aquaria without any special inducement if kept at the right temperature, fed adequately, and are not crowded.

Such fish as swordtails, platys, and guppies will reproduce readily but are inclined to eat their young unless plant cover is available for hiding the young. Plants such as water sprite (*Ceratopteris thalictrodides*), the *Vallisnerias* (*Vallisneria americana* and *spiralis*), and the Amazon sword plant (*Echinodorm rangeria*) are satisfactory for growth.

Water sprite is most versatile and grows equally well in a substratum and totally submerged, or floating at the surface.

The *Vallisnerias* are grasslike in appearance and require a substratum of sand gravel. They reproduce asexually very rapidly if a sufficient amount of light is present.

The Amazon sword plant is broadleafed and requires a substratum.

A pH range of 5.0 to 9.0 is tolerated by fishes. Overfeeding fish results in a polluted condition of the water. Feed fish just what they can consume.

In general, fishes thrive better on live food rather than any other recommended foods. *Daphnia* are excellent food for young fishes as well as for adults of many small fishes. *Daphnia* are especially abundant in temporary pools. Collect by straining fertile, stagnant water through fine mesh silk or nylon cloth. *Daphnia* may be reared in the laboratory by keeping them in aerated aquaria and feeding them small quantities of dried yeast or bone meal.

Brine shrimp (*Artemia salina*) are hatched from eggs obtained from supply houses. The eggs are placed in salt water (eight tablespoons of salt per gallon of water) kept at a temperature of 70° F to 80° F in an aerated aquarium. They hatch in 24 to 48 hours. If you keep several containers hatching in sequence they provide food for small fish. Brine shrimp are attracted to a light so may readily be separated from the eggs in the aquarium.

Tubifex worms *(Tubifex tubifex)* are found in dense populations at the edge of water in ponds and rivers that are polluted with sewage or other concentrated organic matter. These worms are about one inch in length and are excellent food for most fishes. They can be dipped up with mud in which they remain partly buried. The mud can be washed away and the worms fed to fishes.

White worms *(Enchytrae)* may be purchased from pet dealers and cultured in trays of damp, loam soil at a temperature of 55° to 70° F. They thrive on Pablum, oatmeal, or bread. Cover the tray with burlap and the worms will collect on it. Worms tend to collect at spots where the food is added. They can be picked out in masses with tweezers and fed to the fish.

Earthworms *(Lumbricus* spp.) are good for fish food and may be given to fish either whole or chopped, according to the size of the fish. For breeding worms, place in a bed of rich soil that includes some leaf mold. Feed a variety of substances such as cornmeal, oatmeal, bread, and even small amounts of lard. Soil must be damp but not wet. Use no sand in the bed as sand grains damage the alimentary tract of the worm.

Meal worm larvae of beetles in the family Tenebrionidae can be cultured in chick laying mash placed in a one-inch layer over the bottom of a covered container. Laying mash should be in thin layers as the meal tends to sour. Either the larvae or adult beetles may be obtained from the wild or from a biological supply house. Moisture is added, if necessary, on cotton or a small cloth. Larvae are obtained by sifting the material through a screen and picking them out. Fully developed larvae are about an inch long, but any size may be fed to fish.

Prepared foods for fish are generally made of high proteins from plant materials supplemented with meat scraps, minerals, and torula yeast. The mixture may be ground to a fine powder for smaller fish or made into floatable pellets for larger ones. For fish that feed at the bottom of the container, pellets that sink are available.

Dried foods are available from pet shops.

To make a liver-cereal ration that is to be frozen until used, dice a pound of liver, add a small amount of water, and liquidize in a food blender. Add 20 tablespoonfuls of Pablum and 2 teaspoonfuls of salt and heat until the liver is coagulated. Freeze until fed to fish.

Dried shrimp is available from tropical fish dealers. It is very convenient to use but should be supplemented with other foods.

Many diseases and parasites attack fish, but it may be difficult to determine the cause of the ailment. An inadequate diet can weaken a fish to the extent that infection sets in and causes death.

Mechanical damage with even slight injury to the fish may result in fungal and bacterial infections. The more important infection organisms include *Aeromonas* among the bacteria and *Saprolegnia* among the fungi. These are usually present in the aquarium but become a problem only when fish are mechanically damaged or weakened because of an inadequate diet.

Furunculosis, a bacterial disease caused by the bacterium *Aeromonas salmonicida,* causes capillaries of the fish to burst, resulting in red spots, reddening at the base of the fins, and a bloody fluid from the intestines. The kidneys become swollen and boils or furuncules may or may not develop. Chemical treatment recommended is sulfamerazine at the rate of 0.1 g per pound of fish.

Other species of *Aeromonas* cause different pathogenic conditions in fishes. These may include infectious dropsy, involving accumulation of serous fluids in the tissues, and a fatal surface infection of fish following handling of them.

The surface infection is easily controlled by using 20 ppm of water-soluble Terramycin (veterinarian grade) in the water for 24 hours or longer.

For dropsy, use Chloromycetin or Terramycin in the food at the rate of 1.0 mg per 15 g of fish for a period of four days.

The disease columnaris, caused by the bacterium *Chondrococcus columnaris,* produces gray-white areas on the head, gills, body, and frequently in the mouth of the fish. These areas become ulcerated and the fins may become frayed. Infection becomes general on the exterior of the fish and finally moves to the internal organs. For controlling columnaris, use a 1:2000 parts copper sulfate dip two or three times at 12- to 24-hour intervals during the early stages of the infection.

The protozoan parasites *Costia nectris* and *C. pyriformis* are external fish parasites. They cause a blue or gray film to form on the fish's body. Although this is considered to be a fatal disease that develops quickly, control may be successful by dipping fish for one minute in a 1:500 solution of acetic acid, or a 12-hour treatment of a 1:20,000 formalin solution.

Ichthyophthirius multifilies causes uniformly shaped white specks to be scattered over the host fish. The parasites may be controlled in part by placing fish in running water for three or four days. Keeping the water at 85° F for several days will control "ich" on tropical fishes. If fish are confined in smaller tanks, quinine sulfate at the rate of 0.032 g per gallon of water in a tank and left unchanged for three weeks will control it. Since quinine is poisonous to plants, they must be removed from the aquarium.

Gold dust disease is caused by a dinoflagellate *Oodinium climneticum.* Infected fish appear as if covered with a golden dust that is due to a yellow pigment. The attached stage of *Oodinium* is parasitic and causes infected fish to become emaciated. Young fish may suffer high mortality. Control may be effective by using a malachite green dip of 5 ppm for 2 minutes for controlling encysted forms.

Of fungal diseases, water molds of the genus *Saprolegnia* are the most important in the infection of fishes that are damaged, diseased, or suffering from malnutrition. Molds grow well at lower temperatures and when there is an accumulation of food in the aquaria.

The presence of cotton-like growth of mycelium indicates that *Saprolegnia* is involved.

For treating fish for fungal growths, a copper sulfate dip of 1:2000 for 1 to 2 minutes or a malachite green dip of 1:15,000 for 10 to 30 seconds is recommended.

Lymphocystis is a viral disease of fishes. It affects the epidermis of fishes, causing the production of a wartlike growth on the fins. The best thing to do is to discard the fish and obtain new stock.

Any of these cold-blooded vertebrate animals add interest to a classroom. However, they can do much more than this. Students can make habitat studies of them. They can see which kinds of foods these vertebrates seem to prefer. Which type or types of hiding places do they seem to prefer? Do snakes seem to enjoy being handled? How can you determine this? Can they be taught to respond to various stimuli? When are they most active? When least active? How well do they get along with others of their species in the same environment? How often do they eat? Are they continuous or intermittent feeders? Do any attempt to make nests in the cage?

REFERENCES

Conant, Roger. 1958. *A Field Guide to Reptiles and Amphibians.* Houghton Mifflin Co., Boston.

Lewis, William M. 1963. *Maintaining Fishes for Experimental and Instructional Purposes.* Southern Illinois Univ. Press, Carbondale.

PRESERVATIVES AND FIXING MATERIALS

Changes occur in cells when they are killed. Bacteria multiply and destroy them. Autolysis or self-digestion of cells occurs by means of the enzymes in the cells. Enzymes begin to split proteins into amino acids, which diffuse out of the cells.

In order to prevent these changes from occurring, it is necessary to place the killed cells in fixatives or fixing solutions which will penetrate them rapidly. These solutions should also protect cells against shrinkage and distortion during the processes of dehydration, embedding, and sectioning. Fixatives should also permit the parts of the cells to become clearly and selectively visible when stains and dyes are used.

It is necessary, then, to place the killed specimens in fixatives as soon as possible after death occurs. In selecting a fixative consider how the tissue is to be used. Is an all-purpose fixative adequate, or are certain parts to be considered?

Aqueous fixatives dissolve out glycogen, but alcoholic fixatives remove lipids. How rapidly does the fixative penetrate tissues? What size should the pieces be so penetration of tissues by the fixative is adequate? Does the fixative harden tissues excessively? How much fixing fluid should be used?

If in doubt as to how the tissue will be used later, place it in formalin.

Use an amount of fluid equal to at least ten to twelve times the size of the tissue.

Place the tissues in the fixative as soon as possible after removing from the organism.

If an organ is covered by a tough membrane, trim it with a sharp razor blade or scalpel rather than with scissors, which damages tissues by crushing them.

Do not tear or crush tissues while removing them.

Do not allow tissues to become dried before placing them in fixative.

Using normal saline, wash off any accumulation of blood.

Use as small pieces as possible for fixing.

Shake solution gently with tissue in it so fluid reaches all parts.

Place a piece of cotton, sponge, or glass wool in the bottom of the bottle to

274

prevent tissue from sticking to the bottom of the container.

These solutions are commonly used and are not difficult to prepare. They have been tested on many different kinds of tissue.

1. Formalin

This is a good general fixer for all classes of materials, but not preferred for very fine cytological work. It preserves fat and myelin. It is widely used for brain and spinal cord and is an excellent hardener. Tissues placed in it become tough but not brittle.

Formula
Commercial formalin—
    full strength   1 part
    Water   10 parts
(If fixing marine specimens, use seawater.)

It takes about 12 to 24 hours to fix material. For large materials, such as the brain or liver, fix for several weeks to several months.

When ready to stain the material, wash several times in 50% alcohol.

Any stain may follow this.

2. Formalin-Acetic-Alcohol (FAA) Formula No. 1

Commercial formalin 5 parts
Glacial acetic acid   5 parts
Alcohol 50%   90 parts
This is often called the universal fixer. It is good in all general work, but not recommended for fine cytological work. It is good for field collecting because it fixes and preserves materials for months. It will fix an amount of material that is equal to its own weight. This preservative is preferred by botanists.

Fixing time is 24 to 48 hours. It preserves for an indefinite time.

When ready to stain, wash the material several times in 50% alcohol, then use any stain.

3. Formalin-Acetic-Alcohol (FAA) Formula No. 2

Commercial formalin 6 parts
Glacial acetic acid   1 part
Alcohol, 95%   20 parts
Distilled water   40 parts
This FAA formula is preferred by zoologists.

4. Formalin-Acetic-Alcohol (FAA) Formula No. 3

Commercial formalin 6 parts
Glacial acetic acid   1 part
Alcohol, 95%   15 parts
Distilled water   30 parts
This formula is recommended for arthropods, especially for various insect larvae.

5. Formalin-Acetic-Alcohol (FAA) Formula No. 4

This formula is spoken of as the universal preservative by the General Biological Supply House. It consists of:
Commercial
    formalin   6.5 parts
Glacial acetic acid 2.5 parts
Alcohol, 50%   100.0 parts
This is for general fixing, preserving of whole objects, and field collecting.

6. Formalin-Corrosive

Commercial formalin 1 part
Distilled water   10 parts
Saturate the formalin plus water with potassium carbonate.

To 9 parts of the above mixture, add 1 part glacial acetic acid. Mix this just before using.

The formalin-corrosive mixture is very good for fixing materials such as protozoa, embryological material, and nervous tissue.

Fix for 6 to 24 hours.

Wash in running water 6 to 12 hours.

Use any desired stain.

7. Alcohol

Tissues may be fixed and preserved in alcohol of 70% to 80% strength. If they are to remain for several months, it is preferable to preserve them in a mixture of equal parts of glycerin, distilled water, and 95% alcohol. This prevents them from becoming too hard and brittle.

8. For fixing and preserving insect specimens use:

Distilled water	280 parts
Formalin, full strength	40 parts
Glacial acetic acid	20 parts
Glycerine	50 parts
Isopropyl alcohol	175 parts

9. For Bouin's fluid, a general preservative, use:

Formalin, full strength	20 parts
Glacial acetic acid	5 parts
Picric acid saturated aqueous solution	75 parts

10. For preserving flagellates use:

Cupric acetate crystals	2.5 g
Distilled water	100.0 cc
Formalin, full strength	10.0 cc
Glacial acetic acid	1.5 cc
Picric acid crystals	4.0 g

11. For processing plant tissues and protozoa mix:

Potassium carbonate crystals	14.0 g
Distilled water	200.0 parts
Formalin, full strength	22.5 parts
Glacial acetic acid	25.0 parts

12. To preserve green color in plant tissues use:

Carbolic acid crystals	20.0 g
Cupric acetate crystals	0.2 g
Glycerin	40.0 g
Lactic acid	20.0 g
Cupric chloride crystals	0.2 g
Distilled water	20.0 parts

INJECTING ANIMALS

ANIMALS that are embalmed or preserved, then injected with colored latex to show the circulatory system, may be purchased from biological supply houses—or, you may do the injecting, using animals such as crayfish, frogs, salamanders, small snakes and lizards, turtles, cats, rabbits, and dogs.

Embalming fluid preserves specimens, keeps muscles firm, and helps keep their natural color. It also gives specimens a clean odor and helps prevent them from drying out.

Embalming fluid consists of a mixture of:

Carbolic acid, melted crystals	5 parts
Formalin, 40%	5 parts
Glycerine	5 parts
Water	85 parts

For injecting an average-size cat use about one liter of embalming fluid. Before injecting the embalming fluid, expose a femoral artery by making an incision through the skin on the inner surface of the thigh (the inguinal region). Use sharp, pointed forceps and carefully separate the femoral artery from the closely associated vein and nerve. (The artery is smaller in size and of a lighter color than the vein.) Remove the tissue surrounding the artery for about two inches back of the point where it emerges from the body cavity. Cut off an eight-inch piece of number-ten white cotton thread and tie it loosely around the artery. Use sharp, fine-pointed scissors to make a forward oblique cut about halfway through the artery. Insert a sixteen- or eighteen-gauge needle into the artery at the incision point, just far enough so it can be tied in firmly by pulling down the thread that was placed around the artery previously.

Keep embalming solution from escaping by forcing cotton into the throat and mouth of the animal. If a large animal is being embalmed it may be necessary to refill the syringe several times during the process.

After specimens are embalmed they are ready in a few days for injecting the circulatory system with colored latex or starch mass, both of which may be purchased from biological supply companies. Latex comes in red, blue, and yellow colors. Starch mass is made up as indicated. Boil for 10 minutes.

Water	500 cc
Glycerine	100 cc
Formalin, 400%	100 cc
Cornstarch	1 lb
Color (make it as dark or as light as you wish)	

Use finely powdered carmine for the red, Berlin Blue for the blue, and finely powdered lead chromate for the yellow.

In injecting a cat put the latex from behind into the common carotid artery. Expose it by making a longitudinal cut parallel with the fibers of the sternomastoid muscle. Carefully dissect the common carotid away from the internal jugular vein and tie it lightly anteriorly. Posteriorly tie a loose knot around the artery, make an incision in the vessel, insert needle from the rear, and pull knot tight. Inject about 30 cc of the latex, remove needle, and tie the artery.

Fill veins by injecting into the external jugular vein. Use 30 cc to 50 cc of blue starch mass.

Inject the hepatic portal system through an intestinal branch of this system. Cut through the lateral abdominal wall, pull out a coil of the intestine, and inject starch mass into a larger branch of the intestines. When the needle is removed, tie off the vessel and sew up the incision in the body wall, and also in the skin of the leg and neck.

To keep specimens from drying out, place them in plastic bags or wrap them completely in oilcloth and place them on shelves in the laboratory or in tanks.

Some schools will not permit students to embalm and inject laboratory animals. In these cases animals should be purchased from biological supply houses.

General Biological Supply House has a leaflet, *Turtox Service Leaflet No. 21*, that describes the preparation, injection, and care of embalmed specimens. Leaflets are free to teachers. Size of needles and syringes best for injecting different kinds of animals are listed.

WHERE TO INJECT BLOOD VESSELS[1]

MAMMALS

Arteries. Inject caudally in the common carotid.

Veins, systemic. Inject toward the head in the external jugular. To insure filling the posterior veins, inject into the saphenus, a tributary of the femoral on the inner surface of the thigh.

[1] From the leaflet, "How To Use Injection Materials." Ward's Natural Science Establishment, Inc., P.O. Box 1717, Rochester, N.Y. 14603.

Veins, hepatic-portal. Into any tributary of the hepatic-portal, usually the mesenteric.

BIRDS

Arteries. Brachial or femoral.

Veins, anterior. Jugular.

Veins, posterior. Femoral.

REPTILES

Turtles, lizards, snakes

Arteries. Right side of ventricle into aorta or directly into the carotid artery.

Veins, systemic. Caudally in jugular.

Veins, hepatic-portal. Ventral abdominal vein.

Alligators

Arteries. Right ventricle into aorta.

Veins, systemic. Jugular.

Veins, hepatic-portal. Ventral abdominal.

SALAMANDERS

Including Necturus and Cryptobranchus

Arteries. Aorta in thoracic region.

Veins, systemic. Post cava between liver and gonads.

Veins, hepatic-portal. Ventral abdominal.

FROGS

Arteries. Conus arteriosus.

Veins, systemic. Cutaneous vein in region of pectoral limb.

COLLECTING AND PRESERVING SMALL INVERTEBRATES

MANY SMALL INVERTEBRATES can be collected from the surrounding area and maintained in the laboratory. They serve as experimental animals as well as interesting specimens to watch and observe.

Hydra may be collected from areas such as freshwater ponds, slow-moving streams, lakes, and reservoirs. Usually they are found clinging to the lower surface of lily pads or other aquatic plants.

Vegetation should be placed in containers, taken to the laboratory, opened, and permitted to remain undisturbed so the *Hydra* may expand and become visible.

In *Science*, Vol. 117, page 565, May 15, 1953, W. F. Loomis describes a method for cultivating *Hydra* which has proved successful. He uses Littoralis (L) solution instead of pond water and dried eggs of the brine shrimp, *Artemia,* for a living crustacean source.

He grows *Hydra* at room temperature in L. solution in finger bowls or shallow dishes, using 200 ml of culture solution per container.

The following is the composition of the L. solution for growing *Hydra:* For each liter of distilled or demineralized water use 0.35 g NaCl; 0.07 g $CaCl_2$; 0.01 g $NaHCO_3$. Do not use tap water as most tap water is toxic to *Hydra*.

Two days after brine shrimp eggs are put in saline solution, they are about the right size for the *Hydra* to feed upon.

278

To hatch the brine shrimp (eggs may be purchased from biological supply houses or tropical fish shops) dust about ½ g of the dried eggs on the surface of about 500 ml of salt solution (3.5 g salt per liter of distilled water). Algae growing on the sides of a freshwater aquarium, or a minute amount of yeast sprinkled in the live shrimp culture, provide food for the brine shrimp.

To collect the shrimp larvae shine a light at one end of their containers and collect with a fine mesh net that has been dipped in L. solution.

Within a day after the *Hydra* are fed, decant the old culture fluid and replace with fresh L. solution.

Planarians may be found in a variety of habitats, depending upon the species. Some may be located among algae and water plants and under stones and rocks when the water is rather quiet. Other species may be collected from small ponds and brooks that are spring fed, and among algae on the sandy bottom—even when the water is flowing swiftly.

Some people have attracted *Planaria* to small pieces of liver attached to string placed in a stream or near some rocks. *Planaria* seek the liver as food and may be lifted from the stream clinging to the liver.

Planaria may be cultured in enameled pans and on wooden tubs or buckets partially filled with spring or distilled water, and fed worms, pieces of liver, or a hard-boiled egg yolk one or two times a week. For best survival of the *Planaria,* surplus food should be removed and the water changed when the worms finish feeding.

Daphnia or "water fleas" are best collected in a fine-meshed net by sweeping through aquatic plants in shallow water. They may also be collected from

shallow waters of stagnant ponds and ditches.

Best results in rearing *Daphnia* occur if they are cultured in the water in which they are collected. Pond water or water from a well-balanced, established aquarium may also be used.

Crayfish are found in lakes, ponds, and streams. They hide under rocks, water plants, stones, logs and sticks, or they may construct artificial hiding places. They hide during the day and are active at night. Crayfish may be captured by lifting stones or rocks to find the hiding animals. They may also be captured in nets swished around aquatic plants or mud in the bottom of streams and ponds.

Crayfish are interesting to observe in the laboratory, but should be kept in a large tank with a pile of loose rocks and stones at one end. Small specimens are preferable to large specimens. Two or three specimens of a length up to two inches are enough for a six-gallon tank. They may be fed worms or small pieces of raw meat.

If a female with eggs is found (usually in late autumn or early spring) place her in an aquarium and watch the eggs develop into young crayfish.

Some species of *Isopods* (sow bugs) may be found in dead and decaying vegetable matter. Often they are located in greenhouses. They are somewhat flattened and segmented. During the breeding season the female develops lamellae at the base of some thoracic legs which cover and hold the eggs. These may be observed if specimens are kept in wide-mouthed containers of soil and decaying vegetable matter during the breeding season.

Amphipods, somewhat resembling tiny crayfish, live in ponds and slow-flowing streams. Collecting by using small nets is usually satisfactory. A one-to five-gallon aquarium, with sand, growing plants, dead leaves, mud, and sticks from a pond, provides a good growing area for these animals. Fish that would feed upon them should not be included in the tank.

REFERENCES

ANDREWS, E. A. 1904. Breeding Habits of Crayfish. *American Naturalist* 38:165–206.

LOOMIS, W. F. 1953. The Cultivation of *Hydra* Under Controlled Conditions. *Science* 117:565–66.

RICHARDSON, H. 1905. A Monograph on the Isopods of North America. *Bull. U.S. Nat. Mus.* 54:727 pp.

Turtox Service Leaflet No. 27. 1959. Brine Shrimp and Other Crustaceans.

WARD, HENRY BALDWIN; and WHIPPLE, GEORGE CHANDLER. 1945. *Fresh Water Biology.* John Wiley & Sons, Inc., New York.

Books,

Magazines,

Periodicals,

Journals

MAGAZINES AND PERIODICALS

MAGAZINES AND PERIODICALS belong in the classroom library as well as in the school library. They add depth as well as breadth to a biology course.

Articles related to the subject matter on hand may give to students a different perspective from that offered in the text.

The following list includes some magazines and periodicals that should be available for use of both students and teachers.

American Biology Teacher, National Association of Biology Teachers

American Journal of Botany, Botanical Society of America

American Journal of Human Genetics, American Society of Human Genetics

American Midland Naturalist, University of Notre Dame

American Naturalist, American Society of Naturalists

American Zoologist, American Society of Zoologists

Animal Kingdom, New York Zoological Society

Annals of Botany, Oxford University Press

Applied Microbiology, Williams and Wilkins Co.

Audubon, National Audubon Society, Inc.

Audubon Field Notes, National Audubon Society, Inc.

Bacteriological Reviews, American Society for Microbiology

Bausch and Lomb Focus, Bausch and Lomb

Biological Bulletins, Marine Biological Laboratory

BioScience, American Institute of Biological Sciences

Botanical Review, New York Botanical Garden

Canadian Journal of Microbiology, National Research Council of Canada

Developmental Biology, Academic Press, Inc.

Discovery, Professional and Industrial Publishing Co.

Ecology, North Carolina State University

Economic Botany, The Society for Economic Botany

Endeavor, Imperial Chemical Industries

The Explorer, Cleveland Museum of Natural History

Field and Stream, Holt, Rinehart and Winston, Inc.

Iowa Science Teacher's Journal, Iowa Academy of Science

Iowa State Conservationist, Iowa State Conservation Commission

Isaac Walton Magazine, Isaac Walton League of America

Journal of California Teachers Association, California Teachers Association

Journal of Cell Biology, Rockefeller University Press

Journal of Heredity, American Genetic Association

Journal of Research in Science Teaching, John Wiley & Sons, Inc.

Missouri Conservationist, Missouri Conservation Department

National Education Association Journal, National Education Association of the United States

National Geographic, National Geographic Society

National Geographic School Bulletins, National Geographic Society

National Wildlife, National Wildlife Federation

Natural History, American Museum of Natural History

New Scientist, Cromwell House

Newsweek, Newsweek, Inc.

Phytopathology, American Pathological Society

Quarterly Review of Biology, American Institute of Biological Sciences

School Science and Mathematics, Central Association of Science and Mathematics Teachers, Inc.

Science, American Association for the Advancement of Science

Science Education, Science Education, Inc.

Science on the March, Buffalo Society of Natural Science

Science News, Science Service, Inc.

Science Progress, Blackwell Scientific Publications, Ltd.

Science Teacher, National Science Teachers Association

Scientific American, Scientific American, Inc.

Senior Science, Scholastic Magazines, Inc.

Systematic Zoology, Society of Systematic Zoology

Time, Rockefeller Center

Today's Health, American Medical Association

PREPARING A MANUSCRIPT FOR PUBLICATION

MANUSCRIPTS should be neatly typed with double spacing on 8½ x 11-inch sheets of good quality white paper. Typing should be on one side of the paper only, and the sheets numbered consecutively from page 1 to the end. Keep at least one carbon copy of your material.

Do not fasten sheets together in any way. If there are five sheets or less in the manuscript, fold them in thirds and place them in an envelope, but if there are more than five sheets leave them flat and mail them in an envelope.

Put your name and address in the upper left-hand corner of the first page. Place the title about one-third the way down from the top of the page and put your name under it:

TITLE OF ARTICLE

by

Your Name

Start the article or story about half-way down the first page. Be sure you double-space it.

On pages following the first one, put your last name at the top left-hand corner of the page and the number of the page at the top right-hand corner.

Leave a margin of about one and one-half to two inches at the left side, the top, and the bottom of the pages following the first page, and an inch at the right of the sheets.

Use either pica or elite type, but be sure to use a clean ribbon and keep the type faces clean. Black or blue ribbons are preferred over colored ribbons.

In a bibliography, the author's last name should come first, then his first name and middle initial. The date of the book follows this, then the name of the book, and finally the name and address of the publisher, as follows:

Herskowitz, I. H. 1962. *Genetics.* Little, Brown and Co., Boston.

Sinnott, Edmond W., Dunn, L. C., and Dobzhansky, Th. 1958. *Principles of Genetics.* McGraw-Hill, Inc., New York.

Identify footnotes with an asterisk (*) placed by the text material and also at the bottom of the page, or by a number to correspond to the footnote. Indicate as follows: [2] Brian Hocking. *Biology—or Oblivion.* Schenkman Publishing Co., Cambridge, Mass. 1965.

Drawings made as illustrations should have narrow lines densely black and sharply ruled so that when the engraver transfers them photographically as an acid-resisting film to a zinc plate, they will withstand the acid bath that removes the undesirable portions of the plate's surface. He makes letters such as e, p, and b large enough to prevent them from filling with ink and printing as a blur when the book is on the press.

The most satisfactory material available for preparing drawings for reproduction is Strathmore two-ply plate-finish Bristol board. A drawing may be laid out on it in pencil before inking, and as it erases without smudging or roughening, the preliminary pencil sketch can be corrected or changed before finishing. After the drawing is finished in pencil it can be inked with fresh, undiluted Higgins or Keuffel and Esser black waterproof drawing ink. When finished, the drawing may have an overlay of tissue paper on the top and rubber cementing on the back. The overlay is keyed to the drawing at one or two points and the lettering done on it.

Photographs are an excellent tool for communication. They should be sharp and clear and tell a story. Photos may be 5 x 7 inches to 11 x 14 inches in size, usually glossy finished, with a caption attached to the picture as for the drawing. Your name and address should be on the back of each picture either with a gummed name label or printed lightly with a rubber stamp. Any writing shows through when the picture is reproduced.

SUGGESTED TOPICS TEACHERS MAY WRITE ABOUT

1. Evergreens for Christmas—which retain needles longest; trees with best allover shape; best time to buy trees.
2. Honey from your own hive—bee care; flowers for best honey; removal of honey from the hive; care of bees in winter.
3. Ornamental trees for the yard—ones which do well in your section of the country; length of time to produce a full-grown tree; good or poor qualities of the trees.
4. Dutch elm disease—prevention of the disease; signs of progress of disease; treatment.
5. The value of snakes to farmers—worth of a single snake in terms of insects and rodents eaten; recognition of snakes.
6. Fact or fiction—stories, sayings, and tales often told, many times believed. Sort out the basis of a story and see if it has facts or superstitions as its root.
7. Color changes in leaves during autumn —cause of changes; pigments present; rate of color change.
8. Speed at which animals can run and birds can fly—comparison of size to speed, weight, length of legs, or wing span; length of time the speed can be maintained.
9. A camp-out vacation—basic needs for a family; places to go; easily prepared meals.
10. Food habits of various races or tribes— odd animals that make up part of the diet; means of preparing foods; strange habits in eating.
11. Bird migrations—kinds of birds that migrate; migration pathways; pictures of birds migrating, or resting during migration; fall and spring migrations.
12. Tips to photographers—a series of tips that will help photographers get better pictures.
13. Nature photography—ideas for photographing natural phenomena.
14. Care of pets—basic needs of various pets and how best achieved.
15. Poison ivy—how to recognize the plant; what to do if you contact it.
16. Science fairs—their function; how to set up a fair; a fair for the students of a certain school.
17. Odd habits of animals—care of young; preparation of food before eating; ways to fool man.
18. Some of the oldest trees in the world— their location; age; need for protection.
19. Largest animals of the world—where they live; how they survive.
20. Insects affecting your garden—how to

recognize them; damage or help they do; where to find them.

21. Life cycle of a mosquito—how to rid your premises of mosquitoes.

22. Producing miniature trees for your solarium—cutting and maintaining trees that normally grow to a great height so they grow to a foot or less in height.

23. Your aquarium—choice of fish; size of aquarium; care of aquarium; food for fish; plants to use.

24. Eyes of animals—size and shape of pupil; habits of animals depending on whether they see better at night or during the day; distance they see; comparison of mammal eyes and insect eyes.

25. Sounds of animals—kinds of sounds made under varying circumstances such as stress, fear, food-getting, territory rights, mating, care of young.

26. Spring plants—flowers that emerge first in the spring; when they bloom; how to recognize them; photographing them.

27. A wild-flower garden—needs of flowers for your garden; which kinds do well in a wild-flower garden; care they should have.

28. Identifying the trees in your community—list trees of the community; how they can be identified; characteristics of trees.

29. Interesting recipes for unusual foods—foods not normally eaten by the ordinary person such as rattlesnake steak and chocolate-covered grasshoppers.

30. Life cycle of the cicada—17-year and others, with pictures of same.

31. Allergies—common ones; avoiding conditions that cause allergies; what to do about them.

32. Reviews of new books on science for students and adults.

33. Common weeds—how to identify, control, eradicate.

34. Pond "scum" in summer—cause, control.

The list of topics a teacher could write about is limited only by the imagination and the background of the writer. If your background is limited, there are numerous books that help give needed background. By presenting materials for the public, the biology teacher and the subject of biology are constantly kept before the townspeople.

REFERENCES FOR THE BIOLOGY LIBRARY

ABBOTT, R. TUCKER. 1954. *American Seashells*. D. Van Nostrand Co., Inc., Princeton, N.J.

ADAMSON, JAY. 1960. *Born Free*. Pantheon Books, New York.

ALLEN, ARTHUR A. 1961. *The Book of Bird Life*. D. Van Nostrand Co., Inc., Princeton, N.J.

BATES, MARSTON. 1960. *The Forest and the Sea, A Look at the Economy of Nature and the Ecology of Man*. Random House, New York.

BONNER, J. F., and GALSTON, A. W. 1952. *Principles of Plant Physiology*. W. H. Freeman and Co., San Francisco.

BUTCHER, DEVEREUX. 1955. *Seeing America's Wildlife in Our National Refuges*. The Devin-Adair Co., New York.

CARRIGHAR, SALLY. 1959. *Wild Voice of the North*. Doubleday & Co., Inc., New York.

CARSEN, RACHEL. 1951. *The Sea Around Us*. Oxford Univ. Press, New York.

———. 1962. *Silent Spring*. Houghton Mifflin Co., Boston.

CHARLES, BROTHER H. Make Your Own Color Slides. *American Biology Teacher* 12:62 64, 1950.

CORRINGTON, J. D. 1941. *Exploring With Your Microscope*. McGraw-Hill Book Co., Inc., New York.

CREIGHTON, W. S. 1950. *The Ants of North America* (Bulletin of the Museum of Comparative Zoology). Harvard Univ. Press, Cambridge, Mass., Vol. 103.

EDMONSON, W. T. (editor, Ward & Whipple). 1959. *Fresh-Water Biology*. John Wiley & Sons, Inc., New York.

FABRE, J. H. 1950. *The Insect World*. Dodd, Mead & Co., New York.

FABRIEL, M. L., and FOGEL, S. 1955. *Great Experiments in Biology*. Prentice-Hall, Inc., Englewood Cliffs, N.J.

FALE, E. F., and DAVIER, R. 1953. *Adaptation in Micro-organisms*. Cambridge Univ. Press, New York.

GRAUBARD, MARK. 1958. *The Foundation of Life Science.* D. Van Nostrand Co., Inc., Princeton, N.J.

GREEN, IVAN. 1960. *Wildlife in Danger.* Coward-McCann, Inc., New York.

HENDERSON, I. F., and HENDERSON, W. D. (revised by Kenneth, John H.). *Dictionary of Biological Terms.* 8th Ed. D. Van Nostrand Co., Inc., Princeton, N.J.

HICKEY, JOSEPH J. 1943. *A Guide to Bird Watching.* Oxford Univ. Press, New York.

HOLLAND, W. J. 1940. *The Butterfly Book.* Doubleday, Doran & Co., New York.

———. 1941. *The Moth Book.* Doubleday, Doran & Co., New York.

HUMASON, GRETCHEN L. 1962. *Animal Tissue Techniques.* W. H. Freeman & Co., San Francisco.

HUXLEY, JULIAN. 1950. *Ants.* Jonathan Cape and Harrison Smith, Inc., New York.

JACOB, WILLIAM P. What Makes Leaves Fall. *Science American* 193:82–89, 1955.

JAMESON, WILLIAM. 1959. *The Wandering Albatross.* William Morrow Co., New York.

KIPPS, CLARE. 1954. *Clarence, The Life of a Sparrow.* G. P. Putnam's Sons, New York.

KRUTCH, JOSEPH WOOD. 1959. *The Great Chain of Life.* Houghton Mifflin Co., Boston.

———. 1953. *The Twelve Seasons.* William Sloane Associates, New York.

LEOPOLD, A. STARKER, and DARLING, F. FRASER. 1953. *Wildlife in Alaska.* Ronald Press Co., New York.

LEVITT, J. 1956. *The Hardiness of Plants.* Academic Press, New York.

LORENZ, KONRAD. 1952. *King Solomon's Ring.* Thomas E. Crowell, New York.

MACFAYDEN, A. 1957. *Animal Ecology.* Sir Isaac Pitman & Sons, London, England.

MAETERLINCK, MAURICE (translated by Alfred Seetro). 1927. *The Life of the White Ant.* Dodd, Mead & Co., New York.

MILNE, LORUS, and MILNE, MARGERY. 1960. *The Balance of Nature.* Alfred A. Knopf, Inc., New York.

MORGAN, A. H. 1930. *Fieldbook of Pond and Stream.* G. P. Putnam's Sons, New York.

MURIE, O. 1954. *Field Guide to Animal Tracks.* Houghton Mifflin Co., Boston.

NEEDHAM, J. G. 1963. *Guide to the Study of Fresh-water Biology.* 5th Ed. Holden-Day, San Francisco.

OLIVER, JAMES A. 1955. *The Natural History of North American Amphibians and Reptiles.* D. Van Nostrand Co., Inc., Princeton, N.J.

PACKARD, VANA. 1950. *Animal I. Q.* The Dial Press, New York.

PENNACK, R. W. 1953. *Fresh-water Invertebrates of the United States.* Ronald Press Co., New York.

PETERSON, ROGER TORY. 1947. *Field Guide to the Birds.* Rev. Ed. Houghton Mifflin Co., Boston.

PHILLIPS, E. 1959. *Methods of Vegetation Study.* Holt-Dryden Press, New York.

PORTMANN, ADOLPH. 1959. *Animal Camouflage.* Univ. of Michigan Press, Ann Arbor.

SASS, JOHN E. 1951. *Botanical Microtechnique.* Rev. Ed. The Iowa State Univ. Press, Ames.

SCHULTZ, LEONARD, and STERN, ELIZABETH M. 1948. *The Way of Fishes.* D. Van Nostrand Co., Inc., Princeton, N.J.

STEVENS, G. W. 1957. *Microphotography.* John Wiley & Sons, Inc., New York.

SUSSMAN, M. 1960. *Animal Growth and Development.* Prentice-Hall, Englewood Cliffs, N.J.

TEALE, EDWIN WAY. 1951. *Autumn Across America.* Dodd Mead & Co., New York.

———. 1955. *Insect Friends.* Dodd, Mead & Co., New York.

TINBERGEN, NIKO. 1958. *Curious Naturalists.* Basic Books, New York.

———. 1953. *Social Behaviour in Animals.* John Wiley & Sons, Inc., New York.

VON FRISCH, KARL. 1955. *The Dancing Bees.* Harcourt, Brace and Co., New York.

WHITE, P. R. 1954. *The Cultivation of Animal and Plant Cells.* Ronald Press Co., New York.

WINCHESTER, A. M. 1964. *Biology and Its Relation to Mankind.* D. Van Nostrand Co., Inc., Princeton, N.J.

THE "HOW TO KNOW" SERIES

PUBLISHED by the Wm. C. Brown Co., Dubuque, Iowa 52003.

How To Know the Spring Flowers, Cuthbert, 1943, 1949.
How To Know the Mosses and Liverworts, Conard, 1944, 1956.
Living Things—How To Know Them, Jaques, 1946.
How To Know the Trees, Jaques, 1946.
How To Know the Insects, Jaques, 1947.
How To Know the Land Birds, Jaques, 1947.
Plant Families—How To Know Them, Jaques, 1948.
How To Know the Fall Flowers, Cuthbert, 1948.
How To Know the Economic Plants, Jaques, 1948, 1958.
How To Know the Immature Insects, Chu, 1949.
How To Know the Protozoa, Jahn, 1949.
How To Know the Mammals, Booth, 1949.
How To Know the Beetles, Jaques, 1951.
How To Know the Spiders, Kaston, B. J., and Kaston, Elizabeth, 1952.
How To Know the Grasses, Pohl, 1953.
How To Know the Fresh-Water Algae, Prescott, 1954.
How To Know the Western Trees, Baerg, 1955.
How To Know the Seaweeds, Dawson, 1956.
How To Know the Freshwater Fishes, Eddy, 1957.
How To Know the Weeds, Jaques, 1959.
How To Know the Water Birds, Jaques-Ollivier, 1960.
How To Know the Butterflies, Ehrlich, 1961.
How To Know the Eastern Land Snails, Burch, 1962.
How To Know the Grasshoppers, Helfer, 1963.
How To Know the Cacti, Dawson, 1963.
(Other subjects in preparation.)

BSCS PAMPHLETS

THE BSCS Pamphlet Series may be purchased from D. C. Heath & Co., 285 Columbus Ave., Boston, Mass. 02116.

No.	Title	Author
1	*Guideposts of Animal Navigation,* Carr, Archie
2	*Biological Clocks,* Brown, Frank A., Jr.
3	*Courtship in Animals,* Meyerriecks, Andrew J.
4	*Bioelectricity,* Suckling, E. E.
5	*Biomechanics of the Body,* De Brul, Lloyd E.
6	*Present Problems About the Past,* Auffenberg, Walter
7	*Metabolites of the Sea,* Nigrelli, Ross F.
8	*Blood Cell Physiology,* Gordon, Albert S.
9	*Homeostatic Regulation,* Overmire, Thomas G.
10	*Biology of Coral Atolls,* Boolootian, Richard A.
11	*Early Evolution of Life,* Young, Richard S., and Ponnamperuma, Cyril
12	*Population Genetics,* Wallace, Bruce
13	*Slime Molds and Research,* Alexopoulos, C. J., and Koevenig, James
14	*Cell Division,* Mazia, Daniel
15	*Photoperiodism in Animals,* Farner, Donald S.
16	*Growth and Age,* Milne, Lorus J., and Milne, Margery
17	*Biology of Termites,* Miller, E. Morton
18	*Biogeography,* Neill, Wilfred T.
19	*Hibernation,* Mayer, William V.
20	*Animal Language,* Collias, Nicholas E.
21	*Ecology of the African Elephant,* Quick, Horace
22	*Cellulose in Animal Nutrition,* Hungate, R. E.
23	*Plant Systematics,* Raven, Peter H., and Mertens, Thomas R.
24	*Photosynthesis,* Gaffron, Hans

THE CURRENT CONCEPTS IN BIOLOGY SERIES

PUBLISHED by The Macmillan Co., 866 Third Ave., New York, N.Y. 10022

Algae and Fungi—Alexopoulos, C. J., and Bold, H. C., 1967.
Cells and Energy—Golosby, Richard A., 1967.
Development in Flowering Plants—Torrey, John G., 1967.

Integral Animal Behavior—Davis, David E., 1966.

Evolution and Systematics—Solbrig, Otto T., 1966.

Principles of Development and Differentiation—Waddington, C. H., 1966.

Process and Pattern in Evolution—Hamilton, Terrell H., 1966.

Sensory Mechanisms—Case, James, 1966.

Tools of Biology—Lenhoff, Edward, 1966.

Viruses and Molecular Biology—Fraser, Dean, 1967.

MODERN BIOLOGY SERIES

PUBLISHED by Holt, Rinehart and Winston, Inc., 383 Madison Ave., New York, N.Y. 10017

THIS IS a series of paperback books written by experts in the various areas. They are an excellent source of reference for the instructor and the students. However, they are no longer being published. Many libraries have copies of them so they may be available in your library for you and your students to read.

Burnett, Allison L., and Eisner, Thomas—*Animal Adaptation*.

Delevoryas, Theodore—*Plant Diversification*.

Ebert, James D.—*Interacting Systems in Development*.

Griffin, Donald R.—*Animal Structure and Function*.

Levine, Robert Paul—*Genetics*.

Loewy, Ariel G., and Siekevitz, Philip—*Cell Structure and Function*.

Odum, Eugene P.—*Ecology*.

Ray, Peter Martin—*The Living Plant*.

Savage, Jay M.—*Evolution*.

Sistrom, William R.—*Microbial Life*.

REFERENCES FOR PHOTOGRAPHY

Eastman Kodak Publication. 1959. *Adventures in Indoor Color Slides*. Eastman Kodak Co., Rochester, N.Y. 14650.

Eastman Kodak Publication. 1961. *Adventures in Outdoor Color Slides*. Eastman Kodak Co., Rochester, N.Y. 14650.

Eastman Kodak Publication. 1962. *Basic Developing, Printing, Enlarging*. Eastman Kodak Co., Rochester, N.Y. 14650.

Eastman Kodak Publication. 1962. *Copying*. Eastman Kodak Co., Rochester, N.Y. 14650.

Eastman Kodak Publication. 1964. *Here's How* (techniques for outstanding pictures). Eastman Kodak Co., Rochester, N.Y. 14650.

Eastman Kodak Publication. 1952. *How To Make Good Pictures*. Eastman Kodak Co., Rochester, N.Y. 14650.

Eastman Kodak Publication. 1956. *Professional Printing With Kodak Photographic Papers*. Eastman Kodak Co., Rochester, N.Y. 14650.

Embrey, A. Wilson, III. *Catalog of Photography Books* (free), Fredricksburg, Va. 22401.

Haz, Nicholas. 1956. *Image Arrangement*. The Haz Book Co., P.O. Box 10823, Pittsburgh, Pa. 15236.

Howard, Edwin J., and Marchant, John C. A Fast Electronic Eye for Bird Photography. *Audubon Magazine,* Nov.–Dec., 1956, pp. 278–283, 291.

Linton, David. *Photographing Nature* (a handbook for the beginner and the expert). Natural History Press, The American Museum of Natural History, Central Park, West at Seventy-ninth St., New York, N.Y. 10024.

Modern Photography. Published monthly by the Billboard Publishing Co., 165 W. 46th St., New York, N.Y. 10036.

Nathan, Simon. 1957. *Good Photography's Darkroom Guide*. Fawcett Book, Department 332, Greenwich, Conn. 06830.

Popular Photograph Editor, *Photography Directory and Buying Guide* (a new directory is published each year). Ziff-Davis Publishing Co., One Park Ave., New York, N.Y. 10016.

Popular Photography Magazine. Published monthly by Ziff-Davis Publishing Co., One Park Ave., New York, N.Y. 10016.

Turtox Information on Photography:
Dark-Field Microscopy. Turtox News, April 1948; notes on photomicrography, Part II, Turtox News, January 1949.
The Turtox Microscopy Booklet ($1.00). (Deals with basic microscopy, color filters, and 3-D photomicrography.)

Notes on Photomicrography, Part III. Turtox News, March 1949.

Aperture Stop Illumination Control in Photomicrography With Transmitted Light. Turtox News, August 1952.

Electronic Flash Illumination for Photomicrography. Turtox News, January 1954.

Simplified Photomicrography. Turtox Service Leaflet No. 56.

SELECTED REFERENCES FOR TEACHERS

BLOCK, R. J., DURRUM, E. L., and ZWEIG, G. 1958. *A Manual of Paper Chromatography and Paper Electrophoresis.* Academic Press, New York.

CARR, ARCHIE. 1962. *Guidepost of Animal Navigation.* D. C. Heath & Co., Boston.

DEMEREC, M., and KAUFMAN, B. P. 1961. *Drosophila Guide.* Carnegie Institute of Washington, Cold Spring Harbor, N.Y.

GALSTON, ARTHUR. 1961. *Life of the Green Plant.* Prentice-Hall, Englewood Cliffs, N.J.

GREEN, M. M. The Discrimination of Wildtype *Isoalleles* at the White Locus and *Drosophila melanogaster. Proc.,* Natl. Acad. Sci., 45:549–553, 1959.

HARDIN, GARRET. 1966. *Biology: Its Principles and Implications.* W. H. Freeman & Co., San Francisco.

LEDERER, E., and LEDERER, M. 1957. *Chromatography.* D. Van Nostrand Co., New York.

ROE, ANNE, and SIMPSON, GEORGE GAYLORD. 1958. *Behavior and Evolution.* Yale Univ. Press, New Haven, Conn.

SINNOTT, E. W., DUNN, L. C., and DOBZHANSKY, T. 1958. *Principles of Genetics.* McGraw-Hill, Inc., New York.

WEISZ, PAUL B. 1959. *The Science of Biology.* McGraw-Hill, Inc., New York.

———. 1959. *Laboratory Manual in the Science of Biology.* McGraw-Hill, Inc., New York.

———, and FULLER, MELVIN S. 1962. *The Science of Botany.* McGraw-Hill, New York.

APPENDIX F

Sources of

Materials

TRANSPARENCIES

COMMERCIALLY prepared transparencies or master copies from which to make transparencies:

Audio-Visual Library of Science Transparencies, General Aniline and Film Corp., Binghamton, N.Y. 13900.

DCA Educational Products, Inc., 4865 Stenton Ave., Philadelphia, Pa. 19144.

Encyclopaedia Britannica Films, Inc., 1150 Wilmette Ave., Wilmette, Ill. 60091.

General Aniline and Film Corp., 140 W. 51st St., New York, N.Y. 10020.

Harcourt, Brace and World, Inc., 757 Third Ave., New York, N.Y. 10017.

Keuffel and Esser Co., 520 So. Dearborn St., Chicago, Ill. 60604; Hoboken, N.J. 07030.

McGraw-Hill, Inc., 330 W. Forty-second St., New York, N.Y. 10036.

RCA Educational Service, Camden, N.J. 08108.

State University of Iowa, Bureau of Audio-Visual Instruction, Extension Div., Iowa City, Iowa 52240.

Tweedy Transparencies, 321 Central Ave., Newark, N.J. 07103.

United Transparencies, Inc., 57 Glenwood Ave., Binghamton, N.Y. 13905.

Ward's Natural Science Establishment, Inc., P.O. Box 1712, Rochester, N.Y. 14603.

SCHOOL LABORATORY FURNITURE

All-Steel Equipment, Inc., Aurora, Ill. 60507.

American Desk Mfg. Co., Temple, Tex. 76501.

American Seating Co., 173 W. Madison St., Chicago, Ill. 60602.

Art Metal Construction Co., Jamestown, N.Y. 14701.

Beckley-Cardy, 1926 No. Narragansett Ave., Chicago, Ill. 60639.

Borroughs Mfg. Co., 3062 No. Burdick St., Kalamazoo, Mich. 49007.

Brunswick-Balke-Collender Co., School Equipment Div., 69 W. Washington, Chicago, Ill. 60602.

Chicago Hardware Foundry Co., 2500 Commonwealth Ave., North Chicago, Ill. 60064.

Claridge Products and Equipment, Inc., Harrison, Ark. 72601.

Cole Steel Equipment Co., Inc., Aurora, Ill. 60507.

Douglas Fir Plywood Assn., Tacoma, Wash. 98401.

Duralab Equipment Corp., 987 Linwood St., Brooklyn, N.Y. 11208.

Educators Mfg. Co., P.O. Box 1261, Tacoma, Wash. 98401.

Fleetwood Furniture Co., Zeeland, Mich. 49464.

General Electric, 840 So. Canal St., Chicago, Ill. 60607.

Globe-Wernicke Co., 1329 Arlington St., Cincinnati, Ohio 45225.

Griggs Equipment, Inc., P.O. Box 630, Belton, Tex. 76513; or Selma, N.C. 27576.

Heywood-Wakefield School Furniture Div., Menominee, Mich. 49858.

Kewaunee Mfg. Co., 5140 So. Center St., Adrian, Mich. 49221; or Technical Furniture, Inc., Statesville, N.C. 28677.

Laboratory Furniture Co., Inc., Mineola, L.I., N.Y. 11501.

Metalab Equipment Co., Hicksville, L.I., N.Y. 11802.

Michael Art Bronze Co., P.O. Box 668-A5, Covington, Ky. 41012.

Milton Bradley Co., 74 Park St., Springfield, Mass. 01101.

National School Furniture Co., Div. National Store Fixture Co., Inc., Odenton, Md. 21113.

School Equipment Mfg. Corp., 46 Bridge St., Nashua, N.H. 03060.

Shwayde Brothers, Inc., Classroom Furniture Div., Denver, Colo. 80209.

Standard Pressed Steel Co., Hallowell Div., Jenkintown, Pa. 19046.

Standard School Equipment Co., Siler City, N.C. 27344.

The Tolerton Co., 265 No. Freedom Ave., Alliance, Ohio 44601.

Westfield Mfg. Co., Westfield, Mass. 01085.

Yawman and Erbe Mfg. Co., 1099 Jay St., Rochester, N.Y. 14603.

GREENHOUSES AND GREENHOUSE EQUIPMENT

Aluminum Greenhouses, Inc., 14615 Lorain Ave., Cleveland, Ohio 44111.

Griffin Greenhouse Supplies, Inc., 349 Main St., Reading, Mass. 01867.

Ickes-Braun Greenhouses, Inc., 1733 North Western Ave., Chicago, Ill. 60647.

Albert J. Lauer Co., Aluminum Greenhouses, 2450 Lexington Ave. No., St. Paul, Minn. 55113.

Lord & Burnham, Div. of Burnham Corp., Irvington, N.Y. 10533. Also Belmont, Mass. 02178; 1355 Milldale Rd., Cheshire, Conn. 06410; Des Plaines, Ill. 60016; Elyria, Ohio 44035; Malvern, Pa. 19355; and 3118 Depot Rd., Hayward, Calif. 94545.

Ludy Greenhouse Mfg. Corp., P.O. Box 85, New Madison, Ohio 45346.

Metropolitan Greenhouse Mfg. Corp., 155 Allen Blvd., P.O. Box 610, Farmingdale, L.I., N.Y. 11735.

Midwest Greenhouse Mfg. Co., 1228 Harding, Des Plaines, Ill. 60016.

National Greenhouse Co., Pana, Ill. 62557.

J. A. Nearing, Inc., 10788 Tucker St., Beltsville, Md. 28765.

Redfern Prefab Greenhouses, FG-4, Santa Cruz, Calif. 95060.

Rough Brothers, 4229 Spring Grove Ave., Cincinnati, Ohio 45223.

Sherer-Gillett Co., Marshall, Mich. 49068 (mobile greenhouses).

COMMERCIAL CAGES AND TRAPS

AQUARIA and terraria may be converted cages. Students may build cages that meet the specifications you set up. You may wish to use plastic cages for some small animals.

Traps should work easily. This list includes manufacturers who sell cages and traps of various sizes and kinds.

Allcock Mfg. Co., Manufacturers of Havahart Humane Animal Traps, Ossining, N.Y. 13670.

Aloe Scientific, 1831 Olive St., St. Louis, Mo. 63100.

Animal Trap Co. of America, Lititz, Pa. 17543; Pascagoula, Miss. 39567; Niagara Falls, Ontario, Canada.

Econo Scientific Div., c/o Maryland Plastics, Inc., 9 E. 37th St., New York, N.Y. 10016.

Forestry Supplies, Inc., P.O. Box 8397, 205 W. Rankin St., Jackson, Miss. 39202.

General Biological Supply House, 8200 So. Hoyne Ave., Chicago, Ill. 60620.

Hartford Metal Products, Inc., Hartford Cage Systems, Aberdeen, Md. 21001.

National Band and Tag Co., Newport, Ky. 41071.

National Live Trap Corp., Manufacturers of Folding and Rigid Traps, P.O. Box 302, Route 1, Tomahawk, Wis. 54487.

PREPARED MICROSCOPE SLIDES

Cambosa Scientific Co., Inc., 37 Antwerp St., Brighton Sta., Boston, Mass. 02135.

Carolina Biological Supply Co., Burlington, N.C. 27215.

Central Scientific Co., 1700 Irving Park Rd., Chicago, Ill. 60623.

General Biological Supply House, Inc., 8200 So. Hoyne Ave., Chicago, Ill. 60620.

Oregon Biological Supply Co., 303 N.E. Multnomal St., Portland, Oreg. 97232.

Triarch, Inc., George H. Conant, Ripon, Wis. 54971.

Ward's Natural Science Establishment, Inc., P.O. Box 1712, Rochester, N.Y. 14603; also Ward's of California, P.O. Box 1749, Monterey, Calif. 93940.

Will Corp., Rochester, N.Y. 14603.

FROGS

Carolina Biological Supply Co., Elon College, N.C. 27244.

289

General Biological Supply House, Inc. (Turtox), 8200 So. Hoyne Ave., Chicago, Ill. 60620.

E. G. Hoffman, P.O. Box 815, Oshkosh, Wis. 54901.

Lemberger Co., 1222 W. So. Park Ave., P.O. Box 482, Oshkosh, Wis. 54901.

Macalaster Scientific Corp., 60 Arsenal St., Watertown, Mass. 02172.

Quivira Specialties, 4204 W. 21st St., Topeka, Kans. 66600.

Ray Singleton, Interbay Station, Tampa, Fla. 33600.

Southwestern Biological Supply Co., P.O. Box 4084, Dallas, Tex. 75200.

E. G. Steinhilber and Co., Oshkosh, Wis. 54901.

Western Scientific Supply Co., North Sacramento, Calif. 95801.

CHROMOGENIC BACTERIA

ONE OF THE main features of bacteria is their growth pattern on various media. A striking characteristic of certain species is *chromogenesis,* or pigmentation.

In some pigment-producing bacteria the pigment is retained inside the cells, and the entire mass of bacterial cells is colored. In others the pigment is excreted and it colors the medium on which the bacteria live. Intensity of the pigment is influenced by the composition of the medium and the conditions of incubation. It is better to observe pigment produced from growth on solid rather than liquid media.

Some of the commoner bacterial species that produce pigment and retain it within the cells include: *Serratia marcescens* (red color), *Chromobacterium violaceum* (violet), *Staphylococcus aurens* (golden yellow), *Sarcina lutea* (lemon yellow), *Micrococcus flavus* (yellow), *Micrococcus niger* (brown or black).

These as well as other bacteria are frequently available to schools from nearby universities and colleges that maintain colonies of many bacterial species.

Two other places that are good sources of chromogenic bacteria in addition to biological supply houses are: American Type Culture Collection, 12301 Clark Lawn Dr., Rockville, Md. 20852; and Midwest Culture Service, 1924 No. 7th St., Terre Haute, Ind. 47804.

MICROBIOLOGICAL CULTURE MEDIA

Albimi Laboratories, Inc., 16 Clinton St., Brooklyn, N.Y. 11201.

Bacti-Kit, P.O. Box 101, Eugene, Oreg. 97401.

Baltimore Biological Laboratory, Inc., Div. Becton, Dickinson and Co., Baltimore, Md. 21204.

Case Laboratories, Inc., 515 No. Halsted St., Chicago, Ill. 60622.

Derm Medical Co., P.O. Box 78595, W. Adams St., Los Angeles, Calif. 90016.

Difco Laboratories, Detroit, Mich. 48201.

Hach Chemical Co., Ames, Iowa 50010.

Key Scientific Products Co., P.O. Box 66514, Los Angeles, Calif. 90066.

Media, Inc., 89 Lincoln Park, Newark, N.J. 07102.

Microbiological Associates, Inc., P.O. Box 5970, Bethesda, Md. 20014.

Pennsylvania Biological Laboratories, Inc., 5054 Whitaker Ave., Philadelphia, Pa. 19124.

Viobin Corp., Monticello, Ill. 61856.

Yeast Products:

Anheuser-Busch, Inc., 721 Pestalozzi, St. Louis, Mo. 63118.

Fleischmann Laboratories, Std. Brands, Inc., Betts Ave., Stanford, Conn. 06904.

Red Star Yeast and Products Co., Special Yeast Products Div., 221 E. Buffalo St., Milwaukee, Wis. 53202.

ANTIBIOTICS

Abbott Laboratories, Research Div., North Chicago, Ill. 60064.

Beebe Laboratories, Inc., Larpenteur Ave. and E. Beebe Rd., St. Paul, Minn. 55109.

Bristol Laboratories, Inc., P.O. Box 657, Syracuse, N.Y. 13201.

Commercial Solvents Corp., Terre Haute, Ind. 47808.

Eli Lilly and Co., Indianapolis, Ind. 46206.

Hoffman-LaRoche, Inc., Nutley, N.J. 07110.

Lederle Laboratories Div., American Cyanamid Co., North Middleton Rd. Pearl River, N.Y. 10965.

Merck and Co., Inc., Rahway, N.J. 07065.

Chas. Pfizer and Co., Inc., 11 Bartlett St., Brooklyn, N.Y. 11206.

Squibb Institute of Medical Research, New Brunswick, N.J. 08903.

Upjohn Co., Kalamazoo, Mich. 49001.

COMMERCIAL ENZYMES

American Ferment Co., Inc., 1450 Broadway, New York, N.Y. 10018.

Dairyland Food Laboratories, Inc., 419 Frederick St., Waukesha, Wis. 53186.

Dawe's Laboratories, Inc., Chicago, Ill. 60632.

Fermco Chemicals, Inc., 4941 Racine Ave., Chicago, Ill. 60609.

Glogau and Co., 1910 W. Birchwood Ave., Chicago, Ill. 60626.

Grain Processing, Muscatine, Iowa 52761.

Harvest Queen Mill and Elevator Co., 4350 No. Central Expressway, Dallas, Tex. 75206.

George A. Jeffreys and Co., Inc., Salem, Va. 24153.

Paul Lewis Laboratories, Inc., 4253 No. Port Washington Rd., Milwaukee, Wis. 53212.

Merck and Co., Rahway, N.J. 07065.

Miles Chemical Co., Div. Miles Laboratories, Inc., Clifton, N.J. 07015.

Pabst Laboratories Div., Pabst Brewing Co., 1037 W. McKinley Ave., Milwaukee, Wis. 53205.

Red Star Yeast and Products Co., 221 E. Buffalo St., Milwaukee, Wis. 53202.

Rohm and Haas Co., Special Products Dept., Washington Sq., Philadelphia, Pa. 19105.

Royce Chemical Co., Carlton Hill, East Rutherford, N.J. 07073.

Wallerstein Co., Inc., 180 Madison Ave., New York, N.Y. 10016.

J. A. Zurn Mfg., Div. Zurn Industries, Inc., Erie, Pa. 16501.

ENZYMES AND FINE CHEMICALS

Abbott Laboratories, Research Div., North Chicago, Ill. 60064.

Agricultural Biologicals Corp., Lynbrook, N.Y. 11563.

Applied Science Laboratories, Inc., P.O. Box 440, State College, Pa. 16801.

California Corp. for Biochemical Research, 3625 Medford St., Los Angeles, Calif. 90063.

Committee on National Formulary, American Pharmaceutical Assn., 2215 Constitution Ave., N.W., Washington, D.C. 20037.

Delta Chem. Works, 23 W. 60th St., New York, N.Y. 10023.

Dow Chemical Co., Midland, Mich. 48640.

Distillation Products Industries, Div. Eastman Kodak Co., Rochester, N.Y. 14650.

Edwal Laboratories, Inc., 732 Federal St., Chicago, Ill. 60605.

Fine Chemical Sales Div., Midland, Mich. 48640.

General Biochemicals, Inc., 75 Laboratory Pk., Chagrin Falls, Ohio 44022.

General Mills, Inc., Chemical Div., Kankakee, Ill. 60901.

H. and M. Chem. Co., Ltd., 1651–18th St., Santa Monica, Calif. 90404.

Hormel Research Inst., Austin, Minn. 55912.

K and K Laboratories, 29–46 Northern Blvd., Long Island City, N.Y. 11101.

Lederle Laboratories Div., American Cyanamid Co., 30 Rockefeller Plaza, New York, N.Y. 10020.

Mann Research Laboratories, Inc., 136 Liberty St., New York, N.Y. 10006.

Matheson, Coleman and Bell, Matheson Co., 2909 Highland Ave., Norwood, (Cincinnati), Ohio 45212; or P.O. Box 85, East Rutherford, N.J. 07073.

Merck and Co., Rahway, N.J. 07065.

Ortho Research Corp., Raritan, N.J. 08869.

Pabst Laboratories Div., Pabst Brewing Co., 1037 W. McKinley Ave., Milwaukee, Wis. 53205.

Pentex, Inc., 660 No. Schuyler Ave., Kankakee, Ill. 60901.

Schwarz Laboratories, Inc., 230 Washington St., Mount Vernon, N.Y. 10553.

Sigma Chem. Co., 4648 Easton Ave., St. Louis, Mo. 63113.

Supply Co., 3514 Lucas Ave., St. Louis, Mo. 63103.

Western Amino Acids Co., Inc., 7922 Beverly Blvd., Los Angeles, Calif. 90048.

Winthrop Laboratories, Inc., Special Chemicals Div., 1450 Broadway, New York, N.Y. 10018.

MICROSCOPES

American Optical Co., Instrument Div., Buffalo, N.Y. 14215.

Bausch & Lomb, Inc., 78041 Bausch St., Rochester, N.Y. 14602.

Carolina Biological Supply Co., Burlington, N.C. 27215.

Central Scientific Co., 1700 Irving Park Rd., Chicago, Ill. 60623.

Chicago Lens and Instrument Co., 735 No. LaSalle St., Chicago, Ill. 60610.

Cooke, Troughton and Simms, 91 Waite St., Malden, Mass. 02148.

Elgeet Optical Co., Inc., Scientific Instrument & Apparatus Div., 838 Smith St., Rochester, N.Y. 14606.

Faust Scientific Supply, 5108 Gordon Ave., Madison, Wis. 53716.

Pacific Biological Laboratories, P.O. Box 63. Albany Station, Berkeley, Calif. 91716.

Powell Laboratories Div., Gladstone, Oreg. 97027.

Swift Instruments, Inc., 1572 No. Fourth St., San Jose, Calif. 95112.

Technical Instrument Co., 98 Golden Gate Ave., San Francisco, Calif. 94102. (Kyawa compound microscopes.)

Testa Manufacturing Co., 10126 E. Rush St., El Monte, Calif. 91733.

Unitron Instrument Co., Microscope Sales Div., 66 Needham St., Newton Highlands, Mass. 02161.

Ward's Natural Science Establishment, Inc., Biology Div., P.O. Box 1712, Rochester, N.Y. 14603; or P.O. Box 1749, Monterey, Calif. 93940.

NATURE RECORDINGS

RECORDINGS of sounds in nature add interest and enthusiasm for the class and help them become more conscious of the sounds around us. They become better listeners and interpreters and more critical in their evaluation of what they hear. Records may be purchased from a number of places but may also be recorded by students themselves. A portable tape recorder can be taken into the field to capture the sounds of various birds and insects.

Some of the many available recordings include calls of birds, mammals, insects, sea animals, and amphibians; sounds of the human heart, the sea, and the rain forests, as well as other interesting phenomena.

We suggest you check with your local music dealer for catalogs that list available records. Included here are a few places that sell some of these special records:

Cornell University, Laboratory of Ornithology, 124 Robert Place, Ithaca, N.Y. 14850.

Federation of Ontario Naturalists, 1212 Don Mills Rd., Don Mills, Ontario, Canada.

Folkways Records, 117 W. 46th St., New York, N.Y. 10036.

General Biological Supply House, Inc., 8200 So. Hoyne Ave., Chicago, Ill. 60620.

Pierce Book Co., Winthrop, Iowa 50682.

W. Schwann, Inc., 137 Newberry St., Boston, Mass. 02116.

FILM LOOPS

AN INCREASINGLY large number of single-concept films is being made each year as the demand in schools becomes greater. Each film portrays and develops a single idea or theme which can be viewed over and over again without changing the film. Each is contained in a cartridge as a continuous loop.

Teachers and students can produce their own single-concept films very easily, too, in either black and white or color with any 8mm or Super 8 motion picture camera. This can serve as an interesting project for students who wish to keep their results on film to show contrasts or likenesses.

We list some of the educational film producers and distributors, both in the United States and abroad. Catalogs from these places are available.

Audio-Tutorial Systems, 426 So. Sixth St., Minneapolis, Minn. 55415.

Cenco Educational Films, 2600 So. Kostner Ave., Chicago, Ill. 60623.

Communications Films, Inc., 870 Monterey Pass Rd., Monterey Park, Calif. 91754.

The Ealing Corporation, 2225 Massachusetts Ave., Cambridge, Mass. 02140.

Educational Services, Inc., 39 Chapel St., Newton, Mass. 02158.

Encyclopaedia Britannica Educational Corp., 425 No. Michigan Ave., Chicago, Ill. 60611.

Eothen Films, Ltd., 70, Furzehill Rd. Foreham, Wood, Herts, England.

Film Associates Educational Films, Inc., 115559 Santa Monica Blvd., Los Angeles, Calif. 90029.

Gateway Educational Films, Ltd., 470/472 Green Lanes, Palmers Green, London N. 13, England.

Halas & Batchelor Cartoon Films, 3/7 Kean St. Aldwych, London, W. C.2, England.

International Visual Aids Center, 691 Chaussee De Mons, Brussels 7, Belgium.

Jam Hardy Organization, Inc., 2821 E. Grand Blvd., Detroit, Mich., 48211.

Macmillan & Co., Ltd., Little Essex St., London W.C.2, England.

Modern Learning Aids, 1212 Ave. of the Americas, New York, N.Y. 10036.

National Film Board of Canada, P.O. Box 6100, Montreal 3, Canada.

National Instructional Films, 54 East Route

59, Nanuet Professional Center, Nanuet, N.Y. 10954.

The Nuffield Foundation Science Teaching Project, Tavistock Place London, W.C.1, England.

Rank Film Library, 1 Aintree Rd., Perivale, Greenford, Middlesex, England.

Sutherland Educational Films, 201 No. Occidental Blvd., Los Angeles, Calif. 90026.

Thorne Films, Inc., 1229 University Ave., Boulder, Colo. 80302.

Ward's Natural Science Establishment, Inc., P.O. Box 1712, 3000 E. Ridge Rd., Rochester, N.Y. 14603.

Wasp Productions, Palmer Lane West, Pleasantville, N.Y. 10570.

Technicolor, Commercial and Educational Division, 1300 Frawley Drive, Costa Mesa, Calif. 92627, has developed a source directory for educational, single-concept film loops in instant loading magi-cartridges. They list educational film producers and distributors and the kinds of films that can be obtained from each source. These are great teaching aids and should become part of every biology course.

BULLETIN BOARD MATERIALS

Advance Products Co., 2300 E. Douglas St., Wichita, Kans. 67214 (easels).

Beckley-Cardy Co., 1632 Indiana Ave., Chicago, Ill. 60616.

Bulletin Boards and Directory Products, Inc., 724 Broadway, New York, N.Y. 10003.

The Judy Co., 310 No. 2nd St., Minneapolis, Minn. 55401.

Magna-Hold Products Corp., 1750 No. Lindbergh, St. Louis, Mo. 63132.

Magnet Sales Co., 3657 So. Vermont, Los Angeles, Calif. 90007.

Masonite Corp., 111 W. Washington St., Chicago, Ill. 60602 (pegboard display boards).

Match-A-Tach, 26 E. Pearson St., Chicago, Ill. 60611 (magnetic display boards).

294

LETTERING MATERIALS

Carter's Ink Co., Cambridge, Mass. 02142.

Chart-Pak, Inc., 4 Audubon, Leeds, Mass. 01053.

Cushman and Dennison Mfg. Co., 730 Garden, Carlstadt, N.J. 07072.

E. Dietzgen Co., 2455 No. Sheffield Ave., Chicago, Ill. 60614.

Esterbrook Pen Co., Delaware Ave. and Cooper St., Camden, N.J. 08102.

Grace Letter Co., 5 E. 47th St., New York, N.Y. 10017.

Hernard Mfg. Co., 21 Saw Mill River Rd., Yonkers, N.Y. 10701.

C. Howard Hunt Pen Co., 7th and State Sts., Camden, N.J. 08102.

Keuffel and Esser Co., 520 So. Dearborn St., Chicago, Ill. 60604.

Mark-Tex Corp., 453 W. 17th St., New York, N.Y. 10011.

Marsh Co., Bellville, Ill. 62222 (Marsh 77 and Squeez-o-Maker).

Speedry Products, Inc., Richmond Hill, N.Y. (Magic Marker).

BIOLOGICAL STAMPS

Broadway Approvals, Ltd., 50 Denmark Hill, London S.E. 5, England.

Garcelon Stamp Co., Dept. 1NGN, Calais, Maine 04619.

Gray Stamp Co., Dept. N5, Toronto, Canada.

H. E. Harris, Dept. R-127, Boston, Mass. 02117.

Triangle Stamps, Jamestown Dept. H18NG, Jamestown, N.Y. 14701.

Zenith Co., 81 Willoughby St., Brooklyn, N.Y. 11201.

CLOSED-CIRCUIT TELEVISION EQUIPMENT

Ampex Corp., 401 Broadway, Redwood City, Calif. 94063

Concord Electron Corp., 1935 Armacost Ave., Los Angeles, Calif. 90025.

Dage Television Div., Thompson Products, Inc., W. 10th St., Michigan City, Ind. 46360.

General Electric Co., Closed Circuit TV Business Section, Visual Communica-tion Products Dept., Syracuse, N.Y. 13201.

Graybar Electric Co., Graybar Bldg., New York, N.Y. 10017.

Jerrold Electronics Corp., Educational and Communication Systems Div., 401 Walnut St., Philadelphia, Pa. 19105.

Norelco, 100 E. 42nd St., New York, N.Y. 10017.

Packard Bell, 1920 South Zigueron St., Los Angeles, Calif. 90007.

Panasonic, 636 11th Ave., New York, N.Y. 10036.

Radio Corp. of America, Camden, N.J. 08102.

Sarkes Tarzian, Inc., Bloomington, Ind. 47401.

Shibaden Corp. of America, 58–25 Brooklyn-Queens Expressway, Woodside, N.Y. 11377; 21015–23 So. Figueroa St., Torrance, Calif. 90502; 1725 No. 33rd Ave., Melrose Park, Ill. 60165; 100 Martin Ross Ave., Downsview, Ontario, Canada.

Sony Corp. of America, 580 Fifth Ave., New York, N.Y. 10036.

Video Systems, Inc., 6444 No. Ridgeway Ave., Chicago, Ill. 60045.

Westel Corp., 1777 Borol Place, San Mateo, Calif. 94402.

Westinghouse Electric Corp., Commercial-Institutional Products, Route 27, Edison, N.J. 08817.

EDUCATIONAL FILMS

Academy Films, 800 No. Seward St., Hollywood, Calif. 90038.

Association Films, Inc., 347 Madison Ave., New York, N.Y. 10017.

Arthur Barr Productions, 1029 No. Allen St., Pasadena, Calif. 91104.

Coronet Films, 65 E. So. Water St., Chicago, Ill. 60601.

Dowling Pictures, 509 So. Beverly Dr., Beverly Hills, Calif. 90212.

The Ealing Corp., 2225 Massachusetts Ave., Cambridge, Mass. 02140.

Encyclopaedia Britannica Films, Inc., 1150 Wilmette Ave., Wilmette, Ill. 60091.

Family Films, Inc., 5823 Santa Monica Blvd., Hollywood, Calif. 90038.

FOM—Film Strip-of-the-Month Clubs, 355 Lexington Ave., New York, N.Y. 10017.

Jam Hardy Organization, Inc., 2821 E. Grand Blvd., Detroit, Mich. 48211.

Indiana Univ., Audio-Visual Center, Bloomington, Ind. 47401.

International Film Bureau, 332 So. Michigan Ave., Chicago, Ill. 60604.

International Film Foundation, 270 Park Ave., New York, N.Y. 10017.

Johnson Hunt Productions, La Canada, Calif. 91011.

McGraw-Hill, Inc., Text-Film Department, 330 W. 42nd St., New York, N.Y. 10036.

Moody Institute of Science, Educational Film Div., 11428 Santa Monica Blvd., Los Angeles, Calif. 90025.

National Film Board of Canada, 1270 Ave. of the Americas, New York, N.Y. 10020.

Thorne Films, Boulder, Colo. 80302.

United Nations, Film and Visual Information Div., New York, N.Y. 10017.

United World Films, Inc., 221 Park Ave. So., New York, N.Y. 10003.

Visual Sciences, P.O. Box 599B, Suffern, N.Y. 10901.

Walt Disney Productions, 2400 W. Alameda Ave., Burbank, Calif. 91503.

MATERIALS FOR FLANNELBOARDS

American Felt Co., 121 Wacker Dr., Chicago, Ill. 60606.

Educational Supply and Specialty Co., 2833 Gage Ave., Hartington Park, Calif. 90255.

Florez, Inc., 815 Bates St., Detroit, Mich. 48226.

Jacronda Mfg. Co., 5449 Hunter St., Philadelphia, Pa. 19131.

The Judy Co., 310 No. 2nd St., Minneapolis, Minn. 55401.

The Ohio Flock-Cote Co., 5713 Euclid Ave., Cleveland, Ohio 44103.

Oravisual Co., Inc., P.O. Box 11150, St. Petersburg, Fla. 33733.

Visual Specialties Co., 203 No. Saginaw St., Bryon, Mich. 48418.

L. A. Whitney Associates, 250 Daniel's Farm Rd., Trumbull, Conn. 06611.

FREE OR INEXPENSIVE PRINTED MATERIALS

THE Educator's Progress Service in Randolph, Wisconsin, publishes seven volumes each year listing sources of free and inexpensive materials for teaching. They abstract from several thousand sources. Titles of the volumes are:

Elementary Teacher's Guide to Free Curriculum Materials—1,640 listings
Educator's Guide to Free Filmstrips—358 listings
Educator's Guide to Free Science Materials—1,690 listings
Educator's Guide to Free Social Studies Materials—2,357 listings
Educator's Guide to Free Guidance Materials—861 listings
Educator's Guide to Free Tapes, Scripts, Transcriptions—392 listings
Educator's Guide to Free Films—4,943 listings

These listings are in the *1968 Educator's Guide to Free Materials.*

Teachers are encouraged to write and ask for materials available for biology courses. Here is a very short list of addresses. Keep adding to it.

American Cancer Society, 47 Beaver St., New York, N.Y. 10004; or 1405–5th Ave., San Diego, Calif. 92101.

American Dental Association, Order Dept., 222 E. Superior St., Chicago, Ill. 60611.

American Museum of Natural History, 79th St. and Central Park W., New York, N.Y. 10024.

American Nature Association, 1214–16th St., N.W., Washington, D.C. 20036.

Bausch and Lomb Optical Co., 635 St. Paul St., Rochester, N.Y. 14602.

Better Vision Institute, Inc., 630–5th Ave., New York, N.Y. 10020.

Carnegie Institute of Washington, 1530 P. St., N.W., Washington, D.C. 20005.

Carolina Biological Supply Co., Burlington, N.C. 27215, *Carolina Tips;* also Gladstone, Oreg. 97027.

296

Chicago Natural History Museum, Educational Dept., Roosevelt Rd., Chicago, Ill. 60605.

Church and Dwight, Inc., 70 Pine St., New York, N.Y. 10005 (bird pictures).

Conservation Foundation, 30 E. 40th St., New York, N.Y. 10016.

Denoyer-Geppert Co., 5235 N. Ravenswood Ave., Chicago, Ill. 60640.

Eastman Kodak Co., Rochester, N.Y. 14650 (medical radiography and photography).

General Biological Supply House, 8200 So. Hoyne Ave., Chicago, Ill. 60600 (Turtox News and Turtox Service Leaflets).

General Mills, Inc., 400–2nd Ave. So., Minneapolis, Minn. 55401.

Health Information Foundation, 420 Lexington Ave., New York, N.Y. 10017.

H. J. Heinz Co., Public Relations Dept., 1062 Progress St., Pittsburgh, Pa. 15230.

Illinois State Department of Health, 503 State Office Bldg., Springfield, Ill. 62706.

Illinois State Natural History Survey, Natural Resources Bldg., Urbana, Ill. 61801.

Indiana Department of Conservation, Division of Parks, Lands and Waters, 100 No. Senate Ave., Indianapolis, Ind. 46204.

Leitz Optical Co., 468–4th Ave., New York, N.Y. 10016.

Louisiana Department of Wildlife and Fisheries, Visual Education Dept., 126 Civil Courts Bldg., New Orleans, La. 70130.

Mentor Books—New American Library of World Literature, 501 Madison Ave., New York, N.Y. 10022.

Metropolitan Life Insurance Co., School Service Dept., 1 Madison Ave., New York, N.Y. 10010.

National Cancer Association, 1739 H St., N.W., Washington, D.C. 20006.

National Dairy Council, 111 No. Canal St., Chicago, Ill. 60606.

National Institute of Mental Health, Public Health Service, 9000 Wisconsin Ave., Bethesda, Md. 20014.

National Society for the Prevention of Blindness, 1790 Broadway, New York, N.Y. 10019.

National Tuberculosis Association, 1790 Broadway, New York, N.Y. 10019.

National Vitamin Foundation, Inc., 15 E. 58th St., New York, N.Y. 10022.

National Wildlife Federation, Education Dept., 1412–16th St., N.W., Washington, D.C. 20036.

U.S. Department of Agriculture, Agricultural Research Service, Washington, D.C. 20250.

U.S. Fish & Wildlife Service, Department of Interior, Washington, D.C. 20240.

U.S. Government Printing Office, Washington, D.C. 20401.

U.S. Public Health Service, Communicable Disease Center, Atlanta, Ga. 30333.

Upjohn Co., Kalamazoo, Mich. 49001 (vitamins books).

Ward's Natural Science Establishment, Inc., P.O. Box 1712, Rochester, N.Y. 14603; or Ward's of California, P.O. Box 1749, Monterey, Calif. 93940 *(Ward's Bulletin)*.

Welch Manufacturing Co., 1515 No. Sedgwick St., Chicago, Ill. 60610.

Westinghouse Electrical Corp., School Service, 306–4th Ave., Pittsburgh, Pa. 15222 (teaching aids).

G

Projects and

Class Studies

LABORATORY EXERCISES WITH ENZYMES

1. ENZYMES IN CORN KERNELS

WHEN CORN KERNELS germinate, they use food stored inside. Food, in the section of the kernel called endosperm, is largely starch. Of course, starch molecules are very large and not soluble in water. In this form, starch makes excellent stored food. It is not soluble, and therefore remains fixed inside the corn kernel. The starch molecule is large enough to stay inside the cell because it will not pass through the cell membrane.

When corn begins to grow, these starch food reserves must change into sugar. Sugar molecules are smaller and will pass through cell membranes, diffusing from one part of the kernel to another. The starch is changed into sugar by enzymes, which diffuse from the embryo into the starch endosperm.

You can show this digestion of starch into sugar by following the procedure below:

a. Soak corn kernels for about 24 hours in tap water. You may use field corn or hybrid seed corn. Try to use corn that has a rather high germination capability, since the success of the experiment will partly depend on how well the corn embryos grow. Make a cold-water starch-agar paste in the following way:

(1) Weigh out 1 g of common starch, cornstarch, or any other form of starch you have available, for each 100 ml of final starch-agar suspension. Weigh out 1 g of agar for each 100 ml of final suspension.
(2) Mix the starch and agar together, dry.
(3) Make a cold-water paste of the starch-agar mixture.
(4) Dilute the paste with cold water until it is somewhat runny, then pour it into the quantity of boiling water necessary to end up with a 1% suspension.
(5) Bring the starch-agar water mixture to a boil. The final suspension should contain starch-agar and water in the ratio of 1:1:100. This is known as a 1% starch-agar suspension (even though it is really 1:1:98).
(6) After it has cooled slightly below the boiling point, pour into the petri dishes. The size of the petri dishes is not critical and neither is the amount of starch-agar suspension you put into the dishes. Generally a dish filled one fourth to one third full of starch-agar suspension is satisfactory. Cover the dishes and allow them to cool to room temperature. At this point the starch-agar suspension should be solid.

b. Remove the endosperm from a corn kernel embryo with your fingernails. Do not use razor blades as they may damage the living embryos.

c. Carefully press the living embryos into the surface of the starch-agar.

This exercise is quite successful as a class project. You may prepare the starch-agar petri dishes beforehand, and you may have your students remove the endosperm from the corn embryos in class and press the embryos into the surface of the agar.

In 24 to 48 hours the embryos should begin to grow. You will be able to observe growth taking place in the form of the epicotyl, which may turn green. The hypocotyl should also be visible in 24 to 48 hours.

Test the germinated embryos for utilization of starch by using iodine solution. Flood the petri dish containing the embryos with iodine solution made up in the following proportions: Dissolve 1.5 g potassium iodide and .3 g iodine in 1000 ml tap water. The plate should not be flooded deeply with iodine solution, but the surface only moistened. If the exercise has been successful, there will be areas around each corn embryo stained only slightly blue-black. The rest of the petri dish with its agar will be intensely blue-black, indicating that digestion has gone on in the areas close to the corn embryo.

Discussion Questions

1. Why might some of the areas around the corn embryos on the agar plate give a positive starch test while others give a negative starch test?
2. What stored foods are present in the corn kernel?
3. Where did the enzymes come from?
4. What caused the starch in the circular area around the growing corn embryos to disappear?
5. When starch is digested, what are the digestion products?
6. Based on results of this exercise, how could you prove what the digestion products of starch are?
7. What modification of this experiment could you design? For example, what would be the effect if you put some petri dishes in the dark, some in the refrigerator, some in incubator ovens at 110°, or used embryos of plants other than corn?

2. THE EFFECTS OF pH AND TEMPERATURE ON ENZYME ACTION

Not only is the chemical structure of enzymes complex, but their mechanisms of operation are also complex. For example, enzymes seem to work best at specific temperatures, pH, and on specific substrate molecules. In this exercise we will determine the best pH and temperature for the enzyme Taka-Diastase. Taka-Diastase may be purchased from a drugstore. Generally it is in dilute alcohol solution. In this form Taka-Diastase generally digests 200 to 400 times its weight in starch in 5 to 10 minutes.

Procedure

a. To show the catalytic action of Taka-Diastase at room temperature.

   (1) Put two separate drops of 1% starch suspension on the microscope slide. To make up the starch suspension weigh out common starch, cornstarch, or any other type of starch you happen to have on hand. Use 1 g of starch for each 100 ml of water. Make a cold-water paste first, using a few ml of the water and the amount of starch you need. Then pour this paste into a beaker of hot water, bring to a boil, and then let cool to room temperature.
   (2) Place 2 drops of the starch suspension on a slide.
   (3) Add a drop of the enzyme Taka-Diastase to 1 drop of starch suspension but not the other.
   (4) Wait 5 minutes.
   (5) Add a drop of iodine solution to each of the 2 separate drops on the slide.
   (6) Place the slide on top of a piece of white paper so you may observe the results of the reaction.

This test should produce a blue-black reaction between the iodine and starch drop, but the starch that contains the drop of Taka-Diastase will not turn blue-black. The above procedure is simply to show the digestive action of Taka-Diastase on starch.

299

b. (1) Fill the rounded ends of each of 3 test tubes with 1% starch suspension.

(2) Put one test tube in a beaker of ice water, one in a beaker of boiling water, and leave the third at room temperature.

(3) After 2 to 3 minutes add 10 drops of Taka-Diastase to each tube and then keep them at their respective temperatures for 5 more minutes.

(4) To each of the 3 test tubes add a quantity of iodine solution equal to the contents of each test tube.

Taka-Diastase works best at room temperature, and it is in this test tube that the iodine test should be negative. You may have to adjust the quantities of starch suspension, Taka-Diastase, or the amount of iodine you add for the final test. For example, starch suspension that is weaker than 1% may be more satisfactory. A more dilute iodine solution may work better. Generally the iodine solution should be a light tan color like a cup of tea. There may be variation between the iodine test and the hot and cold test tubes, but each of them should be more blue-black than the one that remained at room temperature.

c. To show the effect of pH on enzyme action.

(1) Fill the rounded end of each of 5 test tubes with a 1% starch suspension. Then put an equal quantity of pH solution in each test tube. The pH solutions to add are the following: pH 3, 5, 7. 9, 11. Check the final pH with pH paper, just to make sure. To make the 5 pH solutions, purchase powdered buffered tablets from a chemical or biological supply house. These buffered tablets may be dissolved in water beforehand, and kept from one semester to the next. You may find it convenient to put the buffered solution in small dropping bottles so that the students will use them more readily.

(2) Add 10 drops of the enzyme Taka-Diastase to each of the 5 test tubes.

(3) Wait 5 minutes, and then add an equal quantity of iodine solution to each of the 5 test tubes.

Taka-Diastase works best at a pH between 5 and 7, and the contents of these 2 test tubes should show almost no color reaction with the iodine solution. The test tubes with contents of pH 3, 9, and 11 should produce a definite blue-black reaction with the iodine solution.

Discussion Questions

1. What is meant by optimum pH? by optimum temperature?
2. Where does the enzyme Taka-Diastase come from?
3. What is the substrate upon which Taka-Diastase acts?
4. Of what significance in biology is the fact that Taka-Diastase and other enzymes work at specific pH's and temperatures?
5. What test could you use to determine the digestive products of starch and Taka-Diastase?

3. ENZYMES AND CHOCOLATE-COVERED CHERRIES

There are some interesting biological principles connected with the manufacturing of chocolate-covered cherries. You may use this simple exercise to open your discussion of enzymes in biological systems, or you may use it to summarize the functions of enzymes. Used either way, it will be motivating and interesting for your students to understand how the liquid centers in chocolate-covered cherries get there in the first place.

When the cherries are manufactured, the white fondant obviously must be solid since it has to be coated with chocolate by a dipping process. The fondant is composed of a highly concentrated solution of sucrose which is semi-solid, especially at lower temperatures.

But in addition to the sucrose fondant, the enzyme in this case is mixed with the fondant when it is prepared for molding. After being coated with chocolate, the fondant and its accompanying enzyme react together making use of the moisture in the fondant to allow the hydrolysis (digestion) of the sucrose. While being packaged, shipped, and merchandized in stores, this enzyme digestion of sucrose takes place. When the chocolate-covered cherries are purchased, 2 or 3 weeks after they are manufactured, the invertase has digested sucrose to fructose and glucose. Since the fructose component of sucrose is much more soluble in water, it dissolves in the moisture in the fondant and becomes a sugar syrup. The centers of the chocolate-covered cherries thus change from a solid to a liquid.

Procedure

Pass a box of chocolate-covered cherries around your class. As the students begin eating them, turn to the discussion questions below.

Discussion Questions

1. Were the chocolate-covered cherry centers liquid during the time they were manufactured? If not, what was the form of the centers?
2. What is the white part inside the chocolate-covered cherries, in addition to the liquid?
3. What chemical test could you make to determine the nature of:
   a. the liquid center?
   b. the white residue?
4. What biological processes are similar to this digestion of sucrose with the enzyme invertase?
5. What other products, if any, are produced in this digestive process?
6. Write the chemical equation for the digestive process in this exercise.

SOME BASIC TESTS FOR PLANT FOODS

THERE ARE characteristic chemical tests for the three major classes of foods—the *carbohydrates, lipids,* and *proteins.* Some of the tests are quite simple, and others are very complicated and require specialized apparatus. We will attempt to perform some of the simple tests that give good results without requiring special equipment. The tests, and what they detect, are summarized below:

THE IODINE TEST FOR STARCH

Although iodine is not readily soluble in water, it will dissolve quickly if some potassium iodide is added. Because an iodine solution is made up with potassium iodide, the resulting mixture is often called iodine-potassium iodide solution. However, for simplicity we refer to it as iodine solution.

On the other hand, iodine crystals alone will dissolve readily in various alcohols. This type of solution, called a tincture, is not as useful for these tests as an aqueous solution.

1. Place amount of material the size of a pencil eraser on a microscope slide.
2. Put the slide on a piece of white paper so results will show clearly.
3. Add a drop of iodine solution. For preparing iodine solution dissolve 0.3 g iodine and 1.5 g potassium iodide in 100 ml distilled water.
4. A blue-black color indicates the presence of starch.

BENEDICT'S TEST FOR GLUCOSE

1. Place amount of material the size of a pencil eraser in a test tube.
2. Add enough Benedict's solution to fill rounded end of test tube. To prepare Benedict's solution add 173 g sodium citrate and 100 g anhydrous sodium

carbonate to 600 ml distilled water. Heat until chemicals are dissolved. Filter. Add 17.3 g copper sulfate to 150 ml distilled water. After copper sulfate is completely dissolved, add this solution to the first, stirring constantly. Add enough distilled water to make 1 liter of solution.
3. Heat tube containing material to be tested and Benedict's solution.
4. A brick-red color indicates presence of glucose (or other reducing sugar).

NINHYDRIN TEST FOR PROTEIN

1. Place amount of material the size of a pencil eraser in a test tube.
2. Add enough ninhydrin solution to fill rounded end of test tube. To prepare ninhydrin solution add 1 g ninhydrin powder to 50 ml ethanol (95%) and add to 50 ml distilled water.
3. Heat tube and contents in boiling water about 4 minutes.
4. Purple color shows presence of protein. (Actually the chemical ninhydrin reacts with amino acids rather than protein.)

1. Iodine test—starch
2. Benedict's test—glucose
3. Ninhydrin test—protein

The foods below should be tested using the procedures above.

Record your results using the words *positive, negative,* or *trace.*

In the Benedict's test, if a very small trace of red color appears, or if the original blue color changes to blue-green or light green, report the test as *slight trace.*

FOOD	IODINE TEST	BENEDICT'S TEST	NINHYDRIN TEST
Soybean			
Wheat germ			
Wheat			
Cornstarch			
Karo syrup			
Apple			
Banana			
Powdered sugar			
Oatmeal			

DNA MOLECULE FROM *GROCERICUS STOREII*

ONE PROBLEM in working with biochemicals is that we are often required to believe things about them that we cannot prove. For example, the DNA structure has never been observed directly, even by those scientists who work closely with it. It seems that students have an especially difficult time visualizing the double helix, the bonding, and the purines and pyrimidines. To help them understand the structure of DNA, you may wish to have your class construct a model of the DNA molecule, using soda straws, Cheerios, toothpicks, and string.

In this model the sugar groups are the Cheerios, the phosphate groups are short links of white soda straws, and the toothpicks and string are used to hold the model together. In this model four different colored soda straws represent each of the four bases. For example, adenine might be a red straw, guanine might be yellow, and cytosine and thymine might be blue and pink, respectively.

The model requires about 30 to 40 minutes to complete, if students work in groups of two. This will allow some time either at the beginning of class for introduction and orientation, or at the end of the class for discussion questions. (Figure 8.19 in the text shows a completed model.)

Procedure

1. Working with your laboratory partner, select a small pile of Cheerios, toothpicks, and soda straws of five different colors, and about three feet of string.
2. Cut all the straws into $\frac{1}{2}$-inch lengths.
3. String straws of one color, preferably white, on two strings tied to a ring stand. Alternate the straws on the two

strings with Cheerios so that each string when finished contains a straw-Cheerio sequence.

4. After both sides of the model are completely assembled with straws and Cheerios, tie the top ends of the string to a top ring on the ring stand. At this point you should have two parallel strings about 1½ to 2 inches apart, supported vertically by two rings on a ring stand.

5. Now turn to the toothpicks and short lengths of colored straws. Assign straw colors to each of the four organic bases in the DNA molecule. Fill out the accompanying chart to help you do this.

STRAWS REPRESENTING PURINE AND PYRIMIDINE BASES

Color of Straw	Base
	adenine
	guanine
	cytosine
	thymine
	uracil (if you are making RNA)

6. Put two different colored straws on a toothpick and insert the toothpick through the holes in two opposite Cheerios. At this point you will begin building a ladder. The straw-toothpick assemblies represent the rungs of the ladder. The vertical strings with straws and Cheerios represent the sides of the ladder. Continue building the ladder until all the rungs are in place. Keep in mind that one purine must be matched to a pyrimidine; no two purines or two pyrimidines may form a base pair on a toothpick.

7. When the ladder is complete, remove the top ring from the ring stand and carefully twist it once. This will give the ladder a twist, so that it is no longer straight but resembles a helix. Then replace the top ring on the ring stand. The model is now complete.

Discussion Questions

1. In what ways does this model differ from the real DNA molecule other than in size?

2. If you were to make the purine-pyrimidine base pairs (represented by the straws) more realistic, how would you change their length?

3. In the real DNA molecule, by what means are the sugar phosphate groups held together?

4. In the real DNA molecule by what means are the purine-pyrimidine pairs held together?

5. Attempt to compute how large a man would be if this model of DNA were the actual size as the DNA in each of his cells.

EXERCISES ON PHOTOSYNTHESIS

THERE ARE five exercises in this section. They are designed to show: 1. Light Absorption by Plant Pigments, 2. Necessity of CO_2 for Photosynthesis, 3. Use of CO_2 by an Aquatic Plant, 4. Relationship of Oxygen, CO_2, and pH in Photosynthesis, and 5. Production of Starch in Photosynthesis.

Some of these exercises use common glassware found in a biology laboratory, and some need more elaborate apparatus. As an example, in Exercise 1 you need a spectroscope which can usually be borrowed from the physics department; in Exercise 2, large bell jars; Exercise 3, the indicator bromothymol blue; and in Exercise 4, a special water-testing kit that can be purchased from the Hach Chemical Co., Ames, Iowa 50010, for about $35. Each of these exercises is designed to show a specific aspect of photosynthesis. Some may be done as class demonstrations and others by individual students, or students working in small groups in the laboratory. The results of each vary from time to time, but we will comment on each exercise so that you may be aware of what possible outcomes to expect.

1. LIGHT ABSORPTION BY PLANT PIGMENTS

In this exercise you need a *spectroscope,* with a diffraction grating of about 12,000 to 15,000 lines per inch. The grating acts as a prism, splitting white light into red, orange, yellow, green, blue, indigo, and violet.

By finding which colors in white light are absorbed by an extract of plant pigments, you can estimate which colors of the spectrum are utilized in photosynthesis.

a. Grind a few leaves in a mortar with a small amount of sand and about 25 ml acetone. Pour the extract into a test tube and hold the tube in front of the slot at the rear of the spectroscope.
b. Put the light source about 8 inches away from the tube of chlorophyll so that the light shines through the tube into the slit. An ordinary 25- or 50-watt light bulb will do.
c. Look through the eyepiece and observe the intensity of the bands of light. Record intensities in the chart below, using the terms *bright, dim,* or *absent.*
d. Repeat the steps above, using a test tube full of pure acetone.
e. Repeat, omitting the test tube and acetone, using the light only.

	red	orange	yellow	green	blue	indigo	violet
Light only							
Light & test tube of acetone							
Light, test tube, & pigment extract							

Discussion Questions

1. What control, if any, is used in this exercise?
2. Do you think sunlight has a different spectrum than artificial light? How would you prove it?
3. According to your results, which colors are more nearly completely absorbed by the pigment extract?

4. From this exercise can you predict which colors of light are most necessary for synthesis to occur?

2. NECESSITY OF CO_2 FOR PHOTOSYNTHESIS

Make two complete setups using two *Coleus* plants of nearly equal size. Place each under a bell jar on a glass plate. Under one bell jar put a small beaker of potassium hydroxide and under the other a beaker of water. Apply a thin coat of vaseline to the bottom of each bell jar and slide it onto the glass plate, making an air-tight seal. Observe the two setups over a period of 14 days.

Discussion Questions

1. For what is potassium hydroxide used in these laboratory experiments?
2. How did the control setup differ from the one in which potassium hydroxide was used?
3. What factors other than the loss of CO_2 could kill the plant?
4. Do plants need oxygen to carry on photosynthesis?
5. What would happen if both plants were placed in the dark?

3. USE OF CO_2 BY AN AQUATIC PLANT

Plants that grow on land get CO_2 for photosynthesis from the air. Aquatic plants get CO_2 from the water where it is present as carbonic acid (H_2CO_3).

$$H_2CO_3 \longrightarrow CO_2 + H_2O$$

If water contains a large amount of CO_2, enough H_2CO_3 will form to lower the pH below 7. When CO_2 is removed during photosynthesis, pH rises.

Bromthymol blue is a dye indicator that changes color as pH changes. It is yellow in the presence of an acid and blue in the presence of a base. It can be added to the water in which *Elodea* grows without harming the *Elodea*.

a. Fill three large test tubes ¾ each with water from a tank in which *Elodea* has been growing.
b. Add about 30 drops of bromothymol blue to each tube.
c. Use a straw to gently bubble your breath into two of the three tubes until the solution in each turns a pale yellow.
d. Put a sprig of *Elodea* into one of the yellow-colored solutions.
e. Place a loose-fitting cap of aluminum foil on each tube.
f. Record results in three or four days using the chart provided.

	TUBE 1 no *Elodea*	TUBE 2 no *Elodea*	TUBE 3 *Elodea*
Start	Blue	Yellow	Yellow
Finish			

Discussion Questions

1. What color is bromothymol blue in acid? What color is it in base?
2. As *Elodea* grows in water will the water become more acidic or more basic?
3. Why is the yellow solution with no *Elodea* used in this exercise?
4. Why is the blue solution with no *Elodea* used in this exercise?
5. At the end of the exercise which solution will have the highest oxygen content?

4. RELATIONSHIP OF OXYGEN, CO_2, AND pH IN PHOTOSYNTHESIS

In this exercise you will use the Hach Water Testing Kit. With it you can make simple chemical tests to determine pH, CO_2, and oxygen content of water.

As a land plant photosynthesizes, it uses CO_2 from the air and produces oxygen. When aquatic plants photosynthesize they use CO_2 that is dissolved in water. They also give off oxygen, which dissolves in the water. You can measure the concentration of these two gases before and after photosynthesis. The change in their concentration in the water is an indication of the amount of photosynthesis that has taken place.

Tap water may already contain a high concentration of either of the two gases. To obtain suitable water for beginning the experiment, use tap water that has been boiled to drive off excess gases. You will need about 1 liter of boiled water for this exercise. Let it cool before you begin.

a. Test a liter of boiled water for oxygen, CO_2, and pH, following directions included with the Hach Kit. Record results on the chart.
b. Fill two 400-ml beakers with water and add several sprigs of *Elodea* to one.
c. Set both beakers in light for 12 to 24 hours. Do *not* allow them to heat above room temperature. Cover them with aluminum foil to prevent excessive diffusion of gases in and out of the water.
d. Retest the water in each beaker. Record results on the chart.

	Oxygen in ppm	CO_2 in ppm	pH
Initial test on boiled water			
Final test on water without *Elodea*			
Final test on water with *Elodea*			

Discussion Questions

1. What does "parts per million" mean?
2. Air contains about 20% oxygen. How much would this be in ppm?
3. What kind of acid does CO_2 form when it dissolves in water?
4. Does oxygen form an acid or base when it dissolves in water?
5. Using results from your work just completed, encircle the arrow in each space

in the chart below that summarizes the the results of your experiment.

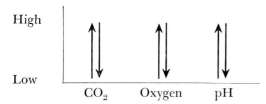

5. PRODUCTION OF STARCH IN PHOTOSYNTHESIS

A large number of plants photosynthesize more food than they can use. They store the excess food in leaves and roots. Plants in the dark cannot photosynthesize, and must use the stored food. In this exercise we will investigate what happens to stored food in leaves kept in the dark.

a. Remove a leaf from a *Coleus* plant that has been growing in the light. Make a sketch of the leaf, labeling the colored areas. Use the boxes below.
b. Put the leaf in a petri dish and cover with alcohol. Boil for about 3 minutes, using a hot plate. Do not boil dry. Replenish the alcohol from time to time. The alcohol removes the chlorophyll. (Caution: Alcohol has a low boiling point, so do not use high heat.)
c. Hold the limp leaf against the bottom of the dish while you drain off the alcohol.
d. Flood the leaf with iodine solution to test for starch.
e. Make another sketch of the leaf, after the test for starch. Use the boxes below.

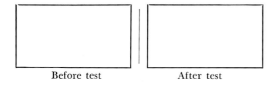

f. Put a *Coleus* plant in total darkness for 24 to 48 hours, then remove a leaf and repeat the 4 steps above. Make sketches in the following boxes.

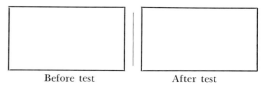

Discussion Questions

1. Photosynthesis produces glucose, yet you tested for starch. Explain.
2. Why do you need to boil the leaf in alcohol to remove the chlorophyll?
3. What is the relationship between the pigmented and nonpigmented areas of the leaves, and photosynthesis?
4. Is there any evidence in this exercise that photosynthesis occurs in one area of a leaf but food storage occurs in another?

FIRE AND RESPIRATION

SOMETIMES students understand respiration better if they can compare it to fire. Fire is easier to investigate, and its products are easier to work with. Heat, light, moisture, and carbon dioxide all are produced when a fire burns. While the heat, light, and moisture are evident, it may not be clear to the students that carbon dioxide is also produced. This exercise is designed to test for the presence of carbon dioxide. The same solution used in this exercise is used in other exercises to test for the presence of carbon dioxide when plants respire.

Procedure

1. Hold a cold test tube inverted at a 45° angle over a burning candle. Keep the test tube in place for about 1 full minute.
2. Set the test tube, open end down, on the table top.
3. While the mouth of the tube is cooling, get a small beaker or dropping bottle of saturated barium hydroxide solution. Be careful not to shake the bottle or

beaker because there may be a white precipitate in the bottom. The barium hydroxide solution will be water-clear.
4. Lift the test tube up slightly from the table and cover the open end with your thumb. Turn it right side up and fill the rounded end of the test tube with barium hydroxide solution.
5. Shake the test tube and observe any changes in the barium hydroxide solution.

Discussion Questions

1. What observable changes took place in the barium hydroxide solution?
2. Write the equation for the chemical reaction between barium hydroxide and carbon dioxide.
3. What other products of combustion were produced in this exercise in addition to those detected by the barium hydroxide solution?
4. How is respiration similar to fire?
5. How do fires differ from respiration?
6. Do you think there is any difference between animal respiration and plant respiration in terms of the products given off?

EXERCISES IN PLANT RESPIRATION

THIS SERIES of exercises can be done as teacher demonstrations, or may be assigned to the students as they work in the laboratory in small groups.

The six exercises in plant respiration are: 1. Oxygen Requirement, 2. Quantitative Determination of Oxygen Requirement, 3. Relationship of CO_2 Produced to Oxygen Consumed, 4. Production of CO_2, 5. Anaerobic Production of CO_2, and 6. Release of Energy.

1. OXYGEN REQUIREMENT

a. Make two small sacks from cheesecloth and put a few soaked corn kernels, oats,

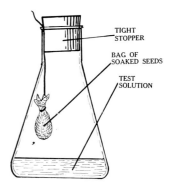

Fig. G.1.

or grass seeds in each. Soak each sack in tap water. (See Figure G.1.)
b. Prepare 50 ml each of a 5% solution of KOH and a 5% solution of pyrogallic acid.
c. Pour the KOH and pyrogallic acid solutions into one 250-ml Erlenmeyer flask. (Caution: Do not let the solution touch the sides of the flask where the sack of seed would absorb the substances.)
d. Quickly put a sack of seeds into the flask, above the solution. Then stopper the flask airtight.
e. Set up a control, using water instead of pyrogallic acid and KOH.

Discussion Questions

1. Why did you use soaked seeds?
2. What does the pyrogallic acid do?
3. What factors could cause the seeds to die in either flask?
4. Why is the flask with water used in this experiment?
5. What part does water play in respiration?

2. QUANTITATIVE DETERMINATION OF OXYGEN REQUIREMENT

a. Pour 350 ml of tap water into each of two gallon jars. (See Figure G.2.)
b. Put 100 g KOH into one jar. Swirl to promote dissolving.
c. Fill each of two glass cylinders ¼ full with soaked peas. Then put in a thin layer of dry cotton to hold the peas in place.

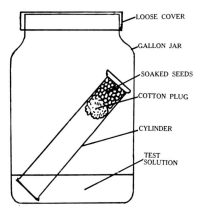

Fig. G.2.

d. Invert the cylinders in the gallon jars. Make sure the open mouth of each cylinder is at least ½ inch below the level of the liquid. Put a lid loosely on each jar.

Discussion Questions

1. Why was KOH added to one jar, but not to the other?
2. Why was cotton used to hold the peas in place, rather than a cork or rubber stopper?
3. What is the KOH solution used for?
4. What is the source of oxygen for the peas in each cylinder?
5. If the experiment worked perfectly, how far up the cylinder should the KOH solution rise?
6. Do you expect the water in the control to rise, fall, or stay at the same level? Why?

3. RELATIONSHIP OF CO_2 PRODUCED TO OXYGEN CONSUMED

a. Fill each of two bottles ¼ full with soaked peas. (See Figure G.3.)
b. Prepare a vial of 20% KOH solution. Cut a small square of paper towel and fold into a fan. Put the fan into the vial. It will absorb KOH from the vial and provide a large surface area for absorbing CO_2 from the atmosphere inside the jar.

c. Hang the vial in one bottle, then stopper *both* bottles with a rubber stopper-glass tubing assembly.
d. Put a bottle of water (colored with a drop of red food coloring) at the end of the glass tube, on both the experiment and the control. Adjust each setup so that the tube extends almost to the bottom of the jar of red-colored water.
e. Check each setup for leaks.

Discussion Questions

1. How does the control differ from the experiment in this case?
2. As the peas respire, what gas will they use up? What gas will they give off?
3. In the equation below what is the relationship between the quantity of O_2 used and CO_2 produced?

$$C_6H_{12}O_6 + 6O_2 \rightarrow 6CO_2 + energy + 6H_2O$$

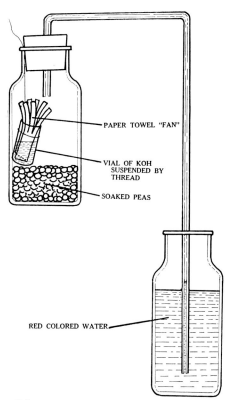

Fig. G.3.

4. List a reason for (a) the red-colored water rising.
   (b) the red-colored water falling.
   (c) the red-colored water staying at the same level.
5. If the experiment worked perfectly, what should happen to the level of the red-colored water?

4. PRODUCTION OF CO_2

a. Set up apparatus as shown in Figure G.4. Use soaked peas and fresh $Ba(OH)_2$ solution.
b. Put some $Ba(OH)_2$ solution in a test tube and blow through the solution, using a soda straw. Write a chemical reaction for the results. What substance in your exhaled breath caused the reaction?
c. Pump air through the train. Observe all the results, then answer the questions below.
d. Repeat, using dry peas. What are the results?

Discussion Questions

1. Complete and balance the equation

$$Ba(OH)_2 + CO_2 \rightarrow$$

2. What is the reason for having the following in the setup:
   bottle 1
   bottle 2
   bottle 4
3. What other factors, if any, could have caused a reaction in bottle 4?

5. ANAEROBIC PRODUCTION OF CO_2

a. Prepare two sets of bottles like the one set shown in Figure G.5.
b. In one bottle put some dough made according to this formula:

   25 g flour
   ½ pkg (or cake) of yeast
   Enough tap water to make a thick dough

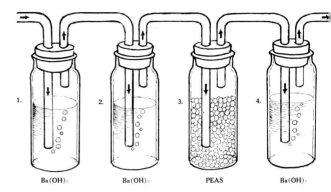

Fig. G.4.

c. Make dough for the second set of bottles in the same way, but leave out the yeast.
d. Fill the carbon dioxide traps with 20% KOH solution and then assemble the apparatus.

Discussion Questions

1. Write the balanced equation for the two types of respiration indicated below, using glucose as the material being respired:
   aerobic respiration:
   anaerobic respiration:
2. Is the respiration in this exercise aerobic or anaerobic?
3. What product of respiration does $Ba(OH)_2$ detect?
4. What products of respiration are *not* detected in this exercise?

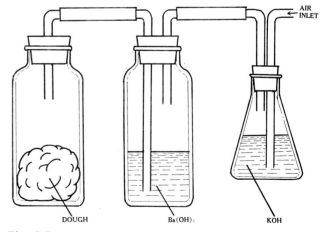

Fig. G.5.

5. Why is Ba(OH)$_2$, rather than KOH, used to detect an end product?
6. How does O$_2$ play a part in this exercise?
7. What happens to the water that is produced by respiration in this exercise?

6. RELEASE OF ENERGY

a. Fill each of three thermos bottles about half full of peas, as shown in Fig. G.6. In bottle 1 put soaked peas; in bottle 2, soaked peas covered with formaldehyde; and in bottle 3, put dry, unsoaked peas.
b. Insert a thermometer in each thermos bottle, then make a loose cotton plug for each. Each thermometer should be in the center of the cotton plug, and within ¼ inch from the bottom of its bottle.
c. Record initial temperature in each thermos and then record temperature every 12 hours for 2 days. Put information on chart.

Discussion Questions

1. Why were cotton stoppers used rather than corks?
2. Why were soaked seeds used? Wouldn't dry seeds in all three bottles work just as well?
3. Why were some seeds covered with formaldehyde?
4. What type of respiratory energy is *not* detected in this exercise?

		Temperatures		
Date	Time	Soaked	Peas + Formaldehyde	Dry

DIFFUSION

USING CARROTS, potatoes, and eggs, your students can do some interesting laboratory work on osmosis. With a few crystals of potassium permanganate they can

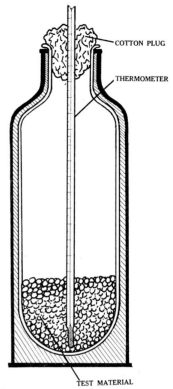

Fig. G.6.

investigate the phenomena associated with diffusion. And with hydrochloric acid and ammonium hydroxide they may do a quantitative determination of the effect of particle size on rate of diffusion.

In the following exercises we will investigate the effects of temperature, type of solvent, and concentration on diffusion.

1. EFFECT OF TEMPERATURE

a. Using two small containers, such as 50-ml beakers, fill one with boiling water and the other with ice water.
b. Select two crystals of equal size of potassium permanganate. (Caution: Causes severe burns when handled.)
c. Working with your laboratory partner, drop a crystal of potassium permanganate into each beaker. Make sure the crystals enter the liquid at the same time.

Discussion Questions

1. In which beaker does diffusion occur more rapidly? Why?
2. Why did you make certain the crystals were of equal size?
3. Why was it necessary that the crystals entered the water at the same time?
4. Would the experiment work the same if both beakers contained water at room temperature?

2. EFFECT OF CONCENTRATION ON DIFFUSION

You can investigate the effect of varying concentrations of a substance on the rate of diffusion by using ammonium hydroxide, a glass tube, and some litmus paper. One problem in this exercise is that the red litmus paper changes to blue through a range of purple colors. It is not always easy to tell when the color change occurs since red changes to blue through the purple range.

Procedure

a. Cut strips of litmus paper the long way so that they are narrow enough to slip into glass tubing about 5 mm in diameter. Slide five or six such narrow strips into the length of the glass tubing with a piece of wire.
b. Place a cotton plug at one end only. Make sure the cotton is very loose and does not form a tight wad in the end of the tube.
c. Put a drop of concentrated ammonium hydroxide on the cotton plug. Using a watch, determine how long it takes for each strip of the litmus paper to change color.
d. Graph your results.

Discussion Questions

1. What color is red litmus in an acid? What color is it in a base?
2. Judging from the results of your experiment, would you say diffusion is directly or inversely proportionate to the con-

centration of the diffusing substance?
3. Was the diffusing substance a solid, liquid, or gas?

3. EFFECT OF PARTICLE SIZE

This exercise uses concentrated ammonium hydroxide and hydrochloric acid. Both chemicals are highly toxic. You may wish to do this as a demonstration for your class for that reason. If so, use a test tube larger than the one recommended here. For a class of about 30 students a glass tube an inch in diameter and 2 to 3 feet long is satisfactory.

If students are doing this exercise in small groups, they can use glass tubing about 8 inches long and 1/4 inch in diameter.

The reaction that occurs is written below.

$$HCl + NH_3 \rightarrow NH_4Cl$$

The ammonium chloride, which is the result of the reaction between hydrochloric acid and gaseous ammonia, is a white salt that precipitates on the inside of the demonstration tube. Because the HCl molecules have a molecular weight of about 36, and the ammonia molecules have a molecular weight of about 17, the ammonia molecules move about twice as fast as the hydrochloric acid molecules. Thus, the white ring will form about one-third the distance from one end of the tube. It is necessary that both the acid and base be applied to the cotton plugs at the ends of the tubes simultaneously. If not, one will get a head start, and cause the ring to form at the wrong position in the closed tube.

Procedure

a. Place a small amount of cotton in each end of a hollow glass tube. The cotton should only form a loose plug on the end and not be a solid wad of cotton.

b. Working with your partner, place a drop of hydrochloric acid on one plug and a drop of concentrated ammonium hydroxide on the other simultaneously. (Caution: These cause severe burns if you touch them.)

c. Watch the two carefully for any evidence of a chemical reaction.

Discussion Questions

1. Write the equation for the reaction between ammonium hydroxide and hydrochloric acid.
2. What would you predict the effect to be if you set the tube on end rather than on its side?
3. What effect would there be if both the acid and base were heated?
4. What factors could cause the ring of ammonium chloride to form at the incorrect distance from the end of the tube?
5. Why do you need to know the molecular weights of the reacting substances to predict how far from the end the white ring of ammonium chloride will form?

OSMOSIS

THESE EXERCISES can be done either as demonstrations or as individual student projects. Some of them use natural materials, such as carrots, potatoes, or eggs. Others make use of synthetic membranes. In several places in these exercises chemicals such as potassium permanganate, hydrochloric acid, and ammonium hydroxide will be used.

1. OSMOSIS WITH AN EGG

If you remove the shell from a fresh egg and soak the shell-less egg in water for a few hours, it will swell up to much greater than its original size. In fact, if you are careful you can bounce this enlarged egg on a table top like a rubber ball.

Procedure

a. Use a dilute solution of hydrochloric acid to remove the eggshell. Acid concentration about one normal to two normal is sufficient. Put the acid in a small beaker and cover the egg. Keep turning the egg until the shell is completely removed. Bubbles of carbon dioxide gas will be given off as evidence of the dissolving of the calcium carbonate shell.

b. Drain off the acid and replace it with water. You may use either tap or distilled water. Measure egg circumference at area halfway between ends. The egg should be completely submerged.

c. After an hour the egg will have become enlarged. Measure it at definite intervals of time.

Be sure to use a beaker large enough to accommodate the swelled-up egg. (Caution: Hydrochloric acid causes severe burns.)

You may wish to have your students calculate the per cent increase in size of the egg. If so, have them measure the volume of the original egg according to how much water it displaces. Measure the amount of water the swelled-up egg displaces and subtract to find the volume of increase.

Discussion Questions

1. Write the chemical equation for the action of hydrochloric acid on the eggshell. Assume the eggshell is calcium carbonate.
2. Would this experiment work if you had used concentrated salt water rather than distilled or tap water?
3. Would this experiment work with a boiled egg?
4. Are you dealing with a living or dead membrane in this case?

2. A CARROT OR POTATO OSMOMETER

We can use plant materials to investigate the action of cell membranes in osmosis. You may do this as a class demonstration or have your students work in small groups to set up the osmometer. The critical factor here seems to be the size hole that is bored into the potato or carrot. The smaller the hole, and the deeper it is, the better the osmometer will work. The hole should be bored with an ordinary cork borer, to a depth nearly equal to the length of the potato or carrot. It should be about one-half inch in diameter. Of course, this means that this one-hole stopper or cork that you use to hold the osmometer tube in place must be about one-half inch in diameter also. A rubber stopper works a little better than cork since the rubber is more flexible and can fit more tightly into the carrot or potato.

Procedure

a. Select a rubber stopper with one hole to receive a glass tube of small diameter. The tube should have an inner bore of about 1 to 2 mm. (See Figure G.7.)
b. Bore a hole in a potato or carrot about the same size as the rubber stopper you have selected. The stopper should pass only half way into the hole. When you insert the stopper, be careful not to split the side of the carrot or potato.
c. Fill the hole about half full of Karo syrup.
d. Insert the glass tubing into the rubber stopper, so that it extends 2 or 3 mm out of the stopper on the other end.
e. Put the rubber stopper-glass tube assembly into the potato or carrot. Support it in a ring stand in a beaker of water.

Discussion Questions

1. What is an osmometer?
2. Would this experiment work if you filled the potato or carrot with distilled water rather than Karo syrup?
3. What would be the effect if you poured Karo syrup into the beaker instead of filling the beaker with water?

4. If the potato or carrot were dead would this have any effect on the experiment?
5. Theoretically, how high up the tube would the liquid rise?

3. A SYNTHETIC-MEMBRANE OSMOMETER

Instead of using a carrot or potato, you can construct an osmometer using a synthetic membrane. Generally, these membranes come from the supply house in the form of flat strips. They may be softened with tap water and formed into a sac.

Procedure

a. Cut a ⅛-inch length of artificial membrane and soften it under tap water so that the two sides may be separated and formed into a cylinder.

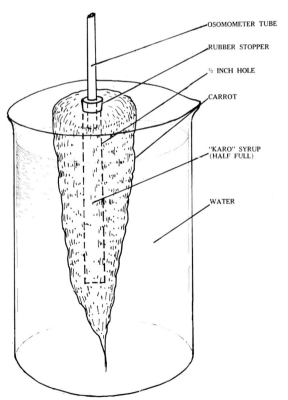

Fig. G.7.

OSOMOMETER TUBE
RUBBER STOPPER
½ INCH HOLE
CARROT
"KARO" SYRUP (HALF FULL)
WATER

b. Tie one end shut and fill with Karo syrup.

c. Select a one-hole stopper and insert it on the appropriate size glass tubing through the small end of the stopper. Thus, when you tie the synthetic membrane sac around the stopper, the sac will not slip off the stopper.

d. Insert the rubber stopper-glass tubing assembly into the open end of the membrane sac. Tie in place. Make sure that there is no air space left between the top surface of the Karo syrup and the bottom surface of the rubber stopper. Put the setup in a beaker of tap water.

Discussion Questions

1. Would this experiment work as well if you used salt water in the beaker rather than tap water?

2. What would happen if you used hot water in the beaker rather than water at room temperature?

3. If you put food coloring in the Karo syrup that was poured into the synthetic membrane sac, what do you predict would happen to the coloring at the end of the experiment?

4. Generally, water will rise in the tube, remain at that level for about a day, and then drop again. Explain.

4. QUANTITATIVE EXPERIMENT IN OSMOSIS

You can use small, equal-sized cylinders of fresh potato to determine the extent of osmosis. In this experiment students bore two- or three-inch lengths of potato with a cork borer, put the bores in salt-water solutions of different concentrations, and check the results each day for two or three days. The diameters of the cylinders change as well as their lengths, depending on the concentration of the salt water.

Procedure

a. Using the cork borer about 1/8 or 3/16 inch in diameter, bore two- to three-inch lengths of potato.

b. Cut the lengths accurately so they are all the same length. Record that length.

c. Prepare four petri dishes of solutions as follows: (1) distilled water, (2) 1% salt water, (3) 5% salt water, and (4) 10% salt water.

d. Lay two lengths of potato in each petri dish, and cover the petri dishes.

e. In 24 and 48 hours measure the potato lengths again. Compare with your original results.

Discussion Questions

1. Did all the potato borings become shorter, or did some become longer?

2. Judging from the results of the experiment, which solution is closest to being isotonic to the contents of a potato?

3. Would this experiment work as well if you had used sugar solutions rather than salt solutions?

4. Why is it helpful to know about isotonic, hypertonic, and hypotonic solutions?

CHROMATOGRAPHY OF INK

CHROMATOGRAPHY—the process of separating a mixture of dyes—can be used to separate mixtures of substances in solution. Often these substances are colored compounds, as in the case of plant pigments. On the other hand, some substances have no color. To separate non-colored compounds, chromatography can also be used. But the colorless compounds must then be identified on a finished chromatogram by spraying with a chemical that stains the separated, colorless compounds.

The principle of chromatography is most easily understood if ink is used. You may either do this as a demonstration for your class, or have the students perform it as a laboratory exercise.

Washable black ink works best. It has a large amount of red, and lesser amounts of green and blue dyes. Ordinary filter paper may be used, or absorbent paper towels, or cleaning tissue. Since the pigments to be separated are already colored, it is not necessary to spray the finished diagram with a developing substance. The exercise outlined below requires only about 20 minutes of class time, and thus will run to completion in an ordinary class period.

Procedure

1. Choose a container, such as a 400-ml beaker or a glass jar about three to four inches in diameter.
2. Fill the container to a depth of about one inch with tap water. The water in this case is the carrier solvent for the dyes.
3. Cut the paper (such as a piece of paper toweling) so that it fits into the container. The edges of the paper should not touch the sides of the container.
4. With a dropper or small brush, put a spot or line of the ink near the bottom of the paper. (See Figure G.8.)

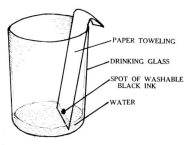

PAPER TOWELING

DRINKING GLASS

SPOT OF WASHABLE BLACK INK

WATER

Fig. G.8.

5. After the ink is dry on the paper, insert the paper, ink end down, into the container of water.
6. Support the sheet of paper so that it does not collapse and fall into the water.
7. In 15 to 20 minutes remove the paper from the water. Different colored inks, making up the black ink you used originally, will be separated on the paper. Generally red ink will move up the paper the farthest distance, followed by blue and green. Traces of yellow may also be present.

Discussion Questions

1. In biology, what substances are generally separated and identified by the processes of chromatography?
2. How do the following types of chromatography differ? (a) paper chromatography, (b) column chromatography, (c) thin-layer chromatography.
3. Does gravity have an effect on chromatography? Test your hypothesis by making a second setup in which the paper is laid in a horizontal plane rather than held vertically above the water.
4. What is meant by: (a) carrier solvent? (b) substrate?

CHROMATOGRAPHY OF AMINO ACIDS

CHROMATOGRAPHY of amino acids is not quite as dramatic as the chromatograph of ink or plant pigments. The amino acids have no color, and therefore are not directly visible on a finished chromatograph.

In the chromatography of colorless substances, such as amino acids, it becomes necessary to spray the finished chromatograph with special dye. In this exercise we will use ninhydrin. Ninhydrin is available commercially in pressurized spray cans. You may make a spraying mixture of ninhydrin by dissolving 1 gm of ninhydrin powder in about 100 ml of ethanol or acetone. This mixture may be sprayed on the chromatograph, using a perfume atomizer.

Procedure

1. Prepare a carrier solvent of:
   30 ml isopropanol

15 ml n-butanol
10 ml water
1 ml glacial acetic acid

Put the solvent in a petri dish bottom and cover with an inverted 1-gallon jar.

2. Spot the amino acids in a line about 1 inch above the bottom of a 9-inch square of Whatman #1 filter paper, according to Figure G.9. Mark the line at the edge of the paper with a pencil. Make about 10 applications of each acid with the broad end of a flat toothpick, allowing each spot to dry before adding more solution. Keep the spots away from the bottom edge of the paper; when the paper is standing in the petri dish *the spots must not touch the solvent.*

3. Roll the paper into a cylinder and staple the edges.

4. Stand the paper cylinder in the petri dish of solvent and cover with the gallon jar. (See Figure G.10.)

5. After 30 to 40 minutes remove the chromatograph, mark the top edge of the solvent with a pencil, and put the chromatograph aside to dry.

6. Spray the chromatograph with a 0.2% solution of ninhydrin in acetone, n-butanol, or 95% ethanol, or use a commercial solution of ninhydrin in a spray can. The purple color will develop within 20 minutes. Encircle the purple spots with a pencil.

7. Using your marked start and stop points of the carrier solvent, and the *centers* of the circled purple spots, compute R_f values for the knowns and unknowns on the chromatographs, using this formula:

$$R_f = \frac{\text{distance unknown moves}}{\text{distance solvent moves}}$$

8. From the R_f values of the known amino acids attempt to identify the unknowns.

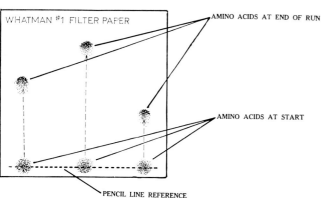

Fig. G.9.

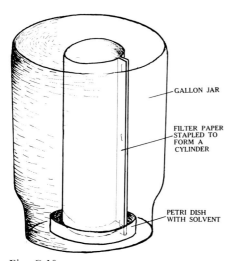

Fig. G.10.

Discussion Questions

1. Is there a relationship between the molecular weights of the amino acids and the height they rise on the finished chromatograph?

2. Do all amino acids react to the same extent as ninhydrin, giving the same intensity of purple spots? Or is the reaction related to the chemical structure of the acids?

3. Why do you think it is necessary to cover the developing chromatogram with a glass jar to provide a closed atmosphere? Do you think the effect would be the same if you did not cover the developing chromatogram with a glass jar?

4. The procedure in this exercise makes use of an ascending chromatogram. Develop and test the chromatograph process, using a descending chromatogram.

5. Does the development time for a chromatograph vary with the molecular weights and complexity of structure of the amino acid molecules?

CHROMATOGRAPHY OF LEAF PIGMENTS

IN THIS EXERCISE we repeat a classic experiment first performed by the Russian botanist, N. S. Tswett. He separated plant pigments by chromatography in 1903.

Since the compounds in a plant pigment extract are colored, it will not be necessary to spray the finished chromatograph with a color-developing solution. The five plant pigments that will separate from this chromatograph are: chlorophyll a, chlorophyll b, carotene, and two of the xanthophylls. Because each of these pigments has a characteristic color, you should be able to identify them on the finished chromatograph. The colors of each pigment are: chlorophyll a—dark green; chlorophyll b—light green; carotene—brilliant yellow; and xanthophylls—pale yellow.

Procedure

1. Put a few leaves, a bit of sand, and about 25 ml acetone in a mortar. Grind until the solvent takes on a dark green color.
2. Allow debris to settle and then pour off the green liquid into a funnel with a piece of tubing attached. *Be sure pinch clamp is on the tubing.*
3. Add 10 ml Skelly B (also called ligroin, or petroleum ether).
4. Slowly add 5-ml portions of distilled water until practically all the pigments are driven into the top layer of Skelly B. Stir slowly to hasten the separation. The top layer should become dark green, and the bottom layer should become yellow-green.
5. Drain off and discard the bottom, yel-low-green fraction.
6. Roll a 9-inch square of Whatman #1 filter paper into a cylinder and staple the edges.
7. With a dropper or small brush apply a thin dense line of the dark green pigment 1 inch up from the bottom of the cylinder. The more dense the line the more vivid the finished chromatograph. Put on several applications, allowing the pigment to dry each time.
8. Put 75 ml Skelly B and 20 drops n-propyl alcohol (normal propanol) in a petri dish bottom. Stand the paper cylinder on end, pigment end down, in the petri dish of solvent. Cover with a 1-gallon widemouthed pickle jar.

Discussion Questions

1. What is the chemical difference in the structure of the chlorophyll a molecule and the chlorophyll b molecule?
2. List some methods of separating a mixture of substances other than by the process of chromatography.
3. List some of the pigments found in leaves that are not separated or identified in this process of chromatography.
4. Why do leaves generally look green, even though they may contain pigments of other colors?
5. How do the results of this exercise help you explain why leaves go through a series of color changes in the fall?

A DEMONSTRATION OF STOMATE ACTION

SOMETIMES it is easier to understand a principle in biology by constructing a model. In this exercise, for use as a demonstration before your class, we suggest a model of two guard cells that surround the stomate. In this model, the guard cells are represented by two elongated balloons. A gain and loss of turgor pressure by the guard cells is an important factor in the stomatal mechanism. In the demonstration you will inflate and deflate the balloons with your breath,

318

simulating the gain and loss of turgor pressure.

The thickened inner walls of guard cells play a part in their opening and closing. We use masking tape in this exercise, attached to the sides of the balloons. The taped balloons represent guard cells.

Procedure

1. Attach about 2 feet of rubber tubing to the single end of a glass "Y" tube. (See Figure G.11.)
2. Attach two elongated balloons to the two open ends of the glass "Y."
3. Inflate the balloons by exhaling into the rubber tube. It may be necessary to tie the balloons to the glass "Y" since they sometimes come off during the inflation process.
4. Hold the two balloons together as shown in the figure. When you inflate and deflate the balloons, as they are held in this position, no space should develop between them.
5. Now partially deflate the balloons and put two strips of masking tape on each balloon.
6. Being careful not to allow the balloons to deflate, turn them so the strips of masking tape are toward the inside of the two balloons.
7. Inflate the balloons, making sure they are held together, with their ends touching. (See Figure G.12.) You may have to shift the position of each balloon slightly in order to keep the masking tape strips on each balloon facing the tape on the opposite balloon. When inflated properly, a space will develop between the two balloons. This space corresponds to the stomatal opening between two guard cells. (See Figure G.13.)

Discussion Questions

1. What role does the thickened wall in

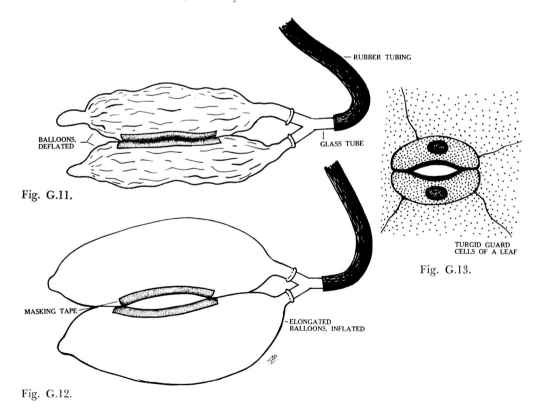

BALLOONS, DEFLATED

RUBBER TUBING

GLASS TUBE

Fig. G.11.

MASKING TAPE

ELONGATED BALLOONS, INFLATED

Fig. G.12.

TURGID GUARD CELLS OF A LEAF

Fig. G.13.

a guard cell play, as illustrated by the use of this model?

2. In living guard cells in leaves, how is the turgor pressure increased and decreased?

3. Of what value to the physiology of a leaf are guard cells and stomates?

MEASURING WITH THE MICROSCOPE

YOU CAN MEASURE with the microscope, using the known diameter of the field of view as a reference. For example, if a *Paramecium* is about one-half as long as the $43\times$ objective field, and you know the objective field diameter is about 400 μ, you can guess that the protist is about 200 μ long. The diameter of the field of view can be calibrated using graph paper printed in 1-mm squares. From the number of squares you can see across the diameter of the field you can determine the diameter in microns.

LABORATORY WORK

1. Mount a slide of graph paper on the stage of your microscope and focus on it with the lowest power objective. Imagine a line across the diameter of the field. Adjust the slide until some of the squares lie on the diameter, with the square at the left edge just touching the left edge of the field.

2. Count the number of squares across the diameter. Record the results on a chart. Also record eyepiece and objective power, and total magnification.

3. Change to the next objective and refocus. (If your microscope has a $97\times$ objective do not use the $97\times$ objective in this exercise.) Again move the slide so that the squares lie on the imaginary diameter of the field, with the left square touching the left edge of the field. This time there may be only a fraction of one square visible and you will have to estimate as accurately as

you can how much of it the field covers. It will be especially difficult to see the lines printed on the paper, so you must focus carefully and adjust the light.

SUGGESTED PROJECTS FOR ECOLOGY

SUGGESTIONS for ecology projects are limited only by the availability of materials and the imagination of the teacher and his students.

They may be carried out on a restricted area far from the school, within the walls of the laboratory itself, or even at the home of a student.

Ecology is such a vital part of biology it should be given an important place in the curriculum. If students can see relationships between certain groups of organisms they begin to understand why things happen as they do.

If a school is located close to a stream, a woods, a bog, or an open meadow, it has unlimited study areas for ecology. For the school located in the center of a large city with some of these facilities not available, the opportunity for ecology studies out-of-doors is more limited. Here, the indoor laboratory must be put into greater use.

The suggestions listed are for both types of schools. Variations of these suggestions may be worked out according to the ingenuity of the persons planning the projects.

1. SMALL RODENTS OF AN AREA

Use an area such as a field, meadow, moist shore area, or woods. Divide the class into pairs. With each pair using 10 (or some given number) mouse traps, the pairs line up in a long line, at intervals of 3 to 6 meters apart. Each pair sights a spot directly at right angles to the line of the pairs of students. Set traps using various kinds of baits (peanut butter mixed with cornmeal, bacon, cheese, no bait, etc.). Alternate the bait so there is a random pattern throughout the area.

After the first traps are set, the pairs move forward 3 to 6 meters and set second traps, the third, etc., until all traps are set.

Mark traps with pieces of cloth tied to weeds or upright markers.

Examine traps at designated times (12- to 24-hour periods) for animals caught, and traps snapped.

Several things can be worked out from this project:

a. Find the percentage of rodents caught to traps set.
b. Identify the kinds of rodents caught.
c. Determine the sex of the animals caught.
d. Open the animals and measure the size of the sex organs and other organs, perhaps the contents of the stomach and the length of the intestines.
e. Determine if any females are pregnant.
f. Examine the teeth to see if they are worn down (old) or in good condition.
g. Make study skins of pelts.
h. Examine the animals for both ectoparasites and endoparasites. Save and identify.
i. Note the variation in color among members of the same species.
j. Which kind of bait seems to be most effective?
k. How does the number of "snapped" traps compare with the number containing caught specimens?
l. How many specimens show any evidence of damage from insects?
m. How do the lengths of tails (or of some other part, such as front or hind legs, or ears) compare in the various specimens?
n. Make a graph or a series of graphs to show your results.
o. Repeat the entire project, using different conditions such as a change in the weather (wet, dry, cold, warm), location of area, or other factors.

Students should note physical conditions such as time of day traps were set, time examined, kind of weather, temperature of soil, temperature of air just above the ground, moonlight or dark night, and other features.

A good reference for identifying the rodents is Booth, *How To Know the Mammals,* part of the Pictured Key Series by Wm. C. Brown Company, 135 South Locust Street, Dubuque, Iowa 52001.

2. ECOLOGY OF AN AREA SET UP IN THE CLASSROOM

Collect things found in an area such as a woods. This could include plants with some of their original soil around them, a rotting log, insects that are available, and any other invertebrates, or vertebrates (frogs, toads). Place them in three or more terraria and keep conditions such as pH of soil, moisture content of soil, temperature, light, placement of plants, logs, and soil as much like the original setting as you can. If you want to observe the effect of a factor such as moisture, change this factor in one terrarium but keep one terrarium as a control with conditions as nearly like the original area as possible.

For each of the other terraria change one condition only. This may be pH of soil, amount of light, or some other condition.

Observe the terraria daily over a period of some set time and note changes as they occur.

Results may be shown in the form of a graph, a write-up, some photographs, or other means devised by the student and his teacher.

Variations may also be shown in a classroom by setting up a bog, meadow, or marsh, with accompanying plants and animals.

3. ECOLOGY IN AN AQUARIUM

An aquarium is more than just an interesting place to keep fish. Its ecology can give students an understanding of principles involved in maintaining a satisfactory one.

Aquaria can be stocked with plants such as *Vallisneria, Sagittaria, Elodea, Myriaphyllum* and *Cabomba*; fish such as native bullheads, sunfish, and minnows; and snails, newts, and aquatic insects. (See Turtox Service Leaflet No. 5, *Starting and Maintaining a Fresh-water Aquarium;* Leaflet No. 23, *Feeding Aquarium and Terrarium Animals;* and Leaflet No. 48, *Aquarium Troubles: Their Prevention and Remedies.*)

The student may follow directions to the letter in setting up the aquarium, but still it may not function properly. Fish may die, plants float to the top of the water and turn brown, and snails multiply too rapidly.

Manipulate physical and chemical conditions so that the environment is right. Keep the correct number of fish for the amount of water. Adjust the pH of the water. Adjust the amount of light.

As a variation for setting up a large class aquarium, students may set up individual aquaria in pickle jars, quart, pint, or even smaller jars. By using correspondingly small organisms the students learn how to vary physical and chemical conditions for these organisms so they live and are maintained in their environment.

Recently teachers inland, far from salt waters, are finding an aquarium stocked with salt-water plants and animals valuable. Again, maintaining a salt-water environment for these creatures makes an interesting project.

4. FIELD SAMPLING TO DETERMINE PLANT SUCCESSION

Students may work in pairs, starting from an area such as the edge of a bog.

By sighting an object 90° from a line where they are standing, they collect samples of the kinds of plants along a straight line from the margin of the bog outward. Plastic bags are excellent for carrying the plants.

After all the pairs have completed collecting, the plants may be laid out to show how they vary from those that grow closest to the bog and those farthest away. Thus plant succession may be observed.

In summarizing the information, photographic record shots of the area may be made as well as pictures of the individual plants that are laid out to show succession.

Other areas, such as a section of a woods or a prairie, may be worked out.

Students may plan individual ways of showing relationships.

5. "BAITS" FOR NIGHT-FLYING INSECTS

Compare the kinds of night-active insects caught under the same climatic conditions, using different kinds of "baits" or traps. Many other suggestions are given in Appendix D 255.

Divide a class into groups and let each group devise a different means for collecting insects.

One group may use a light trap such as a cone with an electric light bulb above it and the small end of the cone placed over a bottle containing a substance such as ethyl acetate on a piece of cotton for killing insects as they drop into the bottle.

A white sheet tacked up between two trees or supports of some sort, with a beam of light turned on the sheet, offers another method for collecting night-flying insects.

Some insects such as mosquitoes and deer flies are attracted directly to the skin of an animal (such as man). This offers still another way of collecting insects.

A comparison of kinds and numbers of insects and ways in which they are caught makes an interesting project. Results may be plotted or made into a graph.

6. POPULATION VARIATION STUDIES

This type of project offers endless possibilities.

Species of snails, mosquitoes, grasshoppers, crayfish, ants, clams, dragonflies, moths, butterflies, flies, deer flies, or other invertebrates may be studied in different habitats. If specimens are collected from different areas such as meadows, woodlands, bogs, and sand shores, differences can be observed in the same specimens.

Divide a class into teams and have each team catch a minimum of 100 specimens of one genus. This might be dragonflies or whatever the group chooses to work on. Each group works in a different habitat. For observation, specimens may be pinned or glued to a cardboard or plaster board according to the differences to be observed such as color, length of wing, legs, body, or some other feature. Comparisons of specimens caught within an area, and also in different areas, make good studies.

Other materials that can be studied include leaves from the same branch or from similar trees, individual pine nee-dles, blades of grass, individual flowers from the same or different plants, and leafhoppers. There is no limit to what can be studied.

This is one type of project that can be set up in the laboratory or in the field.

7. NEST STUDIES FOR PARASITES

Bird nests, or those of mammals, may be observed for parasites that live in them.

Parasites should be collected from each nest and kept in separate containers so determination of the kind and number of parasites may be noted for each kind of nest.

This kind of project may be continued over a period of time.

8. A "LIBRARY" OF HAIRS FROM COMMON MAMMALS OF AN AREA

These may be collected and placed on slides with a mounting medium such as Canada balsam. Distinguishing scale patterns for each species of mammal may be observed using a microscope.

To make sure of getting hairs, mount or tie cockleburs on stakes at the entrance of dens of animals such as woodchucks. Mammals using the dens will leave hair samples on the burs. You can easily collect and mount them.

9. FEATHER TYPES

A display of feathers from birds within an area may be made using cellophane mounts, Saran wrap, or some other transparent cover so the feathers show. These, too, may be observed more closely through the microscope. Feathers indicate the kinds of birds that inhabit a certain area.

10. A COMPARATIVE "BONE BOARD"

A comparison of certain bones of the body, such as the sternum from different birds or mammals, skulls, femurs or humerus bones of various species, or other comparative bones which might be some indication of specialization for a certain type of environment, offers some bases for kinds of "bone boards" that students can make.

11. A BIRD "LOG"

Put a log from a forest area, or one from the woodpile for the fireplace, in a conspicuous place in the room such as a window ledge at the front of the room.

As birds migrate through an area, or as they make their first appearance in the spring, have the student who first reports the appearance of a bird place on a colored picture of it the scientific and common name of the bird, the date first seen, and his own name. Put the picture on a pin and stick it into the log. With a constant reminder before them students are more likely to learn names of birds and to recognize them.

A log is an interesting way to display insects, also.

12. ECOLOGY OF THE DIGESTIVE SYSTEM OF AN AMPHIBIAN

Use either tadpoles or adult stages of frogs or toads. Kill the animal by destroying the brain and spinal cord by pithing (forcing a dissecting needle into the brain, then down the spinal cord).

Open the ventral body wall from the anterior to posterior end.

Different species of intestinal protozoa inhabit rather definite areas within the intestine and rectum.

A flagellate, *Giardia agilis,* may be located in the anterior part.

Endamoeba ranarum may be found in different parts of the intestine and also the rectum.

Three genera of flagellates, *Trichomonas, Hexamitis,* and *Euglena morpha,* and several green flagellates inhabit the rectum.

For observing these individuals alive, the portion of the digestive tract that contains these should be teased out in a drop of 0.7% salt solution (7 g salt in 100 ml water), placed on a glass slide, and covered with a cover glass.

A large ciliate, *Nyctotherus cordiformis,* is often found in the rectum of tadpoles. It seems to act as a scavenger and resembles *Paramecium* in structure. *Opalina ranarum,* another large ciliate, may also be found in the rectum of tadpoles.

13. HABITAT STUDY a la THOREAU

Select an area as far away from the sounds of cars, trucks, and people as possible.

Walk with the class along a path, but every few yards permit one student to step off the path and into the woods a short way.

Explain to the students ahead of time that each one is to sit quietly for a given period (½ to 1 hour as time permits), observe everything in his area—look, listen, smell, touch—and become as much a quiet part of the picture as he can.

At the end of the designated period each comes from his habitat area to a given place—or to the laboratory—and writes an essay on his experiences of the past short period of time.

(We have tried this a number of times with amazing results.)

324

BIOLOGY PROJECTS THAT HAVE BEEN DONE BY HIGH SCHOOL STUDENTS

1. Plaster models of an earthworm, showing both internal and external structures.
2. Skeleton of a squirrel, and a paper discussing adaptations.
3. Skeleton of a duck, and a paper dicussing adaptations.
4. A study of two habitats of Coleoptera.
5. A study of some Iowa prairie insects in October (modify this for your area).
6. A comparison of certain aspects of social insects to nonsocial insects.
7. Adaptations in selected types of bees.
8. A comparative study of the orders of insects found in a bean stubble field, and on Government grounds.
9. Plaster models comparing internal structures of a horseshoe crab and a grasshopper.
10. Plaster model of the digestive system of a grasshopper.
11. Skeleton of a badger or a groundhog, and a paper discussing adaptations as shown by the skeleton.
12. Structural significance of the posterior legs of the frog.
13. Skeleton of a mink, and a paper discussing adaptations.
14. Adaptive and sexual differences as shown by a study of the skeletons of selected mammals.
15. A study of the structural adaptations of prairie insects.
16. A study of insects from a trailer court habitat.
17. Skeleton of the red fox squirrel, and a paper discussing adaptations.
18. A study of the survival of insects of a selected order from one habitat.
19. A study of Arthropods collected at night.
20. Fox skeleton, and a paper discussing adaptations.
21. A comparative study of woods and prairie insects in Orders Coleoptera, Hemiptera, Orthoptera, Diptera, and others.
22. Insects found in houses, gardens, underground, or in woods.

23. The effects of heat and light on insects from various habitats.
24. Design and paint a chart across the front wall of the classroom, showing eras of geological time.
25. Design, model, cast, and paint a three-dimensional model. See the instructor for specific directions. See biological catalogs for pictures of models. Details for making models are shown in Chapter 8.
26. Make a series of detailed biological drawings of selected subjects, such as the anatomy of a vertebrate, the gross morphology of some insects, or a chart designed to illustrate a certain biological principle.
27. Make a skeleton of a vertebrate. Use a dog, cat, pig, rat, mouse, bird (pigeon or chicken), or sheep. See *Turtox Leaflets* for specific instructions.
28. Inject an animal with liquid latex, then remove all tissue, leaving the injected system intact. Arrange in a jar for display.
29. Make an insect collection. See *How To Know the Insects* by Jaques for specific instructions.
30. Arrange for museum display purposes representative animals in one phylum. See the instructor for specific information. If the specimens are dissectible, open them, tie with nylon thread on a glass plate, and insert the plate in a special display jar. Letter 3 x 5 cards with India ink giving information for the specimen on display.
31. Experiment with protozoan cultures to determine population fluctuations and their response to stimuli (electrical, chemical, physical—light, heat, sound).
32. Culture protozoa on artificial growth media.
33. Culture and experiment with fruit flies. See section on Genetics in your text, or genetics references in the library.
34. Experiment with *Dugesia* (brown planaria) and its ability to regenerate.
35. Do a comparative anatomy study of three or four selected vertebrates (frog, bird, fish, pig). Dissect each specimen, remove a complete system, such as the

nervous, circulatory, muscular (in certain areas, such as the legs, or abdomen), excretory, or reproductive. Photograph each system. In write-up, show pictures of your work.

36. Culture brine shrimp (instructor has shrimp eggs for this) and experiment with their response to stimuli. See *Turtox Leaflets* for specific information.

37. Embed biological specimens in plastic. See *Turtox Service Leaflets*.

38. Investigate the protozoan fauna of the intestinal tracts of selected animals; use earthworms, larvae of beetles (called "mealworms" because they live in wheat flour), termites, and other insects.

39. Work with insect blood, preparing stained slides (use Wright's stain, and make up other stains yourself). Make drawings of the blood cells found, and make counts of the relative numbers of cells.

40. Take samples of water and bottom ooze from the *same area* or areas of a stream to determine the changes in plant and animal populations over a period of time.

41. Take samples as indicated in the previous project, but sample the fluctuations of the *same population* over the period of time.

42. Model a series of embryos to represent various stages of development in the early life of an animal. They might be of the same animal, as a man, dog, fish, pig, chick, etc., or they could be of different animals at the same stage of development. The medium for modeling might be a white clay that hardens and can be painted, or it could be modeling clay that could be covered with latex, later peeled off and the rubber mold filled with moulding plaster. After the plaster has hardened and the latex has been removed, the plaster could be painted. Figure 8.17 shows the specific steps in the modeling process.

43. Prepare several one-foot square flats for seeds, using the same kind of soil for each. Soak seeds such as oats, corn, or beans overnight in water to speed germination. Plant a different number of seeds in each flat and determine the effect of crowding upon the plants. Put a different number of seeds in each plot—such as 16, 48, 160, and 1,600. Keep notes on plant growth. You may want to check such things as the number of seeds that germinate; rate of growth in each plot; what happens to plants in dense populations, and in sparse populations. Perhaps there are other conditions you may choose to check.

44. Many variations of the preceding project may be done. Try seeds in different kinds of soils such as sandy, loam, and silt. Plant seeds at different levels in the seed flat and determine whether the plants that come up last are crowded out or not.

45. Plant peas and irradiate them or x-ray them in a doctor's office. Plant their offspring and continue this for several generations to observe any differences they show from the parent generation.

46. Select and mark several bees from a hive and study their activities. Try to determine their duties in a hive, and also in the field. Mark them with tiny dots of quick-drying paint on the dorsal surface of the thorax. Use a hive that is in the classroom and that has a glass plate for observation purposes. Observe the worker bees as they perform the housekeeping duties such as cleaning the brood cells, fanning their wings to keep the temperature at the proper level of 93° F, feeding the larvae (both the older and younger ones), repairing and rebuilding the comb, packing pollen into the cells of the hive, and defending the colony. Try to follow the bees into the field as they carry out job assignments there—scouting for food, collecting nectar and pollen, and returning to the hive to deposit their goods and to signal other bees of the sources of the goodies for the hive.

   An excellent description of the bees and how their work and activities are carried out is given in *Design for Life* by Richard Trump and David Fagle, published by Holt, Rinehart and Winston, Inc. Observations made by some of your students could add materially to information already collected about bees.

17. Make cuttings of stems of a number of plants, including vines. Place these right side up in water to observe formation of roots, stems, and leaves. Use the same plants and make cuttings but place these upside down in water. Find out whether they, too, will produce the plant parts.

18. Lift a cubic foot of soil from an area such as a woods or a prairie and place it in a terrarium. Keep it well watered and observe the succession of plant and animal life that appears.

BEHAVIOR STUDIES

NEARLY ALL STUDENTS express interest and curiosity in the behavior of animals. The few examples listed here hardly scratch the surface of the things that may be observed and done, using behavior as the basis for projects. These may suggest other projects that are of interest to students.

1. Observe the habits of the different living things in one selected area such as a piece of uncultivated land, a bog, a clearing in a woods, a pile of wood, the corner of the school yard, a park, or the edge of a lake or a pond. Note the requirements of each species observed such as food, resting places, activities at various times of the day and year, and the relative number present.

2. In plants, the condition of soil, water, and light might be observed. Relationship between plants and animals in the area might be described. This is a long-term project that may take the combined efforts of several students.

3. A beehive adds an interesting study to the setting of the biology classroom. It is a place to show behavior of bees as they perform "division of labor" within the hive as various members of the colony do the different tasks that need to be done. A close study of members of the hive shows evidence of how some structures such as the stinging apparatus and the

antennae cleaner are specialized for performing definite functions.

If kept inside the room with a pipe leading outside, the bees will come and go as long as the weather is favorable. Large glass sides for the hive provide observation areas.

4. An ant colony provides interest for a class, especially if arranged so the ants can be watched while they are at work doing the jobs that need to be done.

Part of the colony should be exposed so colony members can be observed. Glass sides provide for this.

A colony may be attained by using a shovel to scoop up an area filled with ants. The dirt containing the ants may be placed in a box and the whole thing set on bricks in the middle of a large pan partially filled with water to prevent escape for members of the colony.

After the colony is established, introduce a few ants of another species and observe.

Food habits and food preferences may be studied also.

5. A colony of cockroaches may be kept indefinitely in the classroom, but precautions should be taken to prevent spreading of the colony into all parts of the room and building.

A tall battery jar with sides greased about halfway from the top to near the top keeps the insects from escaping. Small rocks or stones in the bottom of the jar provide space for the colony. Bits of dry biscuit and raw apple provide nutritional needs for the roaches. Other foods may be given the colony to study food preferences.

Personal grooming habits of the roaches are interesting to observe.

Hatching of the young from the capsules of eggs is an interesting observation.

At one school, a colony of special, large cockroaches has been kept in a healthy, breeding condition for a period of over 18 years. They have been used for many different projects, varying from observation studies to physiological experiments.

6. Determine response of *Hydra* to varying conditions of light, to foods, temperature, and of use of nematocysts.

7. Observe courtship and mating activities of birds. In following through with definite pairs of birds the observer may be able to see how the male bird stakes out territorial claims. Make the observations in a zoo or bird sanctuary. Behavior of water birds is of special interest.

8. Observation of black crickets concerning chirping activities and responses might be worked out by some students. Tape recordings of the chirping under various conditions might be made. Record chirping of a male cricket when alone, then again when a female cricket is placed by him, and again when another male cricket is chirping. Variations in the song are measurable.

9. Kittens or puppies (or other mammals) of the same sex and litter, growing up under the same conditions except for training, are interesting to study.

10. Earthworms placed in different buckets of soil may show different responses to light, darkness, moisture, temperature, and food.

11. Responses of insects such as flies, butterflies, and moths to nectar, honey, or sweetened water after the head has been removed from the insect's body may be shown. Length of time the "tongue" responds will vary with different individuals.

12. Goldfish, guppies, or other fish may be trained to respond to certain stimuli such as tapping the glass aquarium in which they reside.

13. Feeding habits can be observed. Place six to ten rabbits, mice, rats, or chickens of the same sex in a large cage or pen. Provide just one feeding station, and observe animals as they feed. Does the same animal always feed first? Is a feeding order established among the group? How does an animal establish itself as a leader?

14. Feeder rights. Based on the previous description, use a bird feeder to determine which species of birds feed first at the feeder. Continue observing to see which members of the species feed first. Is a "peck order" established here?

15. Territory rights. During the nest-building period, observe the singing and fighting by a bird as he establishes his territory. How does he drive away other birds? How great a territory does he establish? What fruits and what trees are included in the area?

16. Effects of colored lights on fish. Use guppies or other tropical fish in a tank of water. Place a colored light close to one end of the tank. Note the effect on the fish. In the next 10 to 15 minutes use a different colored light bulb. Note the effects of red, green, yellow, blue, and white lights. Do fish seem to be attracted by some colors? Are they repelled by others? Are their swimming motions altered?

17. Colored lights for frogs. Place a colored light inside a darkened tank of frogs as was done for fish. Do the frogs move toward the light? Use a series of lights to see the effect of each color.

18. Maze for mammal study. Use white mice or rats in a maze to see how rapidly they learn to follow a maze with a reward of food at the end of one path. The maze should be constructed so that only one pathway leads to food. The others end at a blank wall. Record the results for each animal. After several days discontinue the project, then continue after a few days to see if the learning process persists.

19. Food preference. Use mammals such as kittens, puppies, rabbits, guinea pigs, rats, or mice to determine food preferences. Provide several (4 or 5) different foods for each animal, all placed within equal accessibility. Note the order of preference of each animal. If some foods seem to be entirely ignored, remove them and substitute others. Are the selected foods nutritionally the best? Sample foods might be (1) high protein prepared commercial animal food made especially for pets, (2) bread, (3) lettuce, (4) oatmeal and milk, (5) candy, (6) vitamins. (See Barnett, S. A., *The Rat—A Study in Behavior*, Aldine Publishing Company, Chicago, 1963.)

20. Response to barriers. Provide an area on a table or the floor with several barriers of your choice such as small branches, mounds of dirt or soil, pools of water, and patches of salt. Release animals such as small turtles and chart the pathway each takes.

SUGGESTED FIELD PROJECTS IN MAMMALOGY

1. Compare species composition and numbers of mice in different habitats (woodland versus grassland; prairie versus cultivated crops).
2. Study activity periods of a common mouse by periodic trapping or by the use of captive mice held in cages with activity recorders.
3. Compare the effectiveness of various baits on trap success and species selectivity.
4. Determine sex and age rations for a large sample of a common species (*Peromyscus, Microtus,* or *Sciurus*).
5. Compare success of snap traps, can traps, and other systems for capturing small animals.
6. Make a "bone-board" to show identifying features of selected bones of mice or other small rodents.
7. Study molt patterns of gophers, shrews, or mice.
8. Determine litter size by means of placental scars of the uterus.
9. Evaluate tooth wear, skull size, gonad development, or other criteria of age determination.
10. Make a "library" of hairs of common mammals of your region.
11. Make a collection of plaster or plastic impressions of tracks of common mammals.
12. Study the feeding, drinking, and maintenance behavior of a captive mammal.
13. Collect and identify the parasites of a common species of mammal.
14. Study the field range of a mammal such as a deer or a rabbit. (Capture a rabbit in a cage and mark it with a spot of quick-drying paint, for easy identification.)

REGENERATION AND REPLACEMENT STUDIES

THE LESS COMPLICATED an animal is the more readily will regeneration and replacement of parts occur. Some animals lend themselves very well to studies of this kind.

1. *Earthworms* are most easily collected at night between 10 and 12 o'clock, during or just following a warm rain that has soaked the ground. They come to the surface to feed and may be captured. They feed on dead and decaying leaves. They are easily kept in large boxes filled about 12 inches deep with about equal parts of old leaves and leaf loam taken from the woods. This material should be kept moist but not saturated.

To show regeneration of parts, cut the earthworm in two, just in front of the 20th segment. Keep both parts to see if both regenerate.

Cut other worms in two at segments less than the 20th. Record regeneration efforts.

Keep worms in small amounts of the leaf and loam mixture, and observe. They may be given small bits of bread moistened with milk.

Split the last 10 segments of an earthworm from the dorsal through the ventral surface. Leave both halves attached to the anterior end and observe for appearance of two new "tails."

2. *Crayfish.* Observe the size of the chela of crayfish you collect from streams, or those sent from biological supply houses. Are they the same size or is one larger than the other? If one is smaller, it is likely that it is a replacement for one that was injured or lost earlier in the life of the crayfish.

3. *Starfish.* Observe the size of the arms of starfish collected from the shore, or those sent from a biological laboratory. If one or more is smaller than the rest, this represents replacements of arms formerly lost or broken.

4. *Planaria* may be collected locally

from underneath stones in ponds or streams. They may also be enticed by pieces of liver placed in the water.

In the laboratory they need a clean, darkened habitat. This may be a pan with fresh, clean water, preferably from a well. (Chlorinated water causes planarians to disintegrate.)

Cut the animals into several pieces and observe to see if each piece regenerates a new animal.

Keep the pieces in clean water at all times. As new worms begin to form they can be fed small pieces of liver. However, after the feeding, the water should be changed. Regeneration takes up to two weeks.

5. *Hydra* may show regeneration of tentacles if one or more is cut off. Keep hydra in watch glasses and nonchlorinated water and feed them crushed *Daphnia* or minute bits of liver.

Cut off a tentacle with a tiny "scalpel" made from a dissecting needle that has been flattened and sharpened.

CELL STUDIES

---

1. EFFECTS OF TEMPERATURE ON YEAST CELLS

Crush about one-fourth of a cake of compressed yeast with enough fresh fruit such as grapes or berries to half fill each of three test tubes. Boil contents of tube No. 1, then add a drop of dilute methylene blue to each tube. (Use about 1 g of dry methylene blue powder to saturate 100 ml of 95% alcohol.) Dissolve 1 part potassium carbonate in 10,000 parts water. Add 30 ml of the methylene stain to 100 ml of the potassium carbonate solution. This is to test metabolic rates in the three tubes.

Leave tube No. 1 (that has been boiled) in the laboratory, place No. 2 in the refrigerator, and leave No. 3 in the laboratory.

Examine 24 hours later. Tubes 1 and 2 should still be blue, but No. 3 should be nearly colorless, indicating that the methylene blue has been reduced by the respiring yeast.

To determine whether or not yeast cells have been killed, add Congo red to cells from each tube. (To prepare Congo red, dissolve 0.5 g of Congo red in 10 ml of 95% alcohol, add 90 ml distilled water.)

Congo red stains only the dead yeast cells. Diffusion will not occur through the cell membrane of living cells. Death of the cell results in a breakdown of the membrane, thus permitting the stain to penetrate.

2. COMPARISON OF GUARD CELLS FROM VARIOUS LEAVES

Peel the upper and lower epidermis from various kinds of fresh leaves. Mount on a slide in a drop of water and determine the number of guard cells present per square millimeter.

Compare the number present on upper and lower surfaces of a number of plants such as *Zebrina, Zea maize* (corn), *Poa pratensis* (grass), *Liliacea* spp. (lily), *Pyrus malus* (apple), *Ulmus* spp. (elm), *Quercus* spp. (oak), and *Acer* spp. (maple).

3. EFFECT OF PLANT HORMONES ON CELL GROWTH

The most noticeable effect of gibberellic acid upon plants is the elongation of stems or internodes and a resulting increase in height of the plants. To determine the effect on size of the individual cells, set up a number of kinds of plants in groups of control and test plants. Keep both sets well watered, but spray gibberellic acid on the test group. Use a 5% solution every 3 days for 25 days. Keep records of the height of test and control plants, length of internodes, and number of leaves produced. At intervals of about 5 days examine the cells in the leaves and the stems. Compare in size and shape with those of the controls. (To prepare a solution of gibberellic acid, dissolve 100 mg of the powder in 1 ml of isopropanol to which 10 drops of a detergent such as Tide has been added. Add this mixture to 900 ml of distilled water. Stir vigorously. Add distilled water to make 1 liter of solution. This contains 100 mg gibberellic acid per liter of solution.) This stock solution may be used to prepare test solutions of varying strength for testing on the leaves or plants. For trying out effects of various solutions on cell growth of the plants, it may be diluted 1 to 10 times with a detergent-distilled water mixture.

4. CELL STUDIES FROM INNER AND OUTER CURVES OF PLANT STEMS

Plant a hundred or so seeds such as peas or beans in 2 containers. Grow until seedlings are approximately 5 inches high. Turn them each day so the stems grow straight. The tenth day after they have emerged from the soil, do not turn the container with the experimental plants. Let them grow toward the sun for 2 or 3 days until they are severely bent. Examine several at the area where the bend occurs. Make longitudinal sections of the cells in the area bent toward the sun and also the ones bent away from the sun. What differences are there in size and shape of cells, and in number of chloroplasts?

5. OBSERVE DEVELOPMENT OF FROG EGGS

Cause a female frog to produce eggs by injecting pituitary glands from 5 male frogs into the lymph sacs of her body. (To get a pituitary gland, stun the frog with a blow on the head, decapitate, cut through the angles of each jaw, past the tympanic membranes and across the head to the top of the head at the base of the skull. Carefully cut away the bones of the skull, remove, and lift out the tiny, pink-colored pituitary gland. Mix with enough distilled water to form an emulsion. Inject into female frog.)

Twenty-four hours later press gently on the abdomen of the female frog to remove the eggs. Place them in distilled water with freshly removed sperm cells. (To get sperm cells remove testes from a freshly killed male frog, mash thoroughly, and mix with frog saline—7 g sodium chloride in 1000.0 ml distilled water.) Use only a small amount.

Observe eggs for mitotic development. Tabulate data as to how many eggs go through mitotic stages and what stage of development each reaches.

Note the ones that do not divide and the ones that divide abnormally.

Use the same female frog, remove eggs each day until she produces no more. Add sperm cells as described.

6. BLOOD CORPUSCLES

Compare the kinds of corpuscles found in the blood of several animals such as the frog, the chicken, and man.

Make several blood smears of each kind of blood. Dry, stain with Giemsa stain, and examine under the microscope.

To obtain human blood, sterilize a finger with 70% alcohol and use individual, sterilized lancets to cut through the skin to cause blood to flow. Place a drop or two at one end of a clean slide. Bring a second clean slide just to the blood so it flows evenly across the end of this second slide. Firmly and evenly push the second slide across the first one, to the far end of the slide, pulling the blood along. This should form a thin film of blood across the slide.

Dry the slide in the air.

Fix by standing the slide in a Coplin jar (that has slots along the inside for microscope slides) in 70% methyl alcohol for 5 minutes.

Remove slide and air dry again.

Transfer slide to a second Coplin jar of Giemsa stain for approximately one-half hour.

Wash slide gently in distilled water. Dry, then examine through microscope. (To mix Giemsa stain dissolve 0.5 gm Giemsa powder in 33 ml of glycerine for a period of about 2 hours. Then add 33 ml of acetone-free absolute methyl alcohol. When using, dilute the stain by adding 10 ml of distilled water to each ml of stock solution.)

A library of blood slides may be built up as a permanent teaching unit. If the slides are stained and put away carefully they should last indefinitely. Cover slips may or may not be used. Blood slides make a good instant laboratory.

7. OBSERVING CELLS FROM FROGS AND OTHER ANIMALS

Mount small pieces of the sloughed-off skin from a frog or salamander in Lugol's solution or methylene blue on a clean glass slide. Rim the edge of the cover slip with vaseline and place it over the mount. This will last several hours.

To prepare Lugol's solution dissolve 10 g of potassium iodide in 100 ml of distilled water. Add 5 g of iodine. Stir thoroughly. Use full strength or diluted with distilled water. To prepare methylene blue add 1.48 g of the methylene blue dye to 100 ml of 95% ethyl acetate. Let it stand for two or three days, stirring frequently; filter and store. To use

the stain, dilute 10 ml of the stock solution with 90 ml of distilled water.

Cells from other parts of the body such as the stomach, heart, intestines, lungs, spinal cord, and brain may also be mounted and studied in this way.

USES FOR THE KYMOGRAPH

THE FOLLOWING are experiments using frog muscle tissue for demonstrating:

1. Contraction of gastrocnemius muscle.
2. Response of skeletal muscle, such as the gastrocnemius, to variations in strength of stimuli (all-or-none-law phenomenon).
3. Response of smooth muscle, such as the intestine, to strength of stimuli.
4. Response of smooth muscle to frequency of stimuli.
5. Work performed by skeletal muscle (gastrocnemius).
6. Staircase phenomenon in skeletal muscle.
7. Fatigue in skeletal muscle.
8. Effect of temperature on skeletal muscle.
9. Effect of various salts on action of frog heart.
10. Tracing to represent the normal heartbeat of a frog.
11. Effects of temperature upon the heartbeat.
12. Refractory period of the heart.
13. Cardiac muscle response to various kinds of stimuli.
14. Cardiac muscle response to variations in stimulus strength.
15. Vagal inhibition of heartbeat.
16. Effect of adrenalin, pilocarpine acetylcholine, and other drugs on cardiovascular system.
17. Auricular-ventricular block.

Kymographs may be of the smoke-writing type or the ink-writing kind. The smoke-writing technique is the more sensitive. The price is about the same for the two kinds. The one using smoked paper must have the paper smoked before using, and handled carefully afterwards to avoid smudging. If kept permanently it must be coated with a preservative. On the other hand, the ink kind does not require prior preparation of the paper. The record is permanent without additional treatment. It is less sensitive and less precise than the smoked-paper kymograph.

Kymographs may be obtained from any of the biological supply houses and from instrument companies supplying the basic instruments for use in high schools.

Index

Acetic acid, fixative reagent, 275
Acetone, use of, 315
Acetylcholine, in muscle studies, 331
Actinophrys, 224
Actinosphaerium, 224
Adiantum, 245
Adrenalin, in kymograph studies, 331
Aeromonas, 271–72
Agassiz, Louis, 146
Alcohol
    chromatography solvent, 314–17
    preservative reagent, 274–75
    for slide preparation, 230–31
    staining reactions, 238–39
    in stain preparation, 235
Algae
    collecting, 248–51
    culturing, 248–51
    in field work, 152
    on rocks, 148
Alligators, 268
    injection of, 277
Aloe, 244
Amaryllis, 244
American Biology Teacher, 140, 147, 151
Amino acid
    chromatography of, 314–17
    synthesis, 13
Ammonium chloride, use of, 311
Ammonium hydroxide, use of, 311
Amoeba
    culturing, 221
    feeding, 221
    pseudopodia formation, 226
Amphibians, 268–69, 323
Amphipods, collecting, 278–79
Analogy, effective teaching device, 28
Animal
    behavior studies, 326–28
    bones, 204, 323
    bone staining, 262–63
    cell studies, 329–31
    collecting, 156
    embalming solution, 275, 276
    field studies, mammals, 328
    injecting, 275
    latex injections, 324
    mammal studies, 320
    parasites
        of amphibians, 323
        of rodents, 320
    regeneration, 328–29
    skeleton preparation, 324
    structure, bone boxes, 204

Animals
    alligators, 268, 277
    amphibians, 268–69, 323
    amphipods, 278–79
    for aquaria, 321
    Artemia, 270, 277–78
    bees, individual project, 324, 325
    birds, 85, 262–65, 277, 322, 323, 327
    brine shrimp, 270–71, 277–78
    caimans, 268
    carrion beetles, 154
    chick embryos, 261
    cold-blooded, care, 267–73
    crayfish, 278, 328
    crickets, 269
    Cryptobranchus, 277
    Daphnia, 270, 278
    earthworms, 271
    flukes, in frogs, 120
    frogs, 85–86, 119–21, 252–55, 277, 289–90, 323, 326, 327, 330, 331
    grasshoppers, 269
    Grocericus storeii, 302–3
    hellbenders, 269
    Hydra, 277, 278, 326, 329
    insects, 136–37, 154, 155, 205, 206, 208, 209, 255–59, 275, 321–22, 324, 325
    isopods, collecting, 278
    lizards, 267–68, 277
    Lumbricus, 271
    mammals, 86, 263–66, 276–77, 322, 324, 326–28
    mealworms, 271
    microarthropods, 151
    mudpuppies, 269
    Necturus, 277
    nematodes, 153
    newts, 268
    Planaria, 278, 328–29
    population studies, use in, 322
    protozoa as food, 225
    reptiles, 269, 277
    roaches, 269
    rotifers, 153
    salamanders, 267, 277
    shrimp, dried, 271
    snakes, 267–68, 277
    tadpoles, studies, 323
    taxonomy of, 285
    toads, 268, 323
    tortoises, 269
    Tubifex, 271
    turtles, 267, 268, 269, 270, 277
    water dogs, 269
    white worms, 271
Ant colony, 326
Anthropomorphism, 31
Antibiotics, sources of, 290–91
Apparatus. *See* Equipment
Aquaria, 148
    ecology projects, 321
Arcella, 222–23, 225
Aristotle
    on anatomical relationships, 20

    father of biology, 5
Artemia, 270–71
    collecting and preserving, 277–78
Artery injection, 276–77
Asparagus fern, 244
Aspergillus, 237
Asplenium, 245
Audio-tutorial instruction, 56–59
Auxins
    in plant growth, 31
    plant hormones, 329–30
Avery, MacLeod, and McCarthy, work with DNA, 3–4
Avocado, 243

Bacteria
    Aeromonas, 271–72
    Chondrococcus, 272
    in field work, 152
    flannelgraphs, 205
    sources of, 290
Baiting traps, 320, 321
Balantidium, 228
Barium hydroxide, 309–10
    CO_2 indicator, 306
Basicladia, 132
Beans, 243
Beating net, 154
Beehive studies, 326
Bees, individual project, 324, 325
Benedict's test, 301
Biological concepts, 2–16, 146
    changes in
        journals and texts, 2
        subject matter emphasis, 2
    compartment plan, 6–7
    in curricula, 9
    descriptive to functional, 2, 5, 9
    idea organization, 217–18
    "layer cake," 7
    related books, 280–87
Biological drawing suggestions, 324
Biological prefixes, 201
Biological Sciences Curriculum Study, 142
    pamphlets, 285
    research on curricula, 10
    structure and function inseparable, 32
Biological stamps, sources of, 294
Biological suffixes, 201
Biology
    Concepts in Biology Series, 285–86
    defined, 4–5
    introducing new subjects, 217–18
    laboratory controversy, 173–74
    master plans, 175–76
    night school, 167
    related books, 280–87
    room
        design of, 92–93

room *(cont.)*
  library, 283–87
  map, 112
  short course, 167
  speakers, 202
  themes, 166
Biome, 148
Bioscience, 140
Birds, 85
  banding, 263
  behavior, 327
  "bird log," 323
  bone staining, 262–63
  collecting, 263
  injecting, 277
  skinning, 263–65
Blackboard, 135
Bladderworts, 246–47
Blepharisma, 224
Blood corpuscles, 330–31
Blood typing, 204
Bone
  board, 323
  boxes, 204
  comparisons, 323
  staining, 262–63
Bottom net, 155
Bottom sampler, 260
Bouin's fluid, 275
Brine shrimp, 270–71
  collecting and preserving, 277–78
  individual project, 325
Bryophyllum, 244
Bryophytes, 148, 245
Bulletin board, 142, 218
  materials, sources of, 293
Butterworts, 246–47

Cabinet, display, 89
Cages and traps
  in ecology projects, 319–23
  for frogs, 86
  for insects, 321
  jar, 86
  sources, 289
Caimans, 268
Caladium, 244
Calla, 244
Cameras, 130–32, 142, 149
Canada balsam, 322
Capsella, 11
Carbolic acid, 275
Carbon dioxide
  anaerobic production, 309
  production in respiration, 309
  production in seeds, 308–10
  use in photosynthesis, 304
Carmine, 276
Carnivorous plants, 246–47
Carrel for individualized instruction, 58
Carrion beetles, 154
Castor beans, 243
Cells
  diagrams of, 22, 23

diatoms, 231–34
frog epithelial, 331
protozoa, 220–31
structure of, past and present concepts, 22–23
studies of, 329–31
Centropyxis, 224–25
Ceratopteris, 270
Chambers, electrophoresis, 140
Chara, 249–50
Charts for the laboratory, 209
Chemistry
  "anesthetics" for protozoa, 234
  impact on biology, 12
  ninhydrin protein test, 302
  reagent storage, 208
Chickens, bone staining, 262–63
Chick whole mounts, 260–62
Chilomonas, 221, 222, 225
Chi-square method, 158, 212–14
Chlamydomonas, 248, 250
Chlorella, 249
Chloromycetin, 272
Chlorophyll removal, 238
Chloroplasts, structure and function, 27
Chondrococcus, 272
Chromatography
  amino acids, 315–17
  ink, 314–15
  ninhydrin, 315–16
  paper, 314–17
  plant pigments, 317
  Rf value of amino acids, 316
  solvents, 314–17
  supplies, 140
Circulatory injection, 276–77
Citrus trees, 243
Cladophora, 126, 248–49
Climatarium, 140
Closed-circuit television, 135, 142
  equipment, sources of, 294
Closterium, 248
Club mosses, 246
Cockroaches, behavior, 326
Cold-blooded animals, care of, 267–73
Coleus, 122–23, 244, 245, 304, 306
Collecting
  Actinophrys, 224
  Actinosphaerium, 224
  algae, 248–50
  Amoeba, 221
  amphipods, 278
  Arcella, 222–23
  Chilomonas, 223, 225
  cold-blooded animals, 267–73
  Colpidium, 223, 224
  Colpoda, 223
  crayfish, 278
  Daphnia, 278
  Didinium, 224
  Euglena, 222
  Euplotes, 224
  Halteria, 223
  invertebrates, 277–79
  isopods, 278

  Paramecium, 221–22, 225
  Planarians, 278
  protists, 220–25
  slime mold, 236–37
  Spirostomum, 225
  Stentor, 225
  Tetrahymena, 226–27
  Vorticella, 225
Collecting equipment
  diatometer, 151–52
  gravy baster, 155, 156
  light trap, 154
  nets, 140, 151, 208, 278
    beating, 153–54, 257
    bottom, 155–56, 259
    plankton, 154–56, 259
  plastic bags, 276, 321
  seine, 155
Collecting techniques, 263
  insect light trap use, 154
  for insects, 321–22
  for protists, 153
  sugaring, 154
  tin-can trap use, 154
Colored latex, 276
Colpidium, 222, 225
Colpoda, 223, 225
Community responsibilities
  creating a good public image, 65–67
  service, technical and professional, 69–71
  social participation, 66
  speaking at meetings, 64–65
Compartment plan, 6–7
Congo red in yeast cell studies, 329
Conifers, 243
Conjugation, *Paramecium,* 226
Contractile vacuole, protozoa, 226
Coplin staining jar, 331
Copper sulfate dip, 272
Corn, use of, 243, 298
Cornstarch, 276
Costia, 272
Crayfish
  collecting, 278
  regeneration, 328
Creative biology teacher, qualifications for, 15
Crickets, 269
Crocus, 244
Cryptobranchus, injection of, 277
Culture media
  for algae, 248–51
  for bacteria, 290–91
  for fungi, 290–91
  for microbes, 290–91
  for molds, 290–91
  natural, 207
  proteose-peptone, 227
  for protozoa, 220–31
  for slime mold, 290–91
Culture methods, 220–35
Culturing
  Actinophrys, 224

Actinosphaerium, 224
algae, 248–50
Amoeba, 221
Arcella, 222–23
Blepharisma, 224
brine shrimp, 278
carnivorous plants, 246–47
Centropyxis, 224–25
Chilomonas, 223
Colpidium, 223, 224
Colpoda, 223
Daphnia, 278
Didinium, 224
Euglena, 222
Euplotes, 223, 224
fruit flies, 324
Halteria, 223, 224
Hydra, 277
insectivorous plants, 246–47
Paramecuim, 221–22
Planarians, 278
plants, 240–51
protozoa, 324
slime mold, 236–37
Spirostomum, 225
Stentor, 222, 225
Stylonychia, 224
subculturing protozoa, 220–31
Tetrahymena, 227–28
Vorticella, 225
Cupric acetate crystals, preservative, 275
Cupric chloride crystals, 275
Curricula, 9–10
Biological Sciences Curriculum Study, 10
individual needs, 51
Cutaneous senses, 206
Cuttings of plants, 241
Cystopteris, 245
Cytology
blood corpuscles, 330–31
frog cells, 331
laboratory techniques, 273
muscle cells, 28
plants and animals, 329–31
related books, 283–84
slime molds, 236–37
staining protozoa, 228–31

Dahlia, 244
Daphnia, 270, 278
Darkroom, 142
Darlingtonia, 247
Data presentation
charts, 157–59
displays, 161
graphs, 158–60
DeBakey, Lois, 44
Demonstrations, 45–46
"bird log," 323
"bone board," 323
class openers, 41
coal ball peel, 239–40
osmosis, 312–14
staining protozoa, 228–31

stomate model, 317–19
Descriptive biology, 2, 5, 9, 21
Desmids, 248
Detergents as wetting agents, 330
Diatometer, 151–52
Diatoms, 152, 231–34, 249
Didinium, 223–24, 225
Difflugia, 224
Diffusion, projects and eexrcises, 310–14
Digestive system studies, 323
Dionaea, 247
Disc recordings, 142
Diseases
furunculosis, 272
gold dust, 272
lymphocystis, 272
Display, student project, 60
Dissection, 146
of flowers, 207
of frogs, 252–55
pans, 208
of plants, 237–39
suggestions for comparative, 324–25
DNA
Avery, MacLeod, and McCarty's work, 3–4
experimental evidence, 25
history, 5
models, 26, 81–82, 83, 140, 302–3
structure and function interdependence, 25–27
Wilkins, Watson, and Crick's research, 7, 12
Dredges, 259
Dried shrimp, 271
Droseras, 247
Dryopterius, 245
Dugesia, regeneration, 324

Earthworms, 271
behavior studies, 327
regeneration, 328
Echinodorm, 270
Ecology, 148–49
habital studies, 324
observations, 326–28
population studies, 322
projects, 319–23
quadrat studies, 321
related books, 283–87
transect studies, 321
Educational media. *See* Equipment
Educational television, 134–35, 142
8mm projector, 135
Electrophoresis chamber, 140
Elms, 243
Elodea, 123, 127, 304–5
Embalming solution, animal, 275
Embryology
chick, 260–62
frog egg, 330

Encystment
Chilomonas, 223, 225
Colpoda, 223
Didinium, 224
Euglena, 222, 225
protozoa, 221–23, 225
Endamoeba in amphibians, 323
English ivy, 244
Enzymes
hydrolytic, 301
invertase, use of, 301
lipid digestion, 205
models, 27
plant, 298–302
sources, 291–92
starch digestion, 204–5
structure and function, 27
Equipment. *See also* Materials for teaching; Homemade biology tools
alcohol lamps, 208
aquaria, 148, 321
basic laboratory, 140, 209–10
binocular dissection stereomicroscope, 127
blackboard, 135
bulletin board, 142, 218, 293
cages and traps, 86, 154, 289, 319–23, 321
cameras, 130–32, 142, 149
for chick embryology, 260
chromatography paper, 314–17
chromatography supplies, 140
climatarium, 140
collecting nets, 140, 151, 153–56, 208, 257, 259, 278
Coplin jar, 331
darkroom, 142
for diffusion projects, 309, 310
display cabinet, 89
8mm projector, 135
film-loop projector, 135
glassware, 140, 209
greenhouse, 140, 210, 240–42, 289
homemade, 136–38, 207–9
incubating oven, 140
kymograph, 126, 141, 331–32
micromanipulator, 33
microscopes, 126–28, 140, 141, 292
microtome, 124, 126, 127
opaque projector, 135
osmometer, 312–14
overhead projector, 42–44, 88, 135
for pH, CO_2, O_2 measurement, 148
plastics, 208
pressure cooker, 208
refrigerator, 140
for respiration experiments, 308–10
room plan, 92–93
16mm projector, 135
slide and filmstrip projector, 88, 135

336

Equipment *(cont.)*
  for slide making, 238
  spectroscope, 304
  storage cabinet, 89
  tape recorder, 39–41, 88, 128–30, 142, 292
  for television, educational, 134–35, 142, 294
  terrarium, 247
  torso model, 140
  transfer chamber, 137, 138
  tree trimmer, 140
  video tape recorder, 294
Equisetum, 246
Ethanol, use of, 315
Ethyl acetate, stain preparation, 331
Eudorina, 248, 250
Euglena, 222, 225, 323
Euplotes, 222, 223, 224
Exercise, value of, 100
Exercises. *See* Laboratory investigations
Experiment, nature of, 101–3
Experiments. *See* Laboratory investigations

FAA (Formalin-Acetic acid-Alcohol), fixative, 274
Feather types, 322
Feeding behavior, mammals, 327
Ferns, 245, 246
Fertilizers, greenhouse use, 241
Ficus, 244, 245
Field biology
  concept, 160
  purpose, 147
  summer course, 162
Field projects, mammalogy, 328
Field trips
  indoors, 320
  observation studies, 323
  organization, 161–62
  suggestions for, 210–12
  team approach, 322
  themes, 157
Film loop, 142
  projector, 135
  sources of, 292
Filter paper, Whatman No. 1, 317
Fine chemicals, sources, 291
Fish
  for aquaria, 321
  behavior studies, 327
Fishscopes, 259
Fixatives, 273–75
  alcohol, 275
  FAA solutions, 274
  formalin-corrosive, 274
Flannelgraph, 88, 135, 136, 205, 295
Flukes, in frog lungs, 120
Formaldehyde, use of, 309
Formalin, fixative, 274
Formalin-corrosive, fixative, 274

Free or inexpensive printed materials, sources, 295
Frogs
  behavior studies, 326, 327
  cages for, 85–86
  egg embryology, 330
  injection of, 277
  muscle, kymograph studies, 331
  parasites of, 119–20
    Opalina, 323
  sources of, 289–90
  uses in teaching, 85–86, 119–21, 252–55
Function, new emphasis on, 23
Functional biology, 2, 5, 9
Fungi, 148, 237
Furniture for the biology room, 89
  sources, 288–89
Furunculosis, 272

Genetics
  chi-square test, 158
  emphasis on, 147
Geranium, 244
Giardia, in amphibians, 323
Gibberellic acid, 329
Giemsa stain, 331
Ginkgo, 243
Glassware for the laboratory, 140, 209
Gloeocapsa, 249
Glycerine, as a preservative, 275, 276
Gold dust disease, 272
Gonium, 248–49, 250
Grappling bar, 260
Grappling hook, 260
Grasshoppers, 269
Gravy baster, 155, 156
Greenhouses
  as basic equipment, 140
  equipment for, 210
  fertilizer for, 241
  pests, 242
  plant propagation, 241
  problems of, 240–42
  related books, 283–87
  soils, 240–42
  sources of materials for, 289
  techniques, 240–42
Grocericus storeii, DNA model, 302–3
Guard cells in plants, 329

Habitat studies, 323, 324
Hach water-testing kit, 303, 305
Hair, mammal collection, 322
Halteria, 223
Hellbenders, 269
Hematolechus, 119
Hexamitis, in amphibians, 119, 323
Histology
  alcohol, 274–75

Bouin's fluid, 275
FAA, 274
fixatives, 273–75
formalin, 274
formalin-corrosive, 274
frog tissues, 331
glycerine, 275
Homemade biology tools, 136–38
  bone boxes, 204
  chromatography apparatus, 314–17
  depression slides, 208
  dissecting kit, 140
  DNA model, 140, 302–3
  insect-collecting equipment, 153–56
  laboratory items, 207–9
  osmometer, 312–14
  osmosis demonstrator, 204
  respirometer, 308–10
  styrofoam models, 136
Hormone studies, plants, 329–30
"How To Know" books, 285
Hyacinths, 244
Hydra
  behavior, 326
  collecting and preserving, 277
  feeding, 278
  regeneration, 329
Hydrodictyon, 248

Incubating oven, 140
Individualized instruction
  audio-tutorial, 56–59
  carrel, 58
  a foremost need, 51–53
  linear programming, 55–56
  nature observation, 292, 323
  programmed, 54
  projects, 59–61, 324–26
  single-concept loop films, 58
  teaching machines, 53–56
Indoor insect study display, 206
Injection of circulatory system
  alligators, 277
  birds, 277
  frogs, 277
  mammals, 276
  reptiles, 277
  salamanders, 277
  techniques, 276–77
Ink, chromatography, 314–17
Insectivorous plants, 246–48
Insects
  behavior studies, 326–27
  collecting, 154, 255–59
  flannelgraphs, 205
  habitat studies, 324
  indoor study display, 206
  killing, 155
  log for displaying, 136–37
  nets for collecting, 154, 208, 257
  pinning, 155, 258
  preserving, 275
  rearing, 208
  spreading board, 209

studies of, 324
supplies and equipment, 257–59
traps, 321–22
In-service, for elementary science teachers, 69–70
Instant laboratory. *See* Laboratory equipment
Instruction patterns, 51
Invertase, 301
Invertebrates, collecting and preserving, 277–79
Investigation, art of, 103–4
Investigations. *See* Laboratory investigations
Iodine solution, 299
Iodine test, 301
Isopods, collecting, 278
Ivy, 245

Jar cage, 86
Journals
  list for biology teacher, 280–81
  old and new, 1
  for room library, 140
Jungle, room, 83–84

Kalanchoe, 244
Key, for twigs, 149–51
Klinge, Paul, 147
Koble, Ron, 147
Krebs, Hans, 7
Krutch, Joseph Wood, 146
Kymograph, 126, 141, 331

Laboratory
  exercises, value of, 100
  experiments, nature of, 101–3
  fees, 140, 141
  investigations, art of, 103–4
  for problem solving, 14
  use, 99–100
Laboratory equipment
  basic, 209–10
  charts, 209
  chemicals, 209
  cleaning agents, 208
  culture media, 290–91
  furniture, 288–89
  glassware, 209
  greenhouse, 210
  instant laboratories, 105–6
    blood typing, 204
    bone boxes, 204
    cutaneous senses, 206
    flannelgraphs, 205
    indoor insect study display, 206
    insect flannelgraphs, 205
    lipid digestion, 205
    osmosis demonstration, 204
    protist flannelgraphs, 205
    stains, 206
    starch digestion, 204–5

taste areas, 206
  living materials, 289–90
  microscopes, 292
  microscope slides, 292
  models, 210
  museum, 108–10
  photographic, 210
  plastics, 209
  projection, 210
  skeletons, 210
  stains, 210
Laboratory investigations
  algae, 250, 251
  bone staining, 262–63
  cell studies, 329–31
  cold-blooded vertebrates, 273
  diatoms, 231–34
  DNA model building, 302–3
  ecology, 319–23
  enzymes, 298–302
  frogs, 252–55
  individual projects with plants and animals, 324–28
  insectivorous plants, 246–47
  insects, 255–60
  osmosis and diffusion, 310–14
  photosynthesis, 303–6
  plant hormones, 329–30
  plant use, 245
  protozoa, 220–31
  related books, 283–87
  respiration, 307–10
  staining protozoa, 228–31
  stomate demonstration, 317–19
  study skins, 263–67
  white powder problem, 101–2
Laboratory substitutes, 207–9
  ink for stain, 225
  Perlite, 240
  Vermiculite, 240
Laboratory techniques
  blood corpuscles, 330–31
  chromatography, 314–17
  cleaning slides and cover slips, 208
  clearing plant material, 238–39
  coal ball peel, 239–40
  collecting hatched brine shrimp, 278
  cytology, 273
  diatom slides, 232
  frog cells, 331
  frog embryology, 330
  greenhouse, 240–42
  immobilization of protozoa, 234
  measuring with a microscope, 319
  microdissection, 142
  money saving, 207–9
  ninhydrin protein test, 302
  oilcloth for specimen protection, 276
  osmosis and diffusion, 310–14
  pH paper use, 300
  project suggestions, 298–332
  related books, 283–87

side table demonstrations, 106–8
slides of plant materials, 237–39
staining protozoa, 228–31
tests, 110–11
Lactic acid, preservative, 275
Latex, for injection, 125, 276
Lead chromate, 276
Learning patterns, 51
Leaves, guard cells, 329
Lehninger, Albert L., 27
Lesson plans, 214–15
Lettering materials, sources, 294
Lettuce, 243
Library, room, 283–87
Lichens, 148, 246
Light absorption by plant pigments, 304
Light trap, 154
Ligroin, use of, 317
Linear programming, 55–56
Linnaean system of taxonomy, 5
Litmus paper, use of, 311
Littoralis solution, 277
Liverworts, 245
Living materials, 147
Lizards, 267–68, 277
Lugol's solution, 331
Lumbricus, 271
Lymphocystis, 272

Magazines. *See* Journals
Malachite Green dip, 272
Mammalogy projects, 328
Mammals
  behavior studies, 326–28
  field studies, 328
  hair collections of, 322
  injection of, 276–77
  parasites of, 328
  for science center, 86
  skeleton preparation, 324
  skinning, 265–66
  study skins, 263–65
Manuscript preparation, 281–83
Map, biology room, 112
Maples, 243
Materials for teaching. *See also* Equipment; Homemade biology tools
  algae, 248–51
  aluminum foil, 207
  blood corpuscles, 330–31
  cells from plants, 329–30
  cellulose acetate sheets, 239–40
  coal balls, 239–40
  DNA model, 140, 302–3
  film loops, 58, 142, 292
  films, 41, 294–95
  flannelgraph, 88, 135, 136, 205, 219, 295
  free and inexpensive, 207–9, 295–96
  frogs, 119–21
  frozen specimens, 125–26
  graph paper, 319
  insect log, 136, 137

Materials for teaching *(cont.)*
  insects, 255–59
  iodine solution, 299
  lettering device, 294
  models
    plaster, 137–39, 142–43
    styrofoam, 136
  nature recordings, 292
  picture enlargements, 133
  plants, 122–24, 243–51
  postage stamps, 133–34
  preserved specimens, 124–26
  protists, 121–22, 220–35
  protozoa, 220–31
  recordings, 142, 292, 327
  screens, 88
  single-concept films, 58, 132–33
  16mm films, 294
  slides, 42
  sources of, 288–96
  starch-agar plates, 298
  stomate model, 317–19
  Taka-Diastase, 299
  tape recordings, 41, 292, 327
  teaching file, 215–17
  teaching machines, 53, 54–56
  35mm slides, 157
  transparencies, 288
  water-testing kit, 303
Maze, behavior studies, 327
Mealworms, 271
Measuring
  with microscope, 319
  protozoa, 226
Media. *See* Culture media
Methylene Blue, 235, 331
  in yeast cell studies, 329
Mice, bone-staining, 262–63
Microdissection, 142
Micromanipulator, 33
Microprojector, 140
Microscopes
  as basic equipment, 140, 141
  binocular dissection stereomicroscope, 127
  measuring methods, 319
  monocular, 127
  for projects, 284
  sources of, 292
  uses of, 3, 126–28, 284
Microscope slides
  list, 210
  sources of, 289
Microtome, 124, 126, 127
Microtus, field studies, 328
Mint, 244
Models
  animal dissections, 324
  animal tracks, 328
  for auxin explanation, 31
  DNA, 26, 140, 141, 302–3
  enzymes, 27
  as individual projects, 324–26
  list of essential, 210
  muscle diagrams, 29
  plaster, 137–39, 142–43, 207
  stomate, 317–19

styrofoam, 136
torso, 140
use of, 218
wall, 81
Molds
  Aspergillus, 237
  Penicillium, 237
  Rhizopus, 237
Molecular biology
  emphasis on, 147
  in perspective, 15
Mosses, 148, 245
Mounting media
  Canada balsam, 322
  for chick embryos, 261
  Kleermount, 229
  Permount, 229
  piccolyte, 238
Mudpuppies, 269
Murals
  suggestions for, 324
  wall, 79–81
Muscles
  cells, structure and function, 28
  frog, 252–55, 331
  models, 29
Museum
  display suggestions, 324
  teaching, 108–10

Narcissus, 244
National Science Foundation, support for biology, 7
Nature recordings, sources of, 292
Nature study concept, 146
n-butanol, use of, 316
Necturus, injection of, 277
Nematodes, 153
Nepenthes, 247
Nephrolepis, 245
Nests, parasites, 322
Nets
  aerial, 257
  aquarium, 278
  aquatic, 155–56, 257, 259
  beating, 153–54, 257
  bottom, 155–56, 259
  collecting, 140, 151, 208
  dredge, 259
  insect, 153–54, 208, 257
  plankton, 154–56, 259
  scraper, 257
  seine, 259
  sweeping, 154–55, 257
Newts, 268
Nicotine, action on frog, 255
Ninhydrin
  amino acid chromatography, 315–16
  test, 302
Nitella, 249–50
Nostoc, 249
Nucleic acid, 140
Nyctotherus, 119, 323

Oaks, 243
Observation studies
  animal behavior, 326–28
  crayfish, 278
  nature, 323
  plant behavior, 326–27
  protozoa, 225–31
  Tetrahymena, 228–31
Oedogonium, 250
Oleander, 245
Onoclea, 245
Oodinium, 272
Opalina, 119, 323
Opaque projector, 135
Oscillatoria, 121–22, 126, 248, 249
Osmometer, 312–14
Osmosis, 204, 310–14
Osmunda, 245
Outdoor laboratory, 146
Ovens, incubating, 140
Overhead projector, 42–44, 88, 135
Ovulation, frogs, 330
Oxygen consumption, seeds, 308
Oxygen production, plants, 96

Pandorina, 248, 250
Paramecium
  conjugation, 226
  cultures, 121, 221–22
  as food for *Didinium,* 224
  laboratory use, 225
  staining, 225, 235
  trychocysts, 225
Parasites
  Costia, 272
  of frogs, 120
  of mammals, 328
  in nests, 322
  of rodents, 320
Parsley, 244
Peas, 243
Penicillium, 237
Peperomia, 244
Periodicals, lists for biology teachers, 280–81, 283–84
Perlite, 240
Peromyscus, field studies, 328
Pests, greenhouse, 242
Petroleum ether, use of, 317
pH, effect on enzyme action, 299–300
Philodendron, 244, 245
Photographs
  discussion openers, 41
  enlargement for decoration and study, 133
  in publishing, 282–83
Photography, 130–33
  black and white film, 208
  darkroom, 142
    inexpensive aids, 208
  in ecology, 320
  equipment, 210
  illustrating biology in everyday life, 17
  projection equipment, 210

references for techniques, 286
Photosynthesis, exercises, 303–6
Phylogeny, 10–12
Physarum polycephalum, 236
Piccolyte mounting medium, 238
Picric acid crystals, preservative, 275
Pigments, plant, 304, 317
Pilocarpine acetylcholine, in kymograph studies, 331
Pinguiculas, 247
Pitcher plants, 246, 247
Pituitary gland removal, 330
Planaria
  feeding, 278
  regeneration, 328–29
Plankton net, 155, 156
Plant
  auxins in growth, 31
  behavior studies, 326–28
  books, 285
  cell studies, 329–31
  chromatography, 314–17
  clearing, 237–39
  cuttings, 241
  experiments, 298–318
  grappling bar, 260
  grappling hook, 260
  greenhouse, 240–42
  guard cell studies, 329
  hormone studies, 329–30
  jungle, room, 84
  optimum growth conditions, 241–42
  oxygen production, 96
  pigments, light absorption, 304
  population studies, 322
  propagation, 240–42, 326
  staining, 237–39
  stomate action, 317–18
  succession investigation, 321
  tissue preservative, 275
Plants
  Adiantum, 245
  Aloe, 244
  Amaryllis, 244
  for aquaria, 321
  Asparagus fern, 244
  Asplenium, 245
  avocado, 243
  beans, 243
  bladderworts, 246
  Bryophyllum, 244
  bryophytes, 148
  butterworts, 246
  Caladium, 244
  Calla, 244
  Capsella, 11
  castor beans, 243
  Ceratopteris, 270
  citrus trees, 243
  for the classroom, 243–51
  club mosses, 246
  coal ball peel, 239–40
  coconut, 243
  Coleus, 122, 123, 244, 245, 304, 306

conifers, 243
corn, 243, 298
Crocus, 244
Cystopteris, 245
Dahlia, 244
Darlingtonia, 247
Dionaea, 247
Droseras, 247
Dryopterius, 245
Echinodorm, 270
elephant's ear, 244
elms, 243
Elodea, 123, 127, 304–5
English ivy, 244
enzyme projects, 298–301
Equisetum, 246
ferns, 245
Ficus, 245
Ficus aurem, 244
flowers for dissection, 207
fungi, 236–37
Geranium, 244
Ginkgo, 243
Grocericus storeii, 302, 303
Hyacinths, 244
Kalanchoe, 244
lettuce, 243
lichens, 148, 246
liverworts, 245
maples, 243
mint, 244
mold, 237
mosses, 148, 245
Narcissus, 244
Nepenthes, 247
Nephrolepis, 245
oaks, 243
Oleander, 245
Onoclea, 245
Osmunda, 245
parsley, 244
peas, 243, 325
Peperomia, 244
Philodendron, 244
Physarum polycephalum, 236
Pinguiculas, 247
pitcher plants, 246
Polypodium, 245
Polystichum, 245
Pteretis, 245
Pteridium, 245
Pteridophytes, 246
Sarracenia, 247
Sedum, 244, 245
slime mold, 236
squash, 243
sundews, 246
sweet peas, 243
sweet potatoes, 243
tree seeds, 243
Tulip, 244
Utricularia, 247
Vallisneria, 270
Venus flytrap, 246
yeast, 237
Zebrina, 122–23
Plaster model, 142

Plastic
  embedding specimens, 325
  laboratory material, 209
  specimen bags, 276, 321
Pleurococcus, 250
Polypodium, 245
Polystichum, 245
Population studies, 319–23
Postage stamps depicting nature themes, 133–34
Postlethwait, Samuel, audio-tutorial method, 14
Potassium carbonate crystals, 275
Potassium hydroxide solution, 307–8
Potassium permanganate, 310
Potted plants, care and culture, 244–45
Preservatives, 273–75
Preserved material, 147
Preserving
  Artemia, 277
  brine shrimp, 277
  Hydra, 277
  insects, 275
  methods of, 124–25
  plant tissue, 275
  protozoa, 275
Pressure cooker, laboratory use, 208
Problem solving, plan for, 61
Professional responsibilities
  membership and participation in science organizations, 68, 69
  reading plan, 68
  teacher-student contacts, 67
  workshops, short courses, study groups, 69
  writing and publishing, 68
Programmed instruction, 54
Projects, 298–332. *See also* Laboratory investigations
  amphibian digestive tract, 323
  "bird log," 323
  "bone board," 323
  chick whole mounts, 260–62
  coal ball peel, 239–40
  feather types, 322
  insect collections, 321–22
  insectivorous plants, 247–48
  mammalogy, 328
  ninhydrin, 302
  plant material, 236–37, 245
  population studies, 322
  protozoa, 225–26, 228–31
  student-made, 59–61
  supplies for, 288–96
Protists, 220–35. *See also* Protozoa
  Aeromonas, 271, 272
  algae, 148, 152, 248–51
  bacteria, 152, 237
  Basicladia, 132
  behavior studies, 326–28
  cell studies, 329–31
  Chondrococcus, 272

Protists (cont.)
 Cladophora, 126
 collecting, 152–53, 220–35
 Costia, 272
 culturing, 220–35
 diatoms, 152, 231–34
 encystment, 222, 223, 225
 flannelgraphs, 205
 in frogs, 252
 fungi, 148
 Grocericus storeii model, 302–3
 Hematolechus, 119
 Hexamita, 119
 investigations in animals, 324–26
 Nyctotherus, 119
 Oodinium, 272
 Opalina, 119
 Oscillatoria, 121–22, 126
 parasites, 323
 protozoa, 152, 220–31
 Rhizoclonium, 126
 Saprolegnia, 271, 272
 Serratia marcessens, 122
 Spirogyra, 126
 taxonomy of, 285
 Trichomonas, 119
 used live, 121–22
Protococcus, 250
Protozoa
 Actinophrys, 224
 Actinosphaerium, 224
 Amoeba, 221, 226
 Arcella, 222–23, 225
 Balantidium, 228
 Blepharisma, 224
 Centropyxis, 224–25
 Chilomonas, 222, 225
 collecting, 221
 Colpidium, 222, 223, 224, 225
 Colpoda, 223, 225
 contractile vacuole, 226
 Costia, 272
 culturing, 220–31
 Didinium, 223–24, 225
 Difflugia, 224
 Euglena, 222, 225
 Euplotes, 222, 223, 225
 frog parasites, 323
 growth requirements, 220
 Halteria, 223
 laboratory observations, 225–31
 light conditions, 220
 measuring, 226
 Paramecium, 221–22, 225
 preservatives, 275
 Spirostomum, 225
 staining, 225, 234–35
 Stentor, 222, 225
 Stylonychia, 224
 subculture, 220–21
 succession, 221
 temperature effects, 220
 Tetrahymena, 121, 226–31
 Vorticella, 225
Pteretis, 245
Pteridium, 245

Pteridophytes, 246
Pyrogallic acid as oxygen indicator, 307

Quadrat studies in ecology, 321
Quinine, 272
Quotations, as teaching techniques, 41

Rats, bone staining, 262–63
Reading plan for professional growth, 68
Reagents
 acetic acid, 254, 274–75
 acetone, 315
 acetylcholine, 253, 331
 acid alcohol, 261
 acid for coal ball peel, 239–40
 adrenalin, 253
 amino-acid chromatography, 315–17
 ammonium chloride, 311
 ammonium hydroxide, 310, 311
 antibiotics, sources, 290–91
 barium hydroxide, 306–7, 309–10
 basic laboratory, 209
 Benedict's solution, 301
 borax, 265
 Bouin's fixative, 261
 bromthymol blue, 304–5
 Canada balsam, 322
 carbolic acid, 275
 carmine, 276
 Chloromycetin, 272
 Clorox, 238
 commercial enzymes, sources, 291
 copper sulfate, 272
 cornstarch, 276
 cupric acetate crystals, 275
 cupric chloride crystals, 275
 diatom preparation, 232–33
 enzymes and fine chemicals, sources, 291–92
 ethanol, 315
 ethyl acetate, 257, 331
 FAA, 274
 FAAG, 258
 fixatives, 273
 formaldehyde, 309
 formalin, 272, 274, 276
 formalin-corrosive, 274
 gibberellic acid, 329
 glycerine, 275, 276
 hematoxylin, 230
 hydrogen peroxide, 267
 iodine solution, 299, 301
 iron alum, 230
 Kleermount, 229
 for kymograph studies, 331
 lactic acid, 275
 latex, colored, 276
 lead chromate, 276
 ligroin, 317

 littoralis solution, 277
 Lugol's solution, 331
 Malachite Green, 272
 Methylene Blue, 331
 Meyer's paracarmine, 261
 miticides, 242
 piccolyte, 238
 n-butanol, 316
 nicotine, 255
 ninhydrin test, 302
 Nissenbaum fluid, 230
 Permount, 229
 petroleum ether, 317
 picric acid crystals, 275
 plant material treatment, 238–39
 plaster of Paris, 257
 potassium acetate, 261
 potassium carbonate crystals, 275
 potassium hydroxide, 304, 307–8
 potassium permanganate, 310
 for protozoa staining, 228, 230–31
 pyrogallic acid, 307
 quinine, 272
 Ringer's solution, 253, 261
 Schaudinn fluid, 230
 Skelly B, 317
 sodium hydroxide, 238
 starch-agar, 298
 starch mass, 276
 Taka-Diastase, 299
 Terramycin, 272
 wetting agents, 330
Recording devices, 149
Recordings, disc, 142
Refrigerator, 140
Regeneration
 crayfish, 328
 earthworm, 328
 Hydra, 329
 Planaria, 328–29
 starfish, 328
Report preparation, 61
Reptiles, 269, 277
Respiration, 306–10
Respirometer, 120, 121
Rf value in amino-acid chromatography, 314–17
Rhizoclonium, 126
Rhizopus, 237
Rivularia, 249
Roaches, 269
Rodents, 319–20, 328
Room
 design planning, 76–78, 90–92
 jungle, 82–84
 library, 91, 140–41
Rotifers, 153
Roundworms, 252

Salmanders, 267, 277
Saprolegnia, 271, 272
Sarracenia, 247

Science center
  "bird log" display, 323
  birds, 85
  DNA giant molecule model, 81–82, 83
  essential items, 78
  frogs, 85
  furniture, 89
  jungle, room, 82–84
  mammals, 86
  nonscience equipment, 88–89
  room design planning, 76–78, 90–93
  tanks, 86–87
  wall models, 81
  wall mural, 79–80
Science club
  journals with ideas for, 280
  speakers for, 202
Science methods
  related books, 287
  teaching units, 165–66
Science organizations for teachers, 68, 69
Science Teacher, The, 140
Scientific American, 140
Scientists in biological sciences, chronological list, 184–201
Sciurus, field studies, 328
Sedum, 244, 245
Seeds
  growth experiments, 325
  release of energy, 309
Seine, 155
Serratia marcescens, 122
Short courses, 69
Shrimp, 271
Side table demonstrations, 106–8
Single-concept loop films, 58, 132, 142
  sources, 292–93
16mm projector, 135
Skeletons, 210, 324
Skelly B, 317
Slattery, James L., 4
Slide and filmstrip projector, 88, 135
Slides
  of amphibian parasites, 323
  of blood corpuscles, 330–31
  equipment for making, 238
  Kleermount, 229
  Permount, 229
  piccolyte mounting media, 238
  plant material, 237–38
  preparation
    for diatoms, 232–33
    for protozoa, 228–31
  sources, 289
  stain mountant, 228–29
  as teaching techniques, 42
Slime mold, 236
Snakes, 267–68, 277
Sodium hydroxide, 239
Soil
  ecology, 326
  for greenhouse, 240–42

Solvents, for chromatography, 314–17
Speakers, biology specialists, 202–3
Spirogyra, 126, 248, 250
Spirostomum, 225
Spontaneous generation, 97–98
Squash, 243
Staining
  Paramecium, 225
  protozoa, 225, 228–31, 234–35
  Tetrahymena, 228–31
Stains, 206, 210
  alizarin red S, 262
  Bismarck Brown, 235
  Brilliant Cresyl Blue, 229, 235
  Chlorazol Black E, 238
  Congo red, 329
  Fast Green, 238, 239
  Giemsa, 331
  hematoxylin, 229–31
  Janus Green B, 235
  Methylene Blue, 235, 329
  nigrosin, 225, 228
  preparation of vital, 235
  Safranin O, 238, 239
  substitutes, 208
  Turtox CMC-S, 228
  Wright's, 325
Starch
  agar, 298
  digestion, 204–5
  mass, 276
  production, 306
Starfish, regeneration, 328
Stentor, 222, 225
Stomate demonstration, 317–19
Structure
  comparisons, 324
  diatom, 231–34
  early emphasis on, 21–22
Structure and function, 9, 10, 20–21
  analogy as an effective teaching device, 28
  Biological Sciences Curriculum Study, 32
  of chloroplasts, 27
  complement one another, 10
  DNA, example of interdependence, 25–27
  of enzymes, 27
  in genetics, 24–27
  of muscle cells, 28
  in protozoa, 225–26
  relative emphasis, 23–24
  teaching, 32–33
  in *Tetrahymena,* 226–27
  xylem, 29–31
Study groups, 69
Stylonychia, 224
Succession, investigations in, 321
Sugaring trees, 154
Sundew plants, 246, 247
Sweet peas, 243
Sweet potatoes, 243
Syringe use, 276

Tadpoles, parasite studies, 323
Taka-Diastase, 299–300
Tanks, 86–87
Tape recorder, 39–41, 88, 129–30, 142, 292
Tape recordings, 327
Taste areas, 206
Taxonomy
  dichotomous key, 149–50
  "How To Know" books, 285
  key development, 162
  of plants, 285
Teacher-student contacts, 67
Teaching techniques. *See also* Equipment; Materials for teaching
  analogy, 28
  audio-tutorial instruction, 56–59
  audio-visual, 218
  chalk talk, 42, 43
  chi-square method, 212–14
  class discussion, 41–42
  delivery of lecture material, 37
  demonstration, 41, 45–46
  developing good personality traits, 36–37
  eloquence and logic, 38–39
  eye contact, 39
  "I," "you," "me," 42
  individualized instruction, 51–53
  instruction patterns follow learning patterns, 51
  linear programming, 55–56
  mannerisms, 38
  meeting unexpected situations, 46–47
  models, 26, 27
  nature study, 323
  photography in ecology, 321
  pictures, 41
  programmed instruction, 54
  projects, special student, 59–61
  protozoa in the classroom, 220–31
  questions for laboratory work, 298–332
  quotations, 41
  side table, 33
  simple words, 44
  sketches, 217
  stimulators, 217
  stomate action, 318
  student involvement, 218
  teaching practices, 203–4
  unit method, 164
  voice modulation and volume, 39
Teleology, 31–32
Terramycin, 272
Terraria, 247, 320
Testing sieves, 260
Tests
  evaluation tool, 176
  laboratory, 110–11
  quiz use, 217
  teaching with, 174

Tetrahymena, 121, 226–31
Texts
  authors, 168
  bibliography, 169
  changes in, 1
  emphasis, 168
  glossary, 169
  illustrations, 168–69
  index, 169
  material covered, 167–68
  organization, 168
  paperbacks, 208
  protective book cover, 208
  for room library, 283–87
  scope of, 169
  selection, 167
  style, 168
  survey, 171–72
  in unit building, 172
Tinbergen, Dr. Niko, 159
Tin-can trap for insects, 154
Toads, 268, 323
Tools
  to buy, 138–41
  to make, 136–38
Torso models, 140
Tortoises, 269
Transect studies, ecology, 321
Transfer chamber, 137, 138
Transparencies, sources, 288
Transpiration, 30
Traps and cages

  in ecology projects, 319–23
  for frogs, 86
  for insects, 321
  jar, 86
  sources, 289
Trichocysts, 225
Trichomonas, 119, 323
Tree trimmer, 140
Trump, Richard, text survey, 171–72
Tubifex, 271
Tulip, 244
Turtles, 267, 268, 269, 270, 277
Typing blood, 204

Ulothrix, 248
Units
  audio-visual activities, 179
  course organization, 175–76
  development ideas, 179
  investigative topics, 179
  lesson plans, 177–78, 214–15
Urey-Miller, amino acid research, 12–14
Utricularia, 247

Vallisneria, 270
Vaucheria, 248
Vein injection, 276–77
Venus flytrap, 246, 247

Vermiculite, 242
Video-tape recorders, sources, 294
Visibility disc, 260
Volvox, 248
Vorticella, 225

Wall-mounted screen, 88
Water dogs, 269
Water-testing kit, 303
Wetting agents, 330
White powder problem, 101–3
White worms, 271
Wilkins, Watson, and Crick, research on DNA, 7, 12
Workshops, 69
Writing and publishing, 61, 68, 281–83

X-ray diffraction, in nucleic acid analysis, 25
Xylem, structure and function, 29–31

Yeasts, 237, 309

Zebrina, 122–23
Zygnema, 248